Sedimentation and Sediment Transport

Sedimentation and Sediment Transport

Proceedings of the Symposium held in Monte Verità, Switzerland, from September 2nd – to September 6th, 2002

Edited by

A. GYR
IWH, ETH, Zürich, Switzerland

and

W. KINZELBACH
IWH, ETH, Zürich, Switzerland

KLUWER ACADEMIC PUBLISHERS
DORDRECHT / BOSTON / LONDON

A C.I.P. Catalogue record for this book is available from the Library of Congress.

ISBN 1-4020-1266-7

Published by Kluwer Academic Publishers,
P.O. Box 17, 3300 AA Dordrecht, The Netherlands.

Sold and distributed in North, Central and South America
by Kluwer Academic Publishers,
101 Philip Drive, Norwell, MA 02061, U.S.A.

In all other countries, sold and distributed
by Kluwer Academic Publishers,
P.O. Box 322, 3300 AH Dordrecht, The Netherlands.

Printed on acid-free paper

Printed in the Netherlands.

TABLE OF CONTENTS

III. TURBULENT STRUCTURES AND SEDIMENT TRANSPORT

IV. TWO-PHASE FLOWS

V. SELF ORGANIZATION IN SEDIMENTATION PROCESSES

VI. SELF ORGANIZATION IN SEDIMENT TRANSPORT

VII. PROBLEMS RELATED TO FIELD MEASUREMENTS AND INPUTS NEEDED FOR NUMERICAL MODELS

VIII. MEASURING TECHNIQUES

PREFACE

The Symposium on Sedimentation and Sediment Transport was held from September 2nd to September 6th 2002 on the Monte Verità above Locarno. It was jointly organized by the Institute of Hydromechanics and Water Resources Management and the Institute of Fluid Dynamics under the auspices of the Swiss Federal Institute of Technology Zurich (ETHZ).

The aim of the Monte Verità symposia is to give a platform to well known researchers in a given field of science to discuss new ideas and motivate young researchers working in that field.

It is evident, that for a number of ecological and technical problems in rivers and lakes a better knowledge of sediment transport and sedimentation is needed together with the ability to predict and simulate sediment behavior. On the other hand, a stagnation of research in these topics could be observed in the last decades. At the Symposium an attempt was made to present new results in mathematics and natural sciences relevant for the sediment problem. New strategies were discussed to tackle the complexity of the problem. Basic theoretical research and laboratory experiments alone are incomplete without a feedback from field observations and measurements. For that reason well-known researchers from both basic and engineering sciences were invited.

During the week, 51 scientists from 16 countries participated and discussed problems such as: turbulence, non-local phenomena, stability, interactions, feedback systems, self-organization, two-phase flow and chaotic processes, numerical simulations as well as measurement techniques and field observations.

The intention to bring together scientists from a much broader spectrum than usual was highly appreciated by the participants. Practitioners learned about the problems theoreticians and mathematicians have in describing turbulence and processes strongly influenced by it. On the other hand the theoreticians appreciated from the presentations of field experiments how complex the real world is and where new approaches are needed.

A Symposium is a joint effort of many people. Therefore a few word of thanks are in order. The first thanks go to the participants who communicated their results and even speculations very openly. In general and long lasting discussions of the different themes, they contributed to a very lively and fruitful symposium. Thanks for the administrative help go also to the Centro Stefano Franscini and Veronica Trombini, secretary of the Institute of Hydromechanics and Water Resources Management. In organizing the Symposium we have been helped by Profs. Kleiser and Tsinober. Last but not least we would like to thank Beat Lüthi for his help in formatting the Proceedings for printing.

The conference fee could be kept at a modest level due to financial contributions from the Swiss National Science Foundation and the ETH Zurich. We are especially grateful for the stipends which allowed to invite young scientists and scientists from countries which are in a less fortunate economic situation than Switzerland.

Zurich, November 1st 2002

Albert Gyr & Wolfgang Kinzelbach

I

Turbulence,

mainly its non-local Aspects

SOME MATHEMATICAL CONTRIBUTIONS TO THE UNDERSTANDING OF TURBULENCE

C. BARDOS

Laboratoire Jacques Louis Lions, Université Pierre et Marie Curie, 175 Avenue du Chevaleret 75013, bardos@math.jussieu.fr

1. Introduction

The mathematical analysis of the Navier Stokes equations involved the contributions of such diverse personalities as, Euler, Leray, Kolmogorov, Arnold and others and in view of the needs of practical applications in engineering sciences success have been limited. However the results that have been obtained contribute to our understanding of the physic and the numeric. They may be of some use in applied sciences involving multiscale phenomena and interaction of particles as in the present Monte Verita Symposium. This is what I try do describe with emphasis on the physical relevance of the few available results and of the many conjectures.

2. Hierarchy of Equations

The equations of fluid mechanics are diverse and involve some global or local physical parameters (here are considered the Knudsen number, the Mach number and the Reynolds number). Therefore it is natural to consider a "chain" of equations starting from the Boltzmann equation which describes the evolution of the density of molecules which at some time t have at the point x the velocity v and ending with some models of turbulence used to compute averaged quantities. As it is (hopefully) often the case, the next equation will become relevant when the structure of the phenomena becomes too complicated to be computed by the previous one. The derivation of the macroscopic equations (Euler Navier Stokes etc..) from the Boltzmann equation is at present reasonably well understood [1], [2]. On the other hand there is not rigorous (say from basic principles or Navier Stokes equations) derivation of any turbulence modelling like the ϵk model [3] in spite of the fact that such models are widely used to design the planes that you fly.

Boltzmann Equation

$$Kn \to 0 \Downarrow$$

Compressible Euler Equation

$$Mach \to 0 \Downarrow$$

Incompressible Euler Equation $\quad \partial_t u + \nabla(u \otimes u) + \nabla p = 0, \nabla u = 0$

$$\text{viscosity} \quad \nu \to 0 \Uparrow$$

Incompressible Navier Stokes $\quad \partial_t u + \nabla(u \otimes u) - \nu \Delta u + \nabla p = 0, \nabla u = 0$

Turbulence Modelling

A. Gyr and W. Kinzelbach (eds.), Sedimentation and Sediment Transport, 3–9.

3. Existence Uniqueness theorems for the Euler Equations

Usually mathematicians concerned with time dependent partial differential equations try to prove existence and uniqueness of solutions and use the concept "well posed problems in the sense of Hadamard". This shows that the equation is relevant. An equation with no solution is definitely irrelevant. Then the question of uniqueness of the solution is in general a question of stability. How perturbations in the initial data should affect the phenomena at later time. Quite often the proof contains more information than the result itself.

In the middle of the chain of equations is the compressible Euler equation. It is well posed for small time and smooth initial data however for large time "stable shocks" due to compression appears. These fact was already observed by Euler himself introducing for this purpose forerunners of the modern theory of distributions. As a consequence global in time solutions should be weak . Such result has been obtained in dimension 1 by Glimm [4] in (1965). No serious progress have been obtained since this time. Furthermore one of the tools of Glimm is the use of a special functional space (functions of bounded variation) which is compatible with the presence of shocks. However it has been proven in (1986) [5]) that this concept cannot be used in more than one space variable. For the incompressible Euler equation in 3 space variables the situation is in some sense worse. As above one can show, Lichtenstein (1925) [6] that for smooth initial data there is for "small time" a unique solution. What happens after is "mathematically" unknown in spite of the fact that there is no physical reason for the appearance of a singularity. However it has been observed that the regularity, stability (etc...) of the solutionis driven by a very physical and natural quantity: the size of the vorticity.

Theorem 1: [7] As long as the vorticity $\omega = \nabla \wedge u$ satisfies the relation:

$$\int_0^T \|\omega(t)\|_{L^\infty} dt < \infty \tag{1}$$

the regularity of the solution is governed by the regularity of the initial data.

One of the main reason of the difficulty of the problem is the existence of solutions for which the vorticity may become arbitrarily large in a finite time. Such example can be constructed with a slight modification of the Taylor Green vortex and lead to the following

Theorem 2: [8] In $3d$ and for any p (the enstrophy,L^2 , norm of the vorticity, corresponds to $p = 2$) there is no function

$$\phi(Z,t), \quad \lim_{Z\to 0} \phi(Z,t) = 0$$

such that the vorticity ω of the corresponding solution of the $3d$ incompressible Euler Equation satisfies.

$$\|\omega(t)\|_{L^p} \le \phi(\|\omega(0)\|_{L^p}, t)\,.$$

Considering that with the viscosity ν going to zero, any family of uniformly regular solutions of the incompressible Navier Stokes equation converges to the solution of the above Euler equation, one concludes that the theorem 2 is also true for the Navier Stokes equation with the requirement "ϕ independent of ν".

Therefore there is no chance of detecting a singularity of the 3Euler equation by numerical simulations and the same observation holds for the Navier Stokes equation with small viscosity. For instance it is not possible by numerical simulation to discriminate between a quantity $q(t)$that goes to infinity for $t \to T^*$ and a quantity that in the same region saturates by overflow the computer because it grows like $(exp(expKt)$.

4. More dynamic

The instability results of the previous section leads to the idea that further progress should be based on a better understanding of the dynamical effect of the vorticity. A very naive statement would be as follow: "where the solution is smooth the vorticity is bounded, however when it is not so smooth the vorticity becomes big enough to produce some averaging phenomena that will again stabilize the flow." Along this line some partial but interesting results have been obtained [9]: Start with an initial data of the following form

$$u_\Omega(x,0) = \tilde{u}_0(x) + \Omega e_3 \wedge x\,, \quad \text{in geophysic } \Omega \text{ is the Rossby number}$$

and consider situations where $\Omega \to \infty$ (the authors call it quasi alignment of the initial vorticity.) Then introduce the Poincaré propagator (already used by Sobolev [So])

$$\partial_t v + \Omega \wedge v + \nabla p = 0,\ \ \nabla v = 0, v(x,t) = e^{\Omega J t} v(.,0)$$

and introduce, for the solution of the Euler equation, the change of variables:

$$\tilde{u}_\Omega(x,t) = e^{-\Omega J t} \tilde{v}_\Omega(e^{-\Omega J t} x, t)$$

then the stabilizing effect of the vorticity is illustrated by the following
Theorem 3: The function $\tilde{v}_\Omega$ is smoothly defined on a time interval which increase and goes infinity with$\Omega \to \infty$.

This statement should be compared with a series of results obtained by Constantine, Fefferman and Majda [11] both for the Euler and for the Navier Stokes equation. For instance one can weaken the hypothesis (1) of the theorem 1 with the assumption that the direction varies smoothly. More explicitly one has the
Theorem 4: For a solution of the 3 Euler equation consider the quantity

$$\sup_{|x-y|\le L} \left(\frac{\omega(x,t)}{|\omega(x,t)|} \wedge \frac{\omega(y,t)}{|\omega(y,t)|} \right) \frac{1}{|x-y|} = L(t).$$

Then as long as

$$\int_0^T L(t) \left(\sup_x \int_{|y|\le L} |\omega(x+y,t)| \frac{dy}{|y|^2} \right) dt < \infty$$

the solution remains smooth on $[0,T]$.

5. Regularity of the Leray solution

For the Navier Stokes equation the presence of the viscosity $\nu > 0$stabilize the flow and leads to more results. Formally one has an energy equality.

$$\frac{1}{2}\int |u(x,t)|^2 dx + \nu \int_0^t \int |\nabla \wedge u(x,s)|^2 dx ds = \frac{1}{2}\int |u(x,0)|^2 dx\,.$$

This implies existence of a weak solution u following a construction done by Leray [12]. This construction leaves open the issue of regularity and uniqueness and has never been seriously improved since that time. It has been shown later by Serrin [13],Masuda [14] and others that the hypothesis

$$u \in C(0,T; L^3(\Omega)) \tag{2}$$

implies the regularity of the solution. However the hypothesis (2)is not proven for the Leray solution. On the other hand the result of Serrin and Masuda relies only on the parabolic properties of the equation (and not on the precise structure of the advection or the pressure term). Such proof would hold for any parabolic equation. Moreover the 3 is the critical index for such family of equation:(the space of solutions in L^3 is invariant by the scaling $u(x,t) \to \lambda u(\lambda x, \lambda^2 t))$ and there exist parabolic equations of this type with solutions that do not satisfy (2). As a consequence a more detailed approach should be used and this legitimates the proof of the following results.

Theorem 5: For a weak solution u of the $3d$ incompressible Navier Stokes equation (as constructed by Leray) the following facts are true:

1 The singular support of the solution is as most of Hausdorff dimension 1 in space time (Caffarelli-Kohn-Nirenberg [15], Fanghua Lin [16]).

2 In incompressible fluid the pressure is determined an to an arbitrary function of time. To remove this ambiguity one can replace it by a normalized pressure which goes to zero at infinity. Then one can show that if the normalized pressure is bounded from below (no vacuum) the solution remains smooth (Seregin and Sverak [17]).

3 For to the solution to become singular for $t \to T$ one needs (Constantine and Fefferman [18])

$$\sup_{(x,t),(y,t))\in(R^3\times[0,T[)^2} \left(\frac{\omega(x,t)}{|\omega(x,t)|} \wedge \frac{\omega(y,t)}{|\omega(y,t)|}\right)\frac{1}{|x-y|} = \infty. \tag{3}$$

From the above considerations one can conjecture and in some cases prove the following facts. For the Navier Stokes equation the region where the vorticity is large is very small and this may be a mathematical formulation of the notion of intermittency. To become singular the direction of the vorticity has to vary very much, the flow has to exhibit genuine $3d$ behavior. However then one may conjecture that this turbulent effect will produce averages that may prevent the formation of real singularities and/or justify the introduction of averaged models for numerical computations. Eventually we can speculate that if the normalized pressure becomes very low we encounter the phenomenon of cavitation. Besides its contribution to the theoretical issue of smoothness and uniqueness of the Leray solution the result of [17] means that intermittency, turbulence and high velocity in very small region are related in a measurable way with appearance of very low density of fluid in these regions.

Special $2d$. effect

When the fluid is forced to remain in a plane (by initial and boundary conditions) the direction of the vorticity remains perpendicular to this plane and one has

$$\partial_t \omega + u \cdot \nabla \omega = 0\,. \tag{4}$$

With (3) one obtains a result of existence and uniqueness of the solution for any time under the hypothesis $\omega(x,0) \in L^\infty$. This result has to be completed by the following remarks.

1 The regularity of the initial data propagates. This is not so an easy result it follows from the pair dispersion estimate of Wolibner (1933) [19]. For instance in a bounded domain of diameter the distance between two particles $x(t)$ and $y(t)$ which follows the flow satisfy the estimate.

$$\left(\frac{|x(0)-y(0)|}{D}\right)^{e^{-Ct|\nabla\times u|_{L^\infty()}}} \geq \frac{|x(t)-y(t)|}{D} \geq \left(\frac{|x(0)-y(0)|}{D}\right)^{e^{Ct|\nabla\times u|_{L^\infty()}}}$$

Keeping in mind that

$$\frac{|x(0)-y(0)|}{D} < 1$$

one observes that this estimate deteriorates very fast fort $\rightarrow \infty$.
2 The estimates also deteriorate very fast with loss of regularity of the initial data. The extreme case correspond to some completely unphysical solutions with initial data $u(x,0) \in$ ^{2} and vorticity $\omega(x,0) \notin L^\infty$ constructed by Schaeffer and Shnirelman [20]. Such solutions have compact support in space time. They come from rest and return to rest!
3 The points 2 and 3 are consistent with the appearance and persistency of coherent structures as observed in $2d$ phenomena as diverse as the Jupiter red spot or the " standard" cyclones in the atmosphere. Such solutions are characterized by the relations:

$$\partial_t u = 0 \Rightarrow u\nabla\omega = 0 \Rightarrow \omega = -\Delta\Psi = F(\Psi)\,.$$

Stability in the L^2 norm of the vorticity explains why the structures remains (Arnold and Keyzin [21], Bardos, Guo, Strauss [22] and others). On the other hand since the system is reversible the same remark is valid for negative time. Therefore such structures can be reached only by perturbation in weaker norm (turbulence in the initial data) and large time and this is fully in agreement with the points 1 and 2.

Statistical averaging versus weak convergence

Generally one uses the concept of "statistical theory of turbulence"which implies that the flow is at best described by a random variable$u(x,t\omega)$. Below I intend to pursue the deterministic analysis of the fluid and show that the notion of weak convergence may play the role of statistical average. The basic example of weak convergence is the sequence of oscillating functions $\sin kx$ with $k \rightarrow \infty$ One has (in a weak sense):

$$\lim_{k\rightarrow\infty} \sin kx = 0 \quad \lim_{k\rightarrow\infty} (\sin kx)^2 = \frac{1}{2}\,.$$

Similarly with a weakly converging sequence of solutions of Navier Stokes with viscosity $\nu \rightarrow 0$ one may have:

$$\lim_{\nu\rightarrow 0}(u_\nu \otimes u_\nu) = \lim_{\nu\rightarrow 0} u_\nu \otimes \lim_{\nu\rightarrow 0} u_\nu - R \text{ with } R \neq 0\,. \tag{5}$$

The tensor R which appears in (4) corresponds to the usual Reynolds tensor in the theory of turbulence. However equation for $\lim(u_\nu \otimes u_\nu)$ requires $\lim(u_\nu \otimes u_\nu \otimes u_\nu)$. Equation for $\lim(u_\nu \otimes u_\nu \otimes u_\nu)$ requires $\lim(u_\nu \otimes u_\nu \otimes u_\nu \otimes u_\nu)$ and so on... This is the deterministic counterpart of the closure problem in statistical theory of turbulence. In both cases a natural tool is the Wigner transform :

$$R(x,t,k) = (\frac{1}{2\pi})^n \int e^{-ik\cdot y} < u(x+\frac{y}{2},t) \otimes u(x-\frac{y}{2},t) > dy\,, < u >= 0.$$

which for a random family of flows produces a local spectra of statistic turbulence. At variance for weak convergent sequence one uses the Wigner measure. Observe that the energy estimate gives the relation:

$$\int_0^T \int \nu|\nabla \wedge u_\nu|^2 dxdt \leq Cte\,.$$

Write

$$\tilde{u} = u - \lim_{\nu \to 0} u$$

and introduce

$$R_\nu(x,t,k) = (\frac{1}{2\pi})^n \int e^{-ik\cdot y} < \tilde{u}_\nu(x + \frac{\sqrt{\nu}y}{2}, t) \otimes \tilde{u}_\nu(x - \frac{\sqrt{\nu}y}{2}, t) > dy$$

then with a, by now classical, theorem (proposition 1.7 in [23]) one has

$$\lim_{\nu \to 0} \tilde{u}_\nu(x,t) \otimes \tilde{u}_\nu(x,t) = \int R(x,t,k)dk\, R(x,t,k) = \lim_{\nu \to 0} R_\nu(x,t,k)\,.$$

The "spectra" R is local in (x,t). It depends only on the high frequencies and on the two points correlation. It should be isotropic as this is observed in statistical theory of turbulence [24]

$$R(x,t,k) = R(|k|,x,t)(I - \frac{k \otimes k}{|k|^2})\,.$$

Then if it depends on local properties of the limit fluid u it takes, for instance in 2 dimension [3], the following form

$$\lim_{\nu \to 0} \tilde{u}_\nu(x,t) \otimes \tilde{u}_\nu(x,t) = aI + \mu|(\nabla u + \nabla u^T)|\,.$$

The coefficient μ should be positive to lead to a turbulent diffusion as the one used in turbulence modelling. Of course none of the above statement is proven (they may not be always true) but they are formulated in a explicit and local way which may be an alternative to the classical axioms of the statistical theory of turbulence.

6. References

[1] C. Bardos, F. Golse and D. Levermore: Fluid dynamics limits of kinetic equations I: Formal derivations, J. of Stat. Phys. 63 (1991), 323–344.

[2] F. Golse and L. Saint Raymond (2001), The Navier-Stokes limit for the Boltzmann equation , C.R. Acad. Sci. Paaris 333, 897–902.

[3] B. Mohammadi and O. Pironneau (1994), Analysis of the K-ϵ turbulence model , Research in Applied Math, No 31, J.L.Lions and P. Ciarlet (eds.), Masson-Wiley, Paris.

[4] J. Glimm (1965), Solutions in the large for nonlinear systems of conservation laws, Comm. on Pure and Appl. Math. 18, 685–715.

[5] J. Rauch, BV estimates fail for most quasilinear hyperbolic systems in dimension greater than one (1986), Comm. Math. Phys., 106, 481–484.

[6] L. Lichtenstein: Über einige Existenzproblem der Hydrodynamik homogener unzusammendrückbarer, reibunglosser Flüssikeiten und die Helmholtzschen Wirbelsalitze , Mat. Zeit. Phys. 23 (1925), 89–154; 26 (1927), 193–323; 32 (1930), 608.

[7] J.T. Beale, T. Kato and A. Majda (1984), Remarks on the breakdown of smooth solutions for the 3D Euler equations, Commun. Math. Phys., 94, p. 61-66.

[8] P.L. Lions, Mathematical Topics in Fluid Mechanics Volume 1 Incompressible models (1966), Oxford Lecture Series in Mathematics and its Applications, Oxford.

[9] A. Babin, A. Mahalov, and B. Nicolaenko (1997), Global regularity and integrability of 3D Euler and Navier-Stokes equations for uniformly rotating fluids, Asymptotic Analysis, 15, p. 103–150.

[10] S.L. Sobolev, On a new problem of mathematical physics, Izvestia Akad. Nauk SSSR, Ser. Math 18 (1954), 3-80; DokladyAkad Nauk SSSR 109 (1956), 707; On the motion of a symmetric top with acavity filled by a fluid, Appl. Mech. Tech. Phys. 3 (1960), 20-55.

[11] P. Constantinnd, Ch. Fefferman and A. Majda (1996),Geometric constraints on potential singular solutions for the $3-D$ Euler equation, Comm. in PDEl, 21, p. 559-571.

[12] J. Leray (1934), Sur le mouvement d'un liquide visqueux emplissant l'espace, Acta Math.,63, p. 193-248.

[13] J. Serrin (1962), On the interior regularity of weak solutions of the Navier-Stokes equations, Arch. Rat. Mech. Anal., 9, 187–195.

[14] K. Masuda Weak solutions of the Navier-Stokes equations (1984), Tohoku Math. J., 36, 623–646.

[15] L. Caffarelli, R. Kohn and L. Nirenberg (1982), Partial regularity of suitable weak solutions of the Navier-Stokes equations, Comm. Pure Appl. Math., 35, p. 771-831.

[16] F. Lin (1998), A new proof of the Caffarelli-Kohn-Nirenberg Theorem Comm. Pure App. Math, 51, 241–257.

[17] G. Seregin and V. Sverak (2002), Navier-Stokes equations with lower bound on the pressure, Arch. Rational Mech. Anal., 163, 65–86.

[18] P. Constantin and Ch. Fefferman (1993), Direction of vorticity and the problem of global regularity for the Navier-Stokes equations, Indiana Univ. Math. Journal, 42(3), p. 775-789.

[19] W. Wolibner (1933), Un théorème sur l'existence du mouvement plan d'un fluide parfait, homogène et incompressible, pendant un temps infiniment long, Mat. Z., 37, 698–726.

[20] A. Shnirleman (1997), On the nonuniqueness of weak solution of the Euler equation , Comm. Pure Appl. Math., 50, 1261–1286.

[21] V.I. Arnold and B.A. Khesin (1997), Topological Methods in Hydrodynamics, Applied Mathematical Sciences, 125, Springer.

[22] C. Bardos, Guo and W. Strauss: Stable and unstable ideal plane flows, Chin. Ann. of Math. 23 B (2002), 149-164.

[23] P. Gérard, Microlocal defect measures (1991), Comm. in Partial Differential Equations 16, 1761–1794.

[24] M. Cherkov A. Pumir and B. Shraiman, Lagrangian tetrad and the phenomenology of turbulence. The International Conference on Turbulence, Phys. fluids 11, no 8 (1999), 2394–2410.

NONLOCALITY IN TURBULENCE[1]

ARKADY TSINOBER

Faculty of Engineering, Tel Aviv University, Tel Aviv 69978

1. Nonlocality in 'ordinary' (not particulate) flows

Nonlocality is among the three main reasons[2] why the problem of turbulence is so difficult. The term 'nonlocality' is used here in several related meanings which are clarified in the course of the discussion of the issues throughout this paper.
From the formal mathematical point of view a process is called local if all the terms in the governing equations are differential. If the governing equations contain integral terms or other nonlocal operators then the process is nonlocal. The Navier-Stokes equations are integro-differential for the velocity field in both physical and Fourier space (and any other). Therefore, generally, the Navier-Stokes equations describe nonlocal processes.
From the physical point of view (the above meaning of) nonlocality is associated with or long range or action at a distance. This again is closely related to what is called direct and bidirectional coupling of large and small scales.
Most of the aspects to be discussed here can be put under the common name *direct and bidirectional coupling/interaction between large and small scales.* This, however, contains two ambiguous notions: *scale(s)* and *coupling/interaction.*

1.1. ON SCALES AND THEIR COUPLING

The ambiquity in definition of scales can be avoided by not using any decomposition (D): the small scales (whatever this means) are associated with the field of velocity derivatives (vorticity and strain). Therefore, it is naturally to look at this field as the one objectively (i.e. D-independent) representing the small scales. The velocity fluctuations represent the large scales.
Taking the above position one can state that in homogeneous (not necessarily isotropic) turbulence the large and the small scales are uncorrelated. For example, in a homogeneous turbulent flow the mean Lamb vector $\langle \boldsymbol{\omega} \times \mathbf{u} \rangle = 0$ (and also $\langle (\mathbf{u} \cdot \boldsymbol{\nabla})\mathbf{u} \rangle = 0$). If the flow is statistically reflexionally symmetric then the mean helicity $\langle \boldsymbol{\omega} \cdot \mathbf{u} \rangle$ vanishes too. Similarly, in a homogeneous turbulent flow velocity, u_i, and strain,s_{ij} do not correlate either, $\langle u_i s_{ij} \rangle = 0$. However, vanishing correlations do not necessarily mean absence of dynamically important relations.[3] Indeed, the quantities $(u \cdot \nabla)u \equiv \omega \times u + \nabla(u^2/2)$ and $\omega \times u$, are the main 'guilty' for all we call turbulence. Both contain the large scales (velocity) and small scales (velocity derivatives). Therefore some kind of dynamically essential coupling between the two is unavoidable.

---

[1]This communication is based in part on the material from Tsinober (2001). Therefore mostly references related to the new material are mentioned explicitly.

[2]The three N's: nonlinearity, nonintegrability and nonlocality.

[3]Vorticity and the rate of strain tensor are precisely uncorrelated too $\langle \omega_i s_{ij} \rangle \equiv 0$, but their interaction is in the heart of the physics of any turbulent flow.

A. Gyr and W. Kinzelbach (eds.), Sedimentation and Sediment Transport, 11–22.

1.2. KINEMATIC COUPLING.

Vorticity and strain are not just small scales, since together with boundary conditions the whole flow field is determined entirely by the field of vorticity or strain, so that velocity is a functional of vorticty, $\omega = curl\ u \Rightarrow \nabla^2\mathbf{u} = -curl\omega \Rightarrow \mathbf{u} = \Phi\{(\omega)\}$ and/or strain[4] $2s_{ij} = \partial u_i/\partial x_j + \partial u_j/\partial x_i \Rightarrow \nabla^2 u_i = 2\partial s_{ik}/\partial x_k \Rightarrow u_i = \Psi\{(s_{ij})\}$. Therefore, any altering of vorticity and/or strain results in their 'reacting back' on the velocity field. Likewise the field of a passive scalar is defined by its gradient $\mathbf{G} = \nabla\theta$ and the field of a vector potential $\mathbf{A}$ of a solenoidal passive vector $\mathbf{B}$ ($rot\mathbf{A} = \mathbf{B}$) is defined by $\mathbf{B}$.

The above statement is not that trivial as may seem.

First the purely kinematic reaction back, e.g. of vorticity in the *statitsical* sense, is seen from the following example. Taking a Helmholz decomposition of the most significant part of the nonlinear term in NSE, the lamb vector $\boldsymbol{\omega}\times\mathbf{u} = \nabla\alpha + \nabla\times\boldsymbol{\beta}$ it can be shown that $\langle(\nabla\alpha)^2\rangle \sim \langle(\nabla\times\boldsymbol{\beta})^2\rangle$ if ω and u are random Gaussian and not related, i.e. $\boldsymbol{\omega} \neq rot\mathbf{u}$. However, if $\boldsymbol{\omega} = rot\mathbf{u}$ and $\mathbf{u}$ is random Gaussian, then $\langle(\nabla\alpha)^2\rangle \sim 2\langle(\nabla\times\boldsymbol{\beta})^2\rangle$, i.e. $\boldsymbol{\omega}$ 'reacts back' on velocity for purely kinematic reasons.

1.3. A SIMPLE DYNAMICAL EXAMPLE

We use the above example with the Lamb vector to illustrate the dynamical aspect of the utmost importance of direct and bidirectional coupling between the large and small scales. For this we retreat from homogeneous flows and consider a unidirectional *in the mean* fully developed turbulent flow such as the flow in a plane channel in which all statistical properties depend on the coordinate normal to the channel boundary, x_2, only. In such a flow a simple precise kinematic relation is valid

$$d\langle u_1u_2\rangle/dx_2 \equiv \langle\boldsymbol{\omega}\times\mathbf{u}\rangle_1 = \langle\omega_2u_3 - \omega_3u_2\rangle \neq 0, \tag{1}$$

which is just a consequence of the vector identity $(\mathbf{u}\cdot\nabla)\mathbf{u} \equiv \boldsymbol{\omega}\times\mathbf{u} + \nabla(u^2/2)$, incompressibility and $d\langle\cdots\rangle/dx_{1,3} = 0$, and $\langle\cdots\rangle$ means an average in some sense. Since in these turbulent flows $d\langle u_1u_2\rangle/dx_2$ is essentially different from zero at any arbitrarily large Reynolds number, one can see from (1) that at least some correlations between velocity and vorticity are essentially different from zero too at any arbitrarily large Reynolds number. Hence the relation (1) is a clear indication of a dynamically important statistical dependence between the large ($\mathbf{u}$) and small ($\boldsymbol{\omega}$) scales. Note that both correlation coefficients (betwen ω_2 and u_3, and ω_3u_2) are small ($\sim 10^{-2}$) even at rather small Reynolds numbers and decrease as $Re^{-1/4}$. Nevertheless, in view of the dynamical importance of interaction between velocity and vorticity in turbulent shear flows[5] such 'small' correlation by no means does imply absence of dynamicaly important statistical dependence and direct interaction between large and small scales.

[4]Note that the small scales, generally, are not 'integrated out' in the Bio-Savart-like relations due to the singular nature of the kernel. The simplest example is the potential vortex with 'zero' thickness and (large scale) velocity distribution $v_\theta \sim r^{-1}$.

[5]The relation (1) is approximately valid in many important turbulent flow such as boundary layers, wakes, jets, etc. in which $d\langle\cdots\rangle/dx_{1,3} \ll d\langle\cdots\rangle/dx_2$)

1.4. KINEMATIC COPULING PLUS SELF-AMPLIFICATION OF VORTICITY AND STRAIN.

One of the most basic (and least understood) properties of turbulent flows is the *self*-amplification of the field of velocity derivatives (*both* vorticity and strain). Among other things this is manifesed in that the mean of the corresponding production terms $\omega_i\omega_j s_{ij}$ and $-s_{ij}s_{jk}s_{ki}$ is known to be strictly positive. Most important that $\omega_i\omega_j s_{ij}$ and $-s_{ij}s_{jk}s_{ki}$ (and the corresponding terms in the equations for ω_i and s_{ij}) are dominating in the sense that they are orders of magnitude larger that terms originating from external forcing and/or shear. Similarly the gradient of passive scalar $\mathbf{G} = \nabla\theta$ and a solenoidal passive vector $\mathbf{B}$ are not diffusionless invariants, both are amplified (but not self-) by turbulent (and apparently any random) flow.

The property of self amplification of vorticity and strain is responsible for the fact the neither enstrophy ω^2 nor the total strain s^2 are inviscid invariants as is the kinetic energy u^2. In the context of nonlocality this means that the production of vorticity and strain 'react back' in creating the corresponding velocity field since as mentioned the whole flow field (including velocity, which is mostly a large scale object) is determined entirely by the field of vorticity or strain. In other words the small scales cannot be considered as a kind of passive objects swept by the large scales or just 'slaved' to them. It is noteworthy that due to nonlocality of the relations mostly small scale vorticity and strain are, generally, creating also some large scale velocity. This and other aspects of nonlocality contradict the idea of cascade in physical space, which is local by definition . This means that the common view that *there exists a range of scales (the inertial range) in which the effects of viscosity, boundary conditions, and large-scale structures are unimportant* is basically incorrect.[6]

1.5. THE ROLE OF VISCOSITY/DIFFUSIVITY

It is the right place to stress that independence of some (statistical) parameters or properties of turbulent flows of viscosity at large Reynolds numbers does not mean that viscosity is unimportant. It means *only* that the effect of viscosity is Reynolds number independent.

One cannot switch off the viscosity experimentally and also computationally this switching off is not 'clean' enough. However, it is still possible to have a *direct* impression of the role of diffusivity by comparing the behaviour of material elements (with zero diffusivity) and a passive vector with nonzero diffusivity. This comparison reveals some qualitative differences. For example, in full accordance with intuition the material lines tend to align with the largest positive eigenvector of the rate of strain tensor, whereas passive vector with nonzero diffusivity alignes mostly with the intermediate eigenvector just like does the vorticity.

---

[6]In case of pasive objects the corresponding production terms are $-G_iG_js_{ij}$ for the gradient of passive scalar $\mathbf{G} = \nabla\theta$ and $B_iB_js_{ij}$ for a passive vector $\mathbf{B}$ (more precisely for the production of G^2 and B^2). Both appear to be positive in the mean for real NSE turbulent flows and in artificial random velocity fields, e.g. Gaussian, Tsinober and Galanti (2001). Note that in the case of passive objects the concept of cascade is even less appropriate due to the linearity of the governing equations, Tsinober (2001).

Another aspect of the importance of viscosity is seen from the qualitative differences in the overall flow properties arising from use of hyperviscosity instead of usual viscosity.
So one can claim that in the sense of concern here viscosity/diffusivity are important at all scales.

1.6. PRESSURE

The property of nonlocality of Navier-Stokes equations in physical space is due to pressure ('dynamic' nonlocality). Namely, pressure is a functional of the field of velocity derivatives $p = \Pi\{Q\}$, where $Q = \frac{1}{4}(\omega^2 - 2s_{ij}s_{ij}) = -\frac{1}{4}(\partial^2 u_i u_j/\partial x_i \partial x_j)$ is the second invariant of the velocity gradient tensor $\partial u_i/\partial x_j$. Again (it looks that) pressure too is completely determined by the field of velocity derivatives (not just the field of velocity) in a nonlocal way! The 'but' is that at least in an unbounded fluid flow[7] $p(\mathbf{x}) = -(2\pi)^{-1}\int Q(\mathbf{x})\frac{d\mathbf{y}}{|\mathbf{x}-\mathbf{y}|} = -\frac{1}{3}u^2(\mathbf{x},t)+N(\mathbf{x},t)$, where $N(\mathbf{x},t) = \frac{1}{4\pi}\int_{P.V.}\{3\frac{y_i y_j}{y^2} - \delta_{ij}\}R_{ij}(\mathbf{x}-\mathbf{y},t)\frac{d\mathbf{y}}{|\mathbf{y}|^3}$, and $R_{ij} = u_i u_j$. In other words the pressure field is *directly* defined by the field of velocity itself. However, since the latter is defined by the field of velocity derivatives, so does the pressure.
A related aspect is that the Lagrangian acceleration $a \equiv Du/Dt$ - a kind of small scale quantity – is dominated by the pressure gradient, ∇p. Hence the scaling properties of the acceleration variance do not obey Kolmogorov-like scaling (Hill, 2002) , most probably due to dominating contribution of nonlocality.
The nonlocality due to pressure is strongly associated with essentially non Lagrangian nature of pressure. For example, replacing in the Euler equations the pressure Hessian $\frac{\partial^2 p}{\partial x_i \partial x_j} \equiv \Pi_{ij}$, which is both nonlocal and non Lagrangian, by a local quantity $(1/3)\delta_{ij}\nabla^2 p = (1/6)\rho\{\omega^2 - 2s_{ij}s_{ij}\}$ turns the problem into a local and integrable one and allows to integrate the equations for the invariants of the tensor of velocity derivatives $\partial u_i/\partial x_j$ in terms of a Lagrangian system of coordinates moving with a particle. The reason for disappearance of turbulence (and formation of singularity in finite time) in such models, called restricted Euler models, is that the eigenframe of s_{ij} in these models is fixed in space, whereas in a real turbulent flow it is oriented randomly. This means that nonlocality due to presure is essential for (self-) sustaining turbulence: no pressure Hessian - no turbulence.

1.7. NONLOCALITY OF VORTICITY-STRAIN RELATION

Taking *curl* of the NSE and getting rid of the pressure does not remove the nonlocality. Indeed, the equations for vorticity and enstrophy are nonlocal in vorticity, ω, since they contain the rate of strain tensor, s_{ij}, due to nonlocal relation between vorticity, ω, and the rate of strain tensor, s_{ij} ('kinematic' nonlocality)[8]. The two aspects of nonlocality are related, but are not the same. For example, in compressible flows there is no such relatively simple relation between pressure

[7] A silmilar relation can be obtained for a bounded flow domain using corresponding Green functions.

[8] Nonlocality of the same kind is encountered in problems dealing with the behaviour of vortex filaments in an ideal fluid. Its importance is manifested in the breakdown of the so called localized induction approximation as compared with the full Bio-Savart induction law.

and velocity gradient tensor as above, but the vorticity-strain relation remains the same.

Both aspects of nonlocality are reflected in the equations for the rate of strain tensor and total strain/dissipation, $s^2 \equiv s_{ij}s_{ij}$, and the equations for the third order quantities $\omega_i\omega_j s_{ij}$ and $s_{ij}s_{jk}s_{ki}$. An important aspect is that the latter equations contain invariant (nonlocal) quantities $\omega_i\omega_j\Pi_{ij}$ and $s_{ik}s_{kj}\Pi_{ij}$ $\left(\frac{\partial^2 p}{\partial x_i \partial x_j} \equiv \Pi_{ij}\right)$, reflecting the nonlocal dynamical effects due to pressure and can be interpreted as interaction between vorticity and pressure and between vorticity and and strain. These quantities are among possible candidates preventing formation of finite time singularity in NSE.

1.8. MANIFESTATIONS OF NONLOCALITY

There exist massive evidence not only on direct interaction/coupling between large and small scales but also that this interaction is bidirectional[9] such as the example of turbulent flows in a channel. We mention also the well known effective use of fine honeycombs and screens in reducing large scale turbulence in various experimental facilities. The experimentally observed phenomenon of strong drag reduction in and change of the structure of turbulent flows of dilute polymer solutions and other drag reducing additives is another example of such a 'reacting back' effect of small scales on the large scales.[10] Finally, one can substantially increase the dissipation and the rate of mixing in a turbulent flow by *directly* exciting the small scales.

1.8.1. *Anisotropy*

One of the manifestations of direct interaction between large and small scales is the anisotropy in the small scales. Though local isotropy is believed to be one of the universal properties of high *Re* turbulent flows it appears that it is not so universal: in many situations the small scales do not forget the anisotropy of the large ones. There exist considerable evidence for this which has a long history starting somewhere in the 50-ies. Recently similar observations were made for the velocity increments and velocity derivatives in the direction of the mean shear both numerical and laboratory, see references in Tsinober 2001. It was found that the stastistical properties of velocity increments and velocity derivatives in the direction of the mean shear do not conform with and do not confirm the hypothesis of local isotropy.[11]

---

[9] Note that in case of passive objects there is no such a bidirectional relation – it is only one way.

[10] In two-dimensional (soap) turbulence this effect is observed to be even stronger, Amarouchene and Kellay (2002).

[11] So it is not surprizing that in the 'simlpe' example of turbulent channel flow the flow is neither homogeneous nor isotropic even in the proximity of the midplane, $x_2 \approx 0$, where $dU/dx_2 \approx 0$. Indeed, though $\langle u_1u_2 \rangle \approx 0$ in this region too, $d\langle u_1u_2 \rangle /dx_2$ is essentially nonzero and is *finite* independently of Reynolds number as far as the data allow to make such a claim. This is also a clear indication of nonlocality, since in the bulk of the flow, i.e. *far from the boundaries,* $dU/dx_2 \sim 0$. Therefore the assumption that in the proximity of *the centerline of the channel flow, the local isotropy assumption seem reasonable* is incorrect.

Along with other manifestations of direct interaction between large and small scales the deviations from local isotropy seem to occur due to various external constraints like boundaries, initial conditions, forcing (e.g. as in DNS), mean shear/strain, centrifugal forces (rotation), buoyancy, magnetic field, etc., which usually act as an organizing factor, favoring the formation of coherent structures of different kinds (quasi-two-dimensional, helical, hairpins, etc.). These are as a rule large scale features which depend on the particularities of a given flow and thus are not universal. These structures, especially their edges seem to be responsible for the contamination of the small scales. This 'contamination' is unavoidable even in homogeneous and isotropic turbulence, since there are many ways to produce such a flow, i.e. many ways to produce the large scales. It is the difference in the mechanisms of large scales production which 'contaminates' the small scales. Hence, nonuniversality. This brings us to the next issue.

1.8.2. *Statistical dependence of small scales on large scales.*

An important observation made recently (again see references in Tsinober 2001) is that conditional statistics of small scale quantities (e.g. velocity increments, enstrophy, total strain) conditioned on large scale quantities (velocity) is not independent of the large scale quantities as should be *if* the large scales are *not* coupled directly to the small scales. For example, there is a clear tendency that the conditional averages of the structure functions, enstrophy and total strain increase with the *energy* of fluctuations. Such a tendency, that is the direct coupling, is observed also for the smallest distance of the order of Kolmogorov scale.

1.8.3. *Helicity.*

Helicity $\int \boldsymbol{\omega} \cdot \mathbf{u} d\mathbf{x}$, and its density, $\boldsymbol{\omega} \cdot \mathbf{u}$, deserve here also special mentioning. The formal reason is that if $\langle \mathbf{u} \cdot \boldsymbol{\omega} \rangle \neq 0$, this is a clear indication of direct coupling of large and small scales. So it is not surprising that in flows with nonzero mean helicity the direct coupling between small and large scales is stronger than otherwise. The stronger coupling between the large and small scales in flows with nonzero mean helicity $\langle \mathbf{u} \cdot \boldsymbol{\omega} \rangle$ aids creation of large scale structures out small scale turbuence, as it happens in problems of turbulent dynamo of magnetic fields. As mentioned this does not mean that in case $\langle \mathbf{u} \cdot \boldsymbol{\omega} \rangle = 0$, or even $\mathbf{u} \cdot \boldsymbol{\omega} = 0$ as in two-dimensional flows, such a coupling does not exist.

1.8.4. *Closures and constitutive relations. Memory effects.*

The nonlocality due to the coupling between large and small scales is also manifested (and is a concern) in problems related to various decompositions of turbulent flows and in the so called closure problem. For example, in the Reynolds decomposition of the flow field into the mean and the fluctuations and in similar decompositions associated with large eddy simulations (LES) the relation between the fluctuations and the mean flow (or resolved and unresolved scales in LES, etc.) is a nonlinear *functional.* That is the field of fluctuations at each time/space point depends on the mean (resolved) field in the whole time/space domain. Vice versa the mean (resolved) flow at each time/space point depends on the field of fluctuations (unresolved scales) in the whole time/space domain. This means that in turbulent flows point-wise flow independent 'constitutive' relations analogous to real material constitutive relations for fluids (such as stress/strain relations) can not exist, though the 'eddy viscosity' and 'eddy diffusivity' are used frequently as a crude approximation for taking into account the reaction back of fluctuations (unresolved scales) on the mean flows (resolved scales). The fact that the 'eddy viscosity' and 'eddy diffusivity' are flow (and space/time) dependent is just another expression of the strong coupling between the large and the small scales. The simplest version of this approach with a scalar eddy viscosity leads always to a positive subgrid dissipation (positive energy flux from the resolved to the unresolved scales), whereas *a priory* tests of data from real flows (experiments and DNS) show that there exist considerable regions in the flow with *negative* subgrid-scale dissipation (called backscatter). The exchange of 'information' between the resolved and unresolved scales is pretty reach and is not limited by energy. For example, Bos et al., 2002 (see also references therein) report that the subgrid-scales have a variety of significant effects on the evolution of field of filtered velocity gradients. So it is too optimistic to claim, for examle, that LES *of wall bounded flows ... resolve all the important eddies... has received increased attention, in recent years, as a tool to study the physics of turbulence in flows at higher Reynolds number, or in more complex geometries, than DNS* (Piomelli and Balaras 2002). The qualification large scale (resolved) eddies as the most important ones is too subjective: all eddies are important in view of direct and bidirectional coupling of essentially all eddies. It is doubtful that LES or any other similar approach can be used *as a tool to study the physics of turbulence,* since a vitally important part of physics of turbulence resides in the *unresolved* scales.

1.8.5. *Other related aspects.*

The nonlocality in the sense of concern here is especially strongly manifested in the atmospheric convective boundary layers in which the common downgradient approximation is not satisfactory due to countergradient heat fluxes, Zilitinkevich et al. 1998. We mention also a similar phenomenon in stably stratified turbulent flows - the so called PCG, persistent countregradient fluxes. The essence of PCG is the countergradient transport of momentum and active scalar. It is observed at large scales when stratification is strong, but in small scales it is present with weak stratification as well. Finally, there is a class of flows with the so-called phenomenon of 'negative eddy viscosity', see references in Tsinober (2001). It occurs in the presence of some energy supply other than the mean velocity gradient. In such flows the turbulent transport of momentum occurs against the mean velocity gradient, i.e. from regions with low momentum to regions with high momentum. In other words the Reynolds stresses as one of the agents of coupling the fluctuations with the mean flow act in such flows in the 'opposite' direction as compared to the usual turbulent shear flows. Concomitantly kinetic energy moves in the 'opposite' direction too - from fluctuations to the mean flow.[12] The flows with

[12]There are examples of 'usual' turbulent flows with turbulence induced mean flows. The best known ones are flows in pipes with noncircular crossection. In such flows a mean secondary flow is induced which is absent in purely laminar flow. For example, see Patterson Reif and

negative eddy viscosity are akin to nonturbulent but nonstationary flows in a fluid dominated by its fluctuating components and known (since Rayleigh 1883) under the name (acoustic) steady streaming, Riley, 2001, in the sense that in these flows a mean (time averaged) flow is induced and driven by the fluctuations.
Finally we mention that the far field statistical properties of free shear turbulent flows (mixing layers, wakes, jets) are known to possess strong memory ('nonlocality in time'): they are sensitive to the conditions at their 'start' with some properties not universal in Reynolds number, Bevilaqua and Lykoudis (1978), Dimotakis (2001). These flows develop in space beginning with small scales into the large ones, in apparent contradiction to the Richadrson-Kolmogorov cascade ideas. It is noteworthy that passive tracers in such flows possess even stronger memory, Cimbala et al. (1988). The problem of predictability of turbulent flows involves nonlocality in time: a small scale perturbation (both in time and space) perturb substantially the whole flow including the largest scales within time of the order of integfral time scale. In this sense instability can be seen as nonlocality in time.

2. Particulate flows

The basic interaction of the carrier fluid flow with particulates occurs at the scale of the particle size, i.e. at small scales. However, a number of essential phenomena emerge at much larger scales in a variety of particualate flows: sedimenting suspensions, fluidized beds, formation of bedforms and their interaction with the carrier fluid, preferential concentration of particles/bubles in and modification of turbulent flows, Bernard-Michel et al. (2002), Druzhinin and Elgobashi (2001), Gyr and Kinzelbach (2002), Helland et al. (2002), Ljus et al. (2002), Maxey et al. (1996). These phenomena are treated in terms of large scale instabilities, intrinsic convection in sedimenting suspensions, collective phenomena, long-range multibody hydrodynamic interactions/correlations, clusters.
All these are essentially fluid mediated phenomena/interactions as contrasted with direct particle/particle interactions. Therefore, nonlocality of fluid flows is expected to be significant in these phenomena. For example, an important process in the interaction of the carrier fluid flow with particles (or any other additives) is the production (or more generally modification) of velocity derivatives, i.e. vorticity and strain. The modified field of velocity derivatives reacts back in changing the large scales of the flow (both velocity and pressure). It is tempting to see this process as the one underlying the formation of the mentioned large scale features, though the details in each case are different and in most cases are poorly understood. These processes are modified by specific features such as inertial bias, i.e. inertial response of particles to fluid accelerations and preferential concentration of particles(bubbles) in strain (vorticity) dominated regions. The latter may lead to enhanced bias of strain dominated regions (heavy particles), i.e. regions with large dissipation, or regions with strong enstrophy (bubbles). There are some indications that under certain conditions even in the case in which the fluid and particle density match exactly the particle inertia effects (e.g. due

---

Andersson (2002) for references on such flows in a square duct.

to their finite size) can produce inhomogeneity in particle distribution associated with strain dominated regions, see Cartwright et al. (2002) and references therein.

2.1 NONLOCALITY IN TURBULENCE AND NONLOCAL PHENOMENA IN PARTICULATE FLOWS.

The formation of large scale structures in particulates (uneven particulate distribution in turbulent flows, formation of bedforms) is frequently attributed to the action of and the inreraction with the so called coherent structures in the carrier fluid. This is reasonably clear in free shear flows like mixing layers, and to some extent in wall bounded flows, since both kinds of flows are dominated by large structures. It is still a problem with bed-forms, because the fluid structures (though large) are random in space/ time, so one has to assume a kind of 'locking' mechanism. In quasi-homogeneous and/or quasi-isotropic flows and especially (artificial) structureless flows (such as e.g. Gaussian velocity fields) the situation is even more problematic.

However, large scale features form also in situations where the carrier fluid without particles is just in equilibrium. The well known example is sedimenting suspensions in which under certain conditions the nonlocal processes (or large scale instabilities) lead to formation of large regions with excessive particle content and others regions devoid of particles, Asmolov 2002, Druzhinin 1997. The former descent [13], whereas the latter rise through suspension both leading to large scale convective flow which is called intrinsic convection and which is (most probably) weakly turbulent. The resulting flow has a) long-range particle velocity correlations exceeding 20 interparticulate distances, which at small volume fractions can become larger than half of the cell width, and b) very large distance of influence of the boundaries, which is increasing with decreasing volume fractions and is anisotropic in the sense that horizontal velocity fluctuations are influenced by the wall stronger than the vertical ones.[14]

The next example, is formation of large scale structures in vibrated fluids both vertically (clusters), Voth et al. 2002 and horizontally (ripples, dunes), Andersen et al. 2002. Voth et al. observed short-range repulsion and long-range attraction (as fluid-mediated interactions) between particles, and proposed an explanation based on the phenomenon of mentioned above steady streaming, Riley 2001. It is likely that this approach may be useful (at least qualiatively and/or in part) in more complicated situations such as formation of bedforms.[15]

---

[13]The velocity u_b of such an individual blob at constant volume fraction surrounded by and infinite fluid is increasing with the 'blob' size as $u_b \sim u_s \phi R^2$ (here u_s is the Stokes velocity, ϕ is the local volume fraction, and R the size of the blob). Indeed, theoretical and numerical results show that the variance of the velocity of particles is increasing with the increase of the system size - a clear indication of nonlocality. Experimentally, there is no such a divergence due to the nonlocal effect of side walls. It is noteworthy that $u_b \sim \phi$ in a quasi-homogeneous and isotropic turbulence for particles with Stokes numbers up to ~ 1 and local concentrations up to five time the mean, Aliseda et al., 2002.

[14]For a similar phenomenon of the influence of boundaries on 'regular' turbulent convection see Daya and Ecke 2001.

[15]Note that Papanicolaou, 2000 find the best correlation of the start of the lift of sediments

2.2 WHAT ARE THE SCALES?

Turbulent flows with particles are known to exhibit preferential concentration of, e.g. heavier particles in convergent zones dominated by large strain and regions void of particles dominated by large enstrophy, Druzhinin 2001, As mentioned one of the mechanisms for clustering is the inertial bias, Elperin et al. 1996, 2002, Rouson and Eaton 2001, Maxey et al. 1996 and references therein[16]. Similarly bubbles accumulate preferentially in vorticity dominated regions, Druzhinin and Elgobashi 2001. However, the scales of the 'structures' seen at the snapshots in various particilate flows (Aliseda et al. 2002, Bernard-Michel et al. 2002, Fallon and Rogers 2002, Fessler et al. 1994, Helland et al. 2002, Maxey et al. 1996) do not look small and (at least some of them) are comparable with the integral scale of turbulence.[17]. On the other hand, the existing methods of quantifying the preferential concentration (Hogan and Cuzzi 2001, Kostinski and Shaw 2001, Wang et al. 2000) reveal that the maximum deviation from random distibution (i.e. clustering) occurrs in the scales of the order of ten Komlogorov scales. However, this maximum is not steep, so that the effect is considerable at much larger scales. For example, one can see from figure 3b in Kostinski and Shaw 2001 in which the maximum occurrs at scales $\sim$ 1cm, that the effect is considerable at 1 m and is seen even up to 10 m (see also figure 5 in Rouson and Eaton 2001, in which significant effect is seen at scales of half-width channel). This is why just looking at the snapshots one 'sees' the larger structures, which are one of the manifestations of the nonlocal effects. Another reason is that the particle field is a cumulative result of its (Lagrangian) time history, i.e. nonlocality in time.

2.3. DO STRUCTURES/CLUSTERS EMERGE IN THE LIMIT OF ZERO STOKES NUMBERS?[18]

It is natural to ask whether there is clustering which is caused not by inertial bias. One of the common answers is negative, e.g. Rouson and Eaton 2001: *Low Stokes number particles act as flow tracers. Their mean spacing is fixed by the continuity constraint, so they are precluded from clustering near a point.* However, this is not easy to reconcile with *i*) - the basic rigorously proven result that the distance between two fluid particles is increasing in the mean and *ii*) the behaviour of passive scalars, which exhibit strong intermittency at all known Prandtl/Scmidt numbers. Both *i*) and *ii*) indicate the opposite.[19] For example, a finite number of marked and evenly distributed fluid particles at some time moment will form clusters at later times even in random Gaussian velocity field.

with the instantaneous (not fluctuative) streamwise velocity.

[16]It is noteworthy that this mechanism can be interpreted as excitation of small-scale instability due to particles inertia and compressibility of particles velocity field, Elperin et al. 1996, 2002.

[17]The preferential concentration of heavy particles is thought to be responsible for the substantial increase of the average settling velocity of heavy particles in homogeneous turbulence for particle with Stokes number based on Komlogorov scales $St \sim 1$, observed both in the DNS, Wang and Maxey 1993, and in laboratory, Aliseda et al. 2002.

[18]That is for small neutral particles or simply fluid particles.

[19]Note that in both cases there is no need for real turbulent flow: any random, e.g. Gaussian velocity field produces these effects as well, though somewhat different quantitatively.

There are several reasons, which cause this even distribution (a fixed point in the phase space) to become unstable. First, at least under certain conditions such particles can be regarged as a diffusionless passive scalar. Second, closely related to the first, is the Lagrangian chaos. Finally, both purely kinematic considerations and preliminary DNS of NSE with Lagrangian particle tracking indicate that indeed particles initially located at the grid points form clusters at later times. It seems that clustering (of a finite large enough number of fluid particles) is an intrinsinc property of all Lagrangian chaotic flows, i.e. L-turbulent flows which are not necessarily Euler turbulent and can be pretty simple. For this reason, clustering occurs in a 'purely random' distribution of particles as well even when the particle density match exactly that of the fluid.
Definitely there is more than one reason for clustering just like there are several kinds of clustering all of which exhibit some aspect of nonlocality.

3. References

Aliseda, A., Cartellier, A., Hainaux, F. and Lahseras, J.C., The effect of preferential concentration on the settling velocity of heavy partricles in homogeneous isotropic trubulence, *J. Fluid Mech.,* **468**, (2002), 77-105.

Amarouchene, Y. and Kellay, H., Polymers in 2D turbulence: suppression of large scale fluctuations, *Phys. Rev. Lett.,* **89**, (2002), 104502, 1-4.

Andersen, K.H., Abel, M., Krug, J., Ellegaard, C., Sondergaard, L.R. and Udesen, J., Pattern dynamics of vortex ripples in sand: nonlinear modelling and experimental validation, *Phys. Rev. Lett.,* **88**, (2002), 234301, 1-4.

Asmolov, E.S., 2002 Numerical simulation of the coherent structures in a homogeneous sedimeting suspension, this issue.

Bernard-Michel, G., Monavon, A., Lhuiller, D., Abdo, D. and Simon, H., Particle velocity fluctuations and correlation lenghts in dilute sedimenting suspensions, *Phys. Fluids,* **14**, (2002), 2339–2349.

Bevilaqua, P.M and Lykoudis, P.S., Turbulence memory in self-preserving wakes, *J. Fluid. Mech.*, **89**, (1978), 589–606.

Bos van der, F., Tao, B., Meneveau, C. and Katz, J., Effects of small-scale turbulent motions on the filtered velocity gradieent tensor as deduced from holographic particle image velocimetry, *Phys. Fluids.,* **14** , (2002), 2456-2474.

Cartwright, J.H.E, Magnasco, M.O. and Piro, O., Noise- and inertia induced inhomogeneity in the distribution of small particles, *Chaos*, **12**, (2002), 489–495.

Cimbala, J.M., Nagib, H. M and Roshko, A., Large structures in the far wakes of two-dimensional bluff bodies, *J. Fluid Mech.*, **190**, (1988), 265–298.

Daya, Z.A. and Ecke, R.E., Does turbulent Convection feel the shape of the container, *Phys. Rev. Lett.,* **87**, (2001), 184501–1-4.

Dimotakis, P.E. Recent developments in turbulent mixing, in H., Aref, and J.W. Phillips, editors, *Mechanics for new millenium,* (2001), Kluwer. pp. 327-344.

Druzhinin, O.A., The dynamics of concentration inrerface in a dilute suspension of solid heavy particles, *Phys. Fluids.,* **9**, (1997), 315-324.

Druzhinin, O.A., The influence of inertia on the two-way coupling and modification of isotropic turbulence by microparticles, *Phys. Fluids.,* **13**, (2001), 3738-3755.

Druzhinin, O.A. and Elgobashi, S.E., Dircet numerical simulation of a three-dimensional spatially developing bubble-laden mixing layer wih two-way coupling, *J. Fluid. Mech.*, **429**, (2001), 23 - 61.

Elperin, T., Kleeorin, N. and Rogachevskii, I., Self-excitation of Fluctuations of Inertial Particles Concentration in Turbulent Fluid Flow, *Phys. Rev. Lett.* **77**, (1996), 5373-5376.

Elperin, T., Kleeorin, N., Rogachevskii, I., L'vov, V. and Sokoloff, D., Clustering instability of inertial particles spatial distribution in turbulent flows, *Physical Review,* **E**66, 036302–1-16, (2002).

Fallon, T. and Rogers, C.B., Turbulence-induced preferential concentration of solid particles in microgravity conditions, *Experiments in Fluids,* **33**, (2002), 233-241.

Fessler, J.R., Kulick, J.D. and Eaton, J.K., Preferential concentration of heavy particles in a trubulent channel flow, *Phys. Fluids.,* **6**, (1994), 3742-3749.

Gyr, A. and Kinzelbach, W., Bed Forms in Turbulent Channel Flow, *Appl. Mech. Revs.,* (2002), sub judice.

Helland, E., Occelli, R. and Tadrist, L., Computational study of fluctuating motions and cluster structures is gas-particle flows, *Int. J. Multiphase Flow,* **28**, (2002), 199-223.

Hill, R.,J., Scaling of acceleration in locally isotropic turbulence, *J. Fluid. Mech.*, **452**, (2002), 361 - 370.

Hogan, R.C. and Cuzzi, J.N., Stokes and Reynolds number dependence of preferential particle concentration in simulated three-dimensional turbulence, *Phys. Fluids.,* **13**, (2001), 2938-2945.

Hunt, J.C.R., Kaimal, J.C. and Gaynor, J.E., Eddy structure in the convective boundary layer-new measurements and new concepts, *Q. J. R. Met. Soc.,* **114**, (1988), 827-858.

Kostinski, A.B. and Shaw, R.A., Scale-dependent droplet clustering in turbulent clouds, *J. Fluid Mech.*, **434**, (2001), 389 - 398.

Ljus, C., Johansson, B. and Almstedt, A.-E., Turbulence modification by particles in a horizontal pipe flow, *Int. J. Multiphase Flow*, **28**, (2002), 1075–1090.

Maxey, M., Chang, E.J. and Wang, L.-P., Interaction of particles and microbubbles with turbulence, *Experimental Thermal and Fluid Sci.*, **12**, (1996), 417–25.

Papanicolaou, A.N., The role of near-bed trubulence in the inception of particle motion, *Int. J. Fluid Dyn.,* **4**, (2000), art. 2, 1-15.

Petersson Reif, B.A. and Andersson, H.I., Prediction of turbulence-generated secondary flow in a square duct, *Flow, Turbulence and Combusion,* **68**, (2002), 41-61.

Piomelli, U. and Balaras, E., Wall-layer models for large-eddy simulations, *Annu. Rev. Fluid Mech.*, **34**, (2002), 349-374.

Riley, N., Steady streaming, *Annu Rev. Fluid Mech.,* **33,** (2001), 43-65.

Rouson, D.W.I. and Eaton, J.K. On the preferential concentration of solid particles in turbulent channel flow, *J. Fluid Mech.*, **428**, (2001), 149–169.

Tsinober, A., *An informal introduction to turbulence,* Kluwer, (2001) .

Voth, G., Bigger, B., Buckley, M.R., Kosert, W., Brenner, M.P., Stone, H.A. and Gollub, J.P., Ordered clusters and dynamical states of particles in vibrated fluid, *Phys. Rev. Lett.,* **88**, (2002), 234301, 1-4.

Wang, L.-P., Wexler, A.S. and Zhou, Y., Statistical mechanical description and modelling of turbulent collision of inertial particles, *J. Fluid Mech.*, **415**, (2000), 117–153.

Zilitinkevich, S., Grachev and Hunt J.C.R., Surface frictional processes and non-local heat/mass transfer in the shear-free convective boundary layer, in Plate, E.J., Fedorovich, E.B., Viega, D.X. and Wyngaard, J.C., editors *Buoyant convection in geophysical flows*, Kluwer, (1998), pp. 83-114.

EFFECT OF A ROUGHNESS TRANSITION ON TURBULENT STRUCTURES IN THE OUTER LAYER

GUALA[1] M., CHRISTENSEN[2] K.T., ADRIAN[3] R.J.
[1] *Dipartimento di Ingegneria Ambientale, Università di Genova, via Montallegro 1, Genova, Italy, micche@diam.unige.it*
[2] *Department of Mechanical Engineering, The University of New Mexico, Albuquerque, NM 87131, christensen@me.unm.edu*
[3] *Department of Theoretical and Applied Mechanics, University of Illinois at Urbana Champaign, Urbana, IL 61801, rja@tam.uiuc.edu*

1. Introduction

The physical mechanisms that need to be involved in sediment transport modelling are indissolubly associated to the evolution of coherent structures in a rough boundary layer. In literature many attempt have been found to relate burst frequency and intensity with sediment entrainment or saltation phenomena. Several questions remained partially unanswered concerning the feed back effect of sediment particles on hydrodynamics, but even a deep knowledge of roughness effects on coherent structures has to be involved. The question that may still rise is related to the *undisturbed* spatial scales of the structures responsible for sediment entrainment. Those structures are the very same structures triggered by the vast regime of roughness elements (from sand grains to cobbles or even stones) lying in the bottom of a river. Turbulent flows in a smooth wall boundary layer have been studied extensively over the past decades. Since the early work of Kline (1967), experimental investigations on the detailed structure of the turbulent field in boundary layers have revealed the existence of distinctly organized motions. The present research and the recent investigation of Tomkins (2001) are focused mainly on the effect of a roughness transition over turbulent structures, with the purpose of shedding more light on the variation that may occur on bursting intensity, pattern of dominant structures and representative length scales. In literature, Grass (1993) outlined intense ejection event able to disrupt vortices organization with the consequent reduction of the streamwise integral length scale. Raupach (1991), through quadrant analysis techniques, found an enhanced intensity and frequency of Q_4 events towards the wall, with a higher contribute to Reynolds stress compared to ejections, in the rough wall case. Bandyopadyay & Watson (1988) supported this conclusion inferring that the increased intensity of sweeps event may be associated with the decrease of outward flux caused by the desegregation of hairpin packets and the increase of wall ward flux associated with the very local generation of necklace vortices around the elements. Grass (1993), provide a convincing evidence of the existence of hairpin vortices in rough wall turbulence. Krogstad & Antonia (1994) investigated turbulence statistics, confirming the disruptive effect of roughness element over vortex organization through the decrease of the two point correlation of streamwise velocity and the increased inclination (up to 38° compared to the typical values of smooth wall boundary layers of $10^o \div 15^o$) of the latter contour, believed to represent the averaged *signature* of the random passage of multiple hairpin type vortices. Question that may rise in literature still being a matter of debate, are then related to the effect of the local roughness on the spatial evolution of coherent structures: how they change through a

A. Gyr and W. Kinzelbach (eds.), Sedimentation and Sediment Transport, 23–28.

roughness transition, how they interact, whether they persist downstream a sequence of elements. The techniques that need to be applied have to outline the main characteristics of the turbulent structures without losing too much information in the averaging processes and will be described in the following paragraphs.

2. Experimental set up and PIV setting

In this section we describe the experimental facility that has been used in the present investigations. Experiments have been performed in a wind tunnel Eiffel type, with a working test-section 457 mm high and 914 mm wide. The incoming boundary layer develops over a smooth surface for approximately 45δ (having denoted by δ the boundary layer thickness at the measurements location) before encountering a full array of hemispheres (h^+ =100, h/δ =0.045) mounted on the pre-existent wall and distributed with low roughness density (figure 1) . T label indicates the streamwise location of previous measurement by Tomkins, while A and B, *center* and *lateral*, indicate the image plane location used in the present investigation. PIV measurements in the streamwise wall-normal plane have been performed at several spanwise locations in the smooth wall case and in the rough case. Flow characteristics are presented in table 1.
PIV measurements have been performed by illuminating 0.5 ÷ 2.0 μm diameter oil particles wit a 0.25mm thick light sheet produced by two digitally timed Nd:YAG lasers (Continuum lasers). Digital double-pulsed images are recorded using a 1k by 1k CCD camera ("PIV-CAM 10-30" by TSI, Inc.) focused on the light sheet with a 0.077 magnification, viewing from one side of the wind tunnel. The camera lens was a 50mm focal length. The local particle displacement was determined using the cross correlation PIV Sleugh software (Christensen et al., 2000) over 500 image pairs per spatial location. PIV resolution is 1mm with a field of view of roughly 1.1 δ.

Table 1	Smooth wall	Rough wall
Re_δ	62493	64543
U_∞ [m/s]	11.39	11.39
δ [mm]	82.3	85.0
u_τ [m/s]	0.40	0.44

Tab. 1: Flow parameters

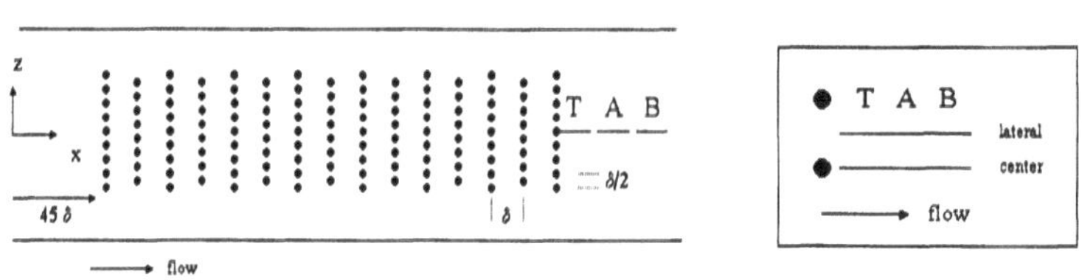

Fig. 1: channel top view

3. Evolution of turbulent structures over the elements: results and discussion

3.1 TWO POINT SPATIAL CORRELATION

The spatial correlation coefficient of the streamwise components has been computed at each spanwise and streamwise location. The two point spatial correlation coefficient has been estimated using fluctuating discrete data obtained subtracting from the measured two dimensional flow fields, the averaged mean velocity profile and assuming in the computation of $\rho_{uu}(r_x,y,y_{ref})$ spatial homogeneity only in the streamwise direction. A first significant experimental evidence is the typical elongated shape of the $\rho_{uu}(r_x,y,y_{ref})$ contour which reveal the imprint of organized motion, in the transition from rough to smooth surface. By comparing result in A and B *center* locations it's evident the shortening of streamwise length-scale as we approach the elements, for $y^+_{ref} = 100 \div 200$ (in fig. 2).

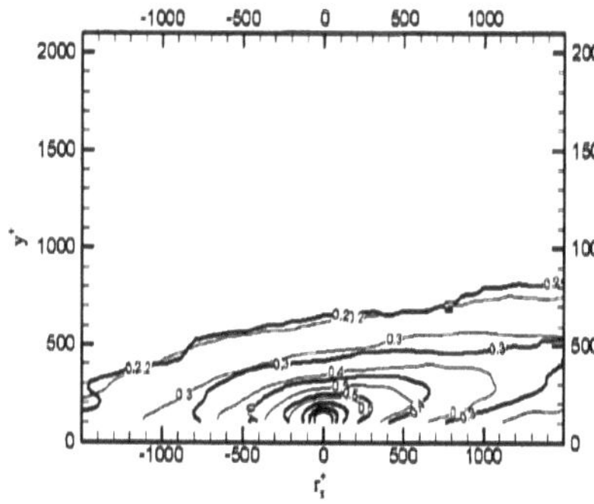

Fig.2: $\rho_{uu}(r_x,y,y_{ref})$ contour **A**, B *center*

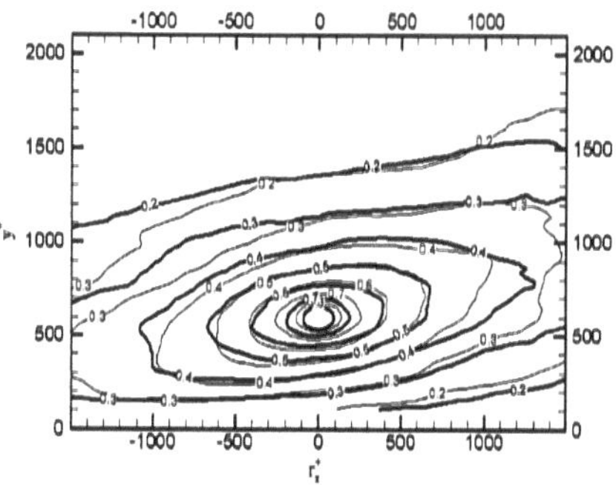

Fig.3: $\rho_{uu}(r_x,y,y_{ref})$ contour **A**, B *center*

This trend appears to be smoothed for $y^+_{ref} = 400$ and even modified for $y^+_{ref} = 600$, where the ρ_{uu} contour appear slightly more compact in the streamwise direction in B rather than in A. It suggests that a lift up mechanism occur in the spatial evolution of the vortices triggered by the elements from A towards B. This results confirm the observations of Tomkins (2001) and Krogstad & Antonia (1999) about the disruptive effect of the roughness elements on the typical hairpin packets found in smooth wall turbulence. As new structures are triggered from the elements, new scales are involved and the average imprints on the statistics denote a loss of coherence of the turbulent structures, even 3 boundary layers downstream the last arrays of hemispheres. Since the two point correlation maintain the shape observed in smooth wall turbulence, outlining the existence of hairpin packets, a question may rise about the nature of the structure triggered by the elements and responsible for the loss of coherence of the packets. Tomkins observed in several instantaneous velocity fields the presence of big vortices, that scale with the roughness elements, and have been identified as hairpin vortices, since they were characterized by the same key features (Christensen & Adrian, 2001). The interaction of these hairpin like structures, periodically triggered from the elements, with the hairpin packets that are reasonably assumed to survive, given the finite length of the matrix of hemispheres, is then believed to be responsible for the variation of $\rho_{uu}(r_x,y,y_{ref})$. A limit case of this trend can be inferred: The total disruption of the typical packet, i.e. the definitive loss of coherence, would force $\rho_{uu}(r_x,y,y_{ref})$ contour to assume the typical round shape of the $\rho_{vv}(r_x,y,y_{ref})$ contour, according to which no significant correlation exist in the streamwise direction. Since the shortening of the length-scale varies with y_{ref}, it's possible to identify where is concentrated in the flow field the interaction between the structures triggered by the elements and the packets. A possible interpretation of the change of behaviour in the B location for $y^+_{ref} > 400$, may be given by assuming the that a lift up mechanism similar to the one observed in the packets and responsible for the ramp like pattern, occur also for the structures triggered by the elements. The imprints of these structures in A at $y^+_{ref} = 100 \div 200$ is then found in B for $y^+_{ref} = 600$. Examining results obtained at different spanwise location may shed more light on the latter assumption of spatial evolution. For a given streamwise location (A or B) we compare the one dimensional spatial correlations: $\rho_{uu}(r_x, y_{ref})$ in the two spanwise location (*center* and *lateral*) .

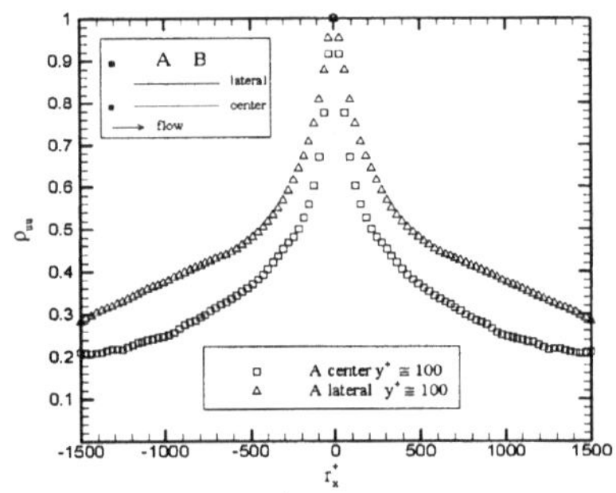

Fig. 4a: A, $y_{ref}^+ = 100$

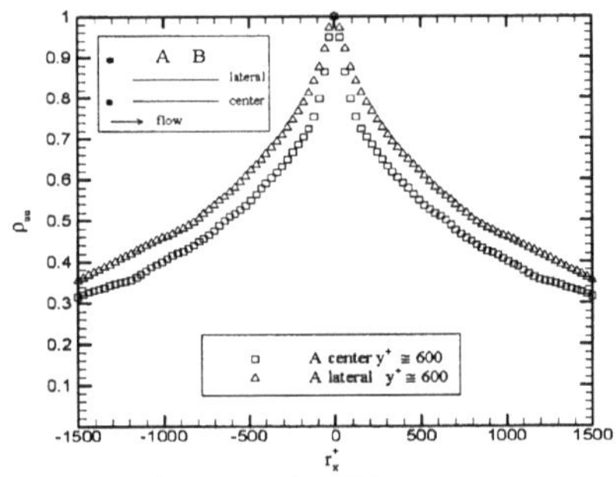

Fig. 4b: A, $y_{ref}^+ = 600$

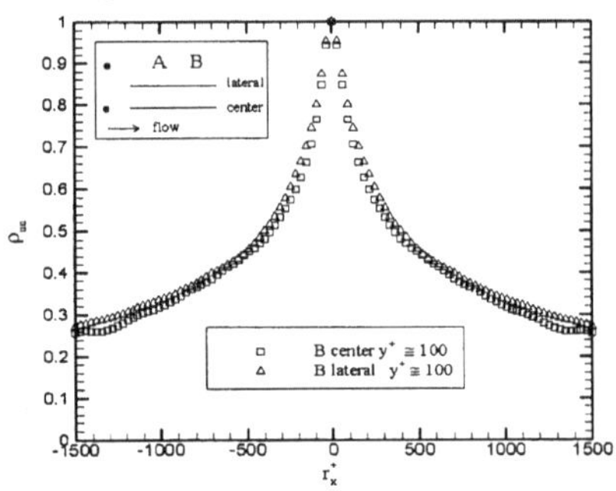

Fig. 4c: B $y_{ref}^+ = 100$

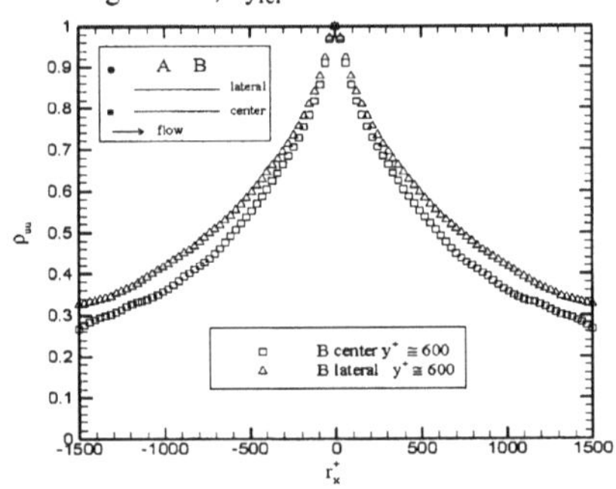

Fig. 4d: B $y_{ref}^+ = 600$

It has to be noted that, according to the regular matrix formed by the elements, the distance between the image plane and the element located at the same spanwise location is the same for *A lateral* and *B center*. Since previous results confirmed the dominant role played by those structures triggered by the hemispheres, we expect to find similar imprints on the statistics. Varying distance from the wall two distinct behaviours are outlined. Close to the roughness, (A) we note a significant difference in the length scales between *center* and *lateral*, that exhibit a maximum close to the wall $y^+_{ref} = 100$ (figure 4a) and attenuate moving further towards the edge of the boundary layer (figure 4b). Further downstream (B), an opposite trend is evidenced: the one-dimensional correlation curves overlap at $y^+_{ref} = 100$ for the two spanwise location (figure 4c). Increasing distance from the wall, the two curves start deviating from each other (figure 4d) up to $y^+_{ref} = 600$, giving rise to the idea that further from the wall vortices tend to be more disorganized, randomly distributed and less frequent. The outlined trend is complete agreement with the precedent idea of organized structures, triggered by the central bump, interacting with the local turbulence, evolving in the streamwise direction and lifting up. The imprints of those modified packets is outlined on the statistics by a difference in the one dimensional correlation of the streamwise fluctuation $\rho_{uu}(r_x,y_{ref})$: the packets is evidenced by a shortened length scale in the A location close to the wall $y^+_{ref} = 100$). As the structures are evolving downstream (B), that difference is found to be located further from the wall. A lift up mechanism is then found to interest those structure as well as typical packets. Another phenomenon has to be noted. Measurement in B location, far from the elements, and close to the wall $y^+_{ref} < 200$ (figure 4c) are not affected any more by varying the spanwise location. This may indicate that a complete merging of the structures triggered by two adjacent elements occur, but in that case a shortening of characteristic length scales should characterized the disorganization and the loss of coherence typical of a interference. Another interpretation is possible: Since the diagonal zone where the structures triggered by the last rows of elements have been identified as a ramp like vortex packets, close to the wall in A and further from the wall in B, a portion of the flow field in B close to the wall may be governed by the local, smooth, regime. In this case we would expect that the imprints of these new structures on the statistics were only a function of the wall normal variable y^+, with no influence between different spanwise locations. Assuming that these structures are typical of smooth turbulence, it has to be noted how surprisingly fast turbulent structures reach a *mature* stage, typical of the fully developed smooth wall regime. This conclusion may require other investigations concerning the formation of these structures: residual turbulence grown upstream the matrix of hemispheres or new structures originated at the smooth wall, downstream the elements. In the latter case a strong instability mechanism has to be assumed, probably due to the spatially persistent evolution of interacting hairpin packets above the smooth wall, able to enhance the classic instabilities that would usually require, at a significantly longer distance, to reach a fully developed regime. Both the two arguments concord to the idea that, besides a short transition zone that strongly depend on the inclination of the ramp like packets, structures in wall turbulence, are controlled by the local roughness.

3.2 STATISTICAL EVIDENCE OF HAIRPIN VORTEX PACKETS

Previous assumptions, concerning the influence of hairpin vortex packets or ramp like vortex on the statistics, require a statistical analysis providing a measure of the effective persistence of these structures in the flow field. Instantaneous realizations indicate some typical features of the vortex organization in a turbulent boundary layer, already mentioned in literature, but not sufficient to determine the patterns that statistically dominate the rich population of events that characterize the flow field. A possible way to obtain in a statistical analysis some information about recurrent patterns of turbulent structures in the flow field is to estimate the average velocity field conditioned to the realization of a given event. Following Christensen & Adrian (2001), we chose as event the RMS value of the swirling strength λ_{ci} . at a given distance from the wall. This vortex marker, defined as the imaginary portion of the complex eigenvalue of the tensor of the two dimensional local velocity gradient, is able to recognize a rotation, or swirling motion, from a shear layer, and, therefore, is more indicated than vorticity to identify unambiguously vortex cores. Basically, we look for the flow field statistically associated to the presence of a vortex core at a given distance from the wall. Stochastic estimation (Adrian, 1989) provide the best linear estimate, in a mean square sense, of conditional averages. The resulting two-dimensional flow field is represented by unitary vectors, for normalization purpose. This procedure consents to obtain information only about persistent pattern of turbulent structures with no distinction on their magnitude. An example is presented in figure 5: The velocity fields are conditioned to an event relatively close to the wall at approximately $y^{+}_{ref} = 400$. The event has been statistically associated to a shear layer and one or more vortex cores (small circles), inclined with respect to the wall at a certain angle, located downstream the given event, indicated by a big circle. As it was found for smooth wall turbulence (Christensen & Adrian 2001), emerges clearly that statistically, vortex cores, or hairpin signatures, are organized in packets that populate the outer layer such densely to give a clear imprints on the statistics, also with elements of roughness. How they differ from packets observed in a smooth boundary layer appears varying streamwise location (from E1, E2 towards A, B) and noting that within the elements the shear layer is found to have a characteristic inclination of roughly $20^{\circ} + 30^{\circ}$ while evolving downstream, tends to an inclination of $10^{\circ} + 15^{\circ}$ characteristic of smooth wall turbulence.

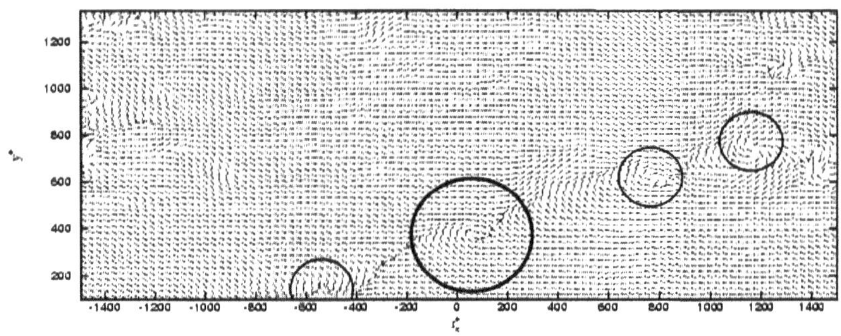

Fig. 5: flow field statistically conditioned to a swirling event (big circle)

References

Adrian R.J., Jones, B.G., Approximation of turbulent conditional averages by stochastic estimation. *Phys. Fluids*, vol. A 1(6): 992-998, 1989.

Bandyopadhyay P.R., Watson R.D., Structure of rough wall turbulent boundary layers. *Phys. Fluids*, vol. 31: 1877-1883, 1988.

Christensen K.T., Adrian, R.J., Statistical evidence of hairpin vortex packets in wall turbulence. *J. Fluid Mech*, vol 431: 433-443, 2000.

Christensen K.T., Soloff S.M., Adrian, R.J., PIV Sleugh: integrated particle image velocimetry (piv) interrogation/validation software. *TAM report, No 943, Department of Theoretical and Applied Mechanics, University of Illinois at Urbana Champaign.* (UIUC) , *2000.*

Grass A.J., Stewart R.J., Mansour-Tehrani M., Common vortical structure of turbulent flows over smooth and rough boundaries. *AIAA Journal*, vol 31(5) : 837-847, 1993.

Krogstad P.A., Antonia R.A., Rough wall turbulent boundary layers, *J Fluid. Mech.*, vol. 277: 1-21, 1994.

Krogstad P.A., Antonia R.A., Surface roughness effects in turbulent boundary layers, *Exp. in fluids*, vol. 27: 450-460, 1994.

Raupach M.R., Antonia R.A., Rajagopalan S., Rough wall turbulent boundary layers *Appl. Mech. Rev.*, vol. 44: 1-25, 1991.

Tomkins, C.D., Structure of turbulence over smooth and rough walls. PhD thesis. TAM 2001

Zhou, J., Adrian, R. J., Balachandar, S. Kendall, T. M, Mechanism for generation of coherent packets of hairpin vortices in channel flow. *J. Fluid. Mech.* , vol. 387, 353-359, 1999.

STOKES FLOWS AT INFINITE RAYLEIGH NUMBER

DAVID G. DRITSCHEL
School of Mathematics and Statistics, University of St Andrews, St Andrews, UK,
dgd@mcs.st-and.ac.uk

1. Introduction

A novel approach is presented for studying viscous Stokes flows at infinite Rayleigh numbers Ra. The flow is driven by a density variations acted upon by gravity, and infinite Ra corresponds to zero density diffusion, i.e. its material conservation. On the other hand, the fluid Reynolds number Re is negligible, allowing one to drop the acceleration in the momentum equations.

Previous work has considered finite Ra Stokes flows, and in particular flows in a narrow gap between two vertical flat plates, i.e. in a Hele-Shaw cell. Most relevant to the current study is that of Graf, Meiburg & Hartel (2002), who examined the linear stability and nonlinear evolution of disturbances to a basic state localised gradient depending only on the vertical coordinate (along which gravity acts). These authors found exponentially growing instabilities, and examined the dependence of the growth rate on the width of the high gradient zone and on Ra.

The infinite Ra limit has so far been inaccessible. Here, we introduce a new approach, based on "contour advection" (Dritschel & Ambaum, 1997), which allows one to study this limit, as well as infinite density gradients — interfaces separating two regions of different uniform density. At $Ra \to \infty$, the interface remains intact, allowing an appealingly simple model of this limit. In particular, the shape of the interface (a two-dimensional surface) determines the entire three-dimensional flow field. A consequence is that the linear stability problem reduces to finding simply the shape of the interface as the "eigenmode".

Below, we review the governing equations in §2, present a consistent way to find the pressure boundary conditions in §3, outline the numerical method developed in §4, discuss the linear stability in §5, and illustrate the nonlinear flow evolution in a few distinct examples in §6. Some conclusions and an outlook for future work round off the paper in §7.

2. The governing equations

Following Graf *et al* (2002), we consider a right-handed system of coordinates with y pointing vertically upward, and with x and z pointing horizontally. The fluid lies between two vertical plates at $z = \pm\frac{1}{2}L_z$, at which the velocity $\boldsymbol{u} = (u, v, w)$ vanishes, and extends periodically over $-\frac{1}{2}L_x < x \leq \frac{1}{2}L_x$. In the vertical y direction, free-slip boundaries (at which $v = 0$) are used at $y = \pm\frac{1}{2}L_y$ to model an infinite domain (valid only for L_y sufficiently large; see below and Graf *et al* (2002)).

The flow is driven by variations in the density ρ, here considered small relative to a uniform background density ρ_0. This allows one to simplify the momentum equations, keeping only terms linear in ρ/ρ_0 — the so-called "Boussinesq approximation". The pressure is divided into a mean hydrostatic part $\bar{p}(y) = p_0 - g\rho_0 y$, where g is gravity and p_0 is an arbitrary reference pressure, and a perturbation denoted p.

A. Gyr and W. Kinzelbach (eds.), Sedimentation and Sediment Transport, 29–42.

Neglecting the fluid acceleration, the momentum equations become

$$0 = -\frac{\boldsymbol{\nabla} p}{\rho_0} - g\frac{\rho}{\rho_0}\hat{\boldsymbol{e}}_y + \nu\nabla^2\boldsymbol{u}\,, \tag{1}$$

where $\hat{\boldsymbol{e}_y}$ is a unit vector in the y direction, and ν is the kinematic viscosity coefficient. In addition, the flow is taken to be incompressible (a very good approximation considering the flow speeds involved),

$$\boldsymbol{\nabla}\cdot\boldsymbol{u} = 0 \tag{2}$$

and to have negligible density diffusion κ so that

$$\frac{\mathrm{D}\rho}{\mathrm{D}t} = \frac{\partial\rho}{\partial t} + \boldsymbol{u}\cdot\boldsymbol{\nabla}\rho = \kappa\nabla^2\rho \approx 0\,. \tag{3}$$

The equations are next made non-dimensional on a length scale $L = L_z$, the gap width, a pressure scale $P = g\Delta\rho L_z$, where $\Delta\rho$ is the given constant increment in density between $y = -\frac{1}{2}L_y$ and $y = +\frac{1}{2}L_y$, and a time scale $T = \rho_0\nu/(g\Delta\rho L_z)$. Velocities are thus scaled by $U = L/T$. With these scalings, (1) may be written free of constants (all variables are henceforth dimensionless):

$$0 = -\boldsymbol{\nabla} p - c\hat{\boldsymbol{e}}_y + \nabla^2\boldsymbol{u}\,, \tag{4}$$

where $c \equiv \rho/\Delta\rho$ is called the "concentration field" in Graf *et al* (2002). Note that c increases by unity from $y = -\frac{1}{2}L_y$ to $y = +\frac{1}{2}L_y$. Without loss of generality, we may take $c = \pm\frac{1}{2}$ at $y = \pm\frac{1}{2}L_y$. Equation (2) remains unchanged, and (3) is replaced by

$$\frac{\mathrm{D}c}{\mathrm{D}t} = \frac{\partial c}{\partial t} + \boldsymbol{u}\cdot\boldsymbol{\nabla} c = Ra^{-1}\nabla^2 c \approx 0\,. \tag{5}$$

where $Ra \equiv UL/\kappa = g\Delta\rho L_z^3/(\rho_0\nu\kappa) \gg 1$ so that the diffusion of density may be neglected. On the other hand, to neglect the fluid acceleration in the momentum equations requires $Re \equiv UL/\nu = g\Delta\rho L_z^3/(\rho_0\nu^2) \ll 1$. A consequence is that the Prandtl number $Pr \equiv \nu/\kappa$ must be large (note: $Ra = Pr\,Re$). These conditions can be achieved in real flows — consult Graf *et al* (2002) & references therein.

The equations are next transformed to enforce $\boldsymbol{\nabla}\cdot\boldsymbol{u} = 0$ directly. We depart from Graf *et al* (2002), who chose to eliminate the pressure by taking the curl of (4) and work with the vorticity $\boldsymbol{\omega} = \boldsymbol{\nabla}\times\boldsymbol{u}$ and c as primary variables. Their transformation avoids having to deal with the implicit pressure boundary condition discussed in the next section. Instead, boundary conditions are prescribed for $\boldsymbol{\omega}$, in addition to those already required for $\boldsymbol{u}$. However, it turns out that the boundary conditions for $\boldsymbol{\omega}$ are not independent of those for p, and in order to prescribe the correct boundary conditions, one must consider p. As a result, the boundary conditions imposed by Graf *et al* (2002), and others, are not exact.

To circumvent this problem, we introduce two scalar potentials ψ and χ, in terms of which

$$u = -\frac{\partial\psi}{\partial y} + \frac{\partial\chi}{\partial x} \quad \& \quad v = \frac{\partial\psi}{\partial x} + \frac{\partial\chi}{\partial y}\,. \tag{6}$$

Then, since the vertical vorticity $\zeta = \partial v/\partial x - \partial u/\partial y$, we have

$$\nabla_h^2 \psi = \zeta \quad \& \quad \nabla_h^2 \chi = -\frac{\partial w}{\partial z} \tag{7}$$

and $\nabla_h^2 \equiv \partial^2/\partial x^2 + \partial^2/\partial y^2$. This shows that u and v can be determined from ζ and $\partial w/\partial z$ by inverting a two-dimensional Laplacian in planes of constant z. Notably, no z boundary conditions need to be imposed to solve for ψ and χ. Moreover, ζ, w and $\partial w/\partial z$ all vanish at the vertical plates $z = \pm\frac{1}{2}$ (note $L_z = 1$; this is the gap width after the scaling above).

It therefore proves convenient to work with the following primary variables: p, w, ζ and c. They satisfy the following linear set of equations

$$\nabla^2 p = -\frac{\partial c}{\partial y} \quad , \quad \nabla^2 w = \frac{\partial p}{\partial z} \quad , \quad \nabla^2 \zeta = \frac{\partial c}{\partial x} \tag{8}$$

together with the nonlinear advection equation (5). The z boundary conditions for w and for ζ are homogeneous as just noted; those for p are more complicated and so are deferred to the next section. In y, the no-slip boundary condition $v = 0$ at $y = \pm\frac{1}{2}L_y$ is enforced by expressing ψ as a series in $\sin \ell\tilde{y}$, where $\ell = 0, 1, 2, \ldots$ and $\tilde{y} \equiv (y + \frac{1}{2}L_y)\pi/L_y$, and by expressing χ as a series in $\cos \ell\tilde{y}$. Then for consistency, u and w are expressed as cosine series, while ζ and $c' \equiv c - y/L_y$ are expressed as sine series. Note that c' is used since $c' = 0$ at $y = \pm\frac{1}{2}L_y$, unlike c. Likewise, it is necessary to use $p' \equiv p - \frac{1}{2}y^2/L_y$ and a cosine series for p' like for w, to be consistent with (8). In x, all fields are periodic, so there are no boundary conditions to impose. Fields are expressed as Fourier series in $\exp(ik\tilde{x})$, where $k = 0, 1, 2, \ldots$ and $\tilde{x} \equiv (x + \frac{1}{2}L_x)2\pi/L_x$.

3. The pressure boundary condition

Henceforth, we deal only with the Fourier transformed fields in x and y and consider each wavenumber (k, l) individually. Depending on the field, $\partial/\partial x \to ik\alpha$ and $\partial/\partial y \to \pm\ell\beta$, where $\alpha \equiv 2\pi/L_x$ and $\beta \equiv \pi/L_y$. Define also $\lambda^2 = (k\alpha)^2 + (\ell\beta)^2$; note that $\nabla_h^2 \to -\lambda^2$. We also drop the primes on c and p; the linear part of c and the corresponding quadratic part of p do not induce any flow and need not be considered further.

In what follows, the implied subscripts k, ℓ are suppressed. The variables p, w, etc... are henceforth considered spectral coefficients depending only on z.

The equations satisfied by the primary variables p, w and ζ are

$$\mathcal{L}p = -\ell\beta c \quad , \quad \mathcal{L}w = \frac{\mathrm{d}p}{\mathrm{d}z} \quad , \quad \mathcal{L}\zeta = ik\alpha c \tag{9}$$

where c is the concentration field, and $\mathcal{L} \equiv \mathrm{d}^2/\mathrm{d}z^2 - \lambda^2$ is the Fourier-transformed 3D Laplace operator. The homogeneous boundary conditions for w and ζ at $z = \pm\frac{1}{2}$ are denoted $w|_\pm = 0$ and $\zeta|_\pm = 0$. As yet, the boundary conditions are unknown for p. However, the boundary conditions for the derived variables u and v are also homogeneous since (cf. (6))

$$\begin{aligned} u &= -\ell\beta\psi + ik\alpha\chi = +\frac{\ell\beta}{\lambda^2}\zeta - \frac{ik\alpha}{\lambda^2}\frac{\mathrm{d}w}{\mathrm{d}z} \\ v &= +ik\alpha\psi - \ell\beta\chi = -\frac{ik\alpha}{\lambda^2}\zeta + \frac{\ell\beta}{\lambda^2}\frac{\mathrm{d}w}{\mathrm{d}z} \end{aligned} \tag{10}$$

and since $\zeta|_\pm = 0$, then $u|_\pm = v|_\pm = 0$ can be satisfied only if $\mathrm{d}w/\mathrm{d}z|_\pm = 0$. (We may exclude $\lambda = 0$, which occurs only for $k = \ell = 0$.) Hence, apparently w has more boundary conditions than needed. However, we have not yet specified the boundary conditions for p, and p appears on the right-hand side of the w equation in (9). In fact, there are just enough boundary conditions now to solve the problem completely, as shown next.

Divide p into two parts, p^i and p^b, where $\mathcal{L}p^i = -\ell\beta c$ with *homogeneous* boundary conditions $p^i|_\pm = 0$, and $\mathcal{L}p^b = 0$ with as yet undetermined boundary conditions. p^b must satisfy

$$p^b = P_+ e^{\lambda z} + P_- e^{-\lambda z} \tag{11}$$

for constants $P_\pm$ to be determined (for each wavenumber pair). On the other hand, p^i can be obtained uniquely, since all boundary conditions are specified. So, at this stage, p^i and ζ are completely determined.

Next divide w into two parts, w^i and w^b, where $\mathcal{L}w^i = \mathrm{d}p^i/\mathrm{d}z$ with $w^i|_\pm = 0$, and $\mathcal{L}w^b = \mathrm{d}p^b/\mathrm{d}z$ with $w^b|_\pm = 0$. Again, w^i can be obtained uniquely since p^i is known. For w^b, we can work out the exact solution since p^b in (11) is a known function of z, apart from the constants $P_\pm$. The result is

$$w^b = A_+ e^{\lambda z} + A_- e^{-\lambda z} + \tfrac{1}{2} z p^b \tag{12}$$

where $A_\pm = \mp(a_1 P_\pm + a_2 P_\mp)$ and

$$a_1 = \tfrac{1}{4}\frac{e_+^2 + e_-^2}{e_+^2 - e_-^2} \quad \& \quad a_2 = \tfrac{1}{4}\frac{2}{e_+^2 - e_-^2} \tag{13}$$

where $e_\pm \equiv \exp(\pm\tfrac{1}{2}\lambda)$. The first two terms in (12) are required to enforce the boundary conditions $w^b|_\pm = 0$. Equation (12) may be also written

$$w^b = -(a_1 P_+ + a_2 P_-)e^{\lambda z} + (a_1 P_- + a_2 P_+)e^{-\lambda z} + \tfrac{1}{2} z (P_+ e^{\lambda z} + P_- e^{-\lambda z})\,. \tag{14}$$

We next apply the second boundary condition $\mathrm{d}w/\mathrm{d}z|_\pm = 0$ to determine $P_\pm$. Since w^i is already known, we need to solve the pair of equations

$$\mathrm{d}w^b/\mathrm{d}z|_\pm = -\mathrm{d}w^i/\mathrm{d}z|_\pm\,. \tag{15}$$

Now, since

$$\begin{aligned}\mathrm{d}w^b/\mathrm{d}z \quad = \quad & [(\tfrac{1}{2} + \lambda(\tfrac{1}{2}z - a_1))e^{\lambda z} - \lambda a_2 e^{-\lambda z}]P_+ \\ & +[(\tfrac{1}{2} - \lambda(\tfrac{1}{2}z + a_1))e^{-\lambda z} - \lambda a_2 e^{\lambda z}]P_-\end{aligned} \tag{16}$$

equation (15) leads to

$$b_+ P_\pm + b_- P_\mp = -\mathrm{d}w^i/\mathrm{d}z|_\pm \tag{17}$$

where

$$\begin{aligned} b_\pm \quad & = \quad (\tfrac{1}{2} - \lambda(a_1 \mp \tfrac{1}{4}))e_\pm - \lambda a_2 e_\mp \\ & = \quad \tfrac{1}{2}e_\pm - \lambda\frac{e_\mp}{e_+^2 - e_-^2}\end{aligned} \tag{18}$$

The solution for $P_\pm$ is therefore

$$P_\pm = \frac{b_- \mathrm{d}w^i/\mathrm{d}z|_\mp - b_+ \mathrm{d}w^i/\mathrm{d}z|_\pm}{b_+^2 - b_-^2} . \tag{19}$$

Note that $b_+^2 - b_-^2 = \frac{1}{4}(e_+^2 - e_-^2) - \lambda^2/(e_+^2 - e_-^2) > 0$ for all $\lambda > 0$ and vanishes only when $\lambda = 0$, which is here excluded. Hence, this is a permissible solution for all wavenumber pairs.

With the above values of $P_\pm$ now determined, the exact form of p^b and hence p is known, as is w^b from (14) and hence w. Thus, p, w and ζ are fully determined in terms of the instantaneous distribution of c, and we can now obtain the remaining velocity components u and v from (10).

Finally, note that the above results give us the form of the boundary conditions satisfied by p. Using

$$b_+e_+ - b_-e_- = \tfrac{1}{2}(e_+^2 - e_-^2) \quad \& \quad b_+e_- - b_-e_+ = \lambda \tag{20}$$

we have from $p|_\pm = p^b|_\pm$ and (11)

$$p|_\pm = -\frac{\frac{1}{2}(e_+^2 - e_-^2)\,\mathrm{d}w^i/\mathrm{d}z|_\pm + \lambda\,\mathrm{d}w^i/\mathrm{d}z|_\mp}{b_+^2 - b_-^2} . \tag{21}$$

4. Numerical method

A numerical method has been developed to solve the above equations on a regular grid, using code adapted from that used by Graf *et al* (2002), in particular the 4th-order compact difference routines used for calculating $\mathrm{d}/\mathrm{d}z$ and $\mathrm{d}^2/\mathrm{d}^2z$, and the penta-diagonal solvers used in inverting $\mathcal{L} \equiv \mathrm{d}^2/\mathrm{d}z^2 - \lambda^2$ in (9). For the y dependence, sine and cosine transform routines were adapted from Numerical Recipes codes, with constants extended to ensure double precision results. For the x dependence, NAG routines were used. The code is written in a flexible way to allow simple substitutions for these routines, if desired.

The use of different primary variables leads to greater computational efficiency since there are only half as many inversions of $\mathcal{L}$ to perform than in the code of Graf *et al* (2002). But the truly novel aspect of the code is the use of "contour advection" (Dritschel & Ambaum 1997) to solve the nonlinear advection equation (5) for c. In contour advection, c is represented by material *contours* $\boldsymbol{x} = \boldsymbol{X}_s(a,t)$, where a is a Lagrangian label, lying on material *surfaces* $z = h_s(x,y,t)$, $s = 1, 2 \ldots, n_{\mathrm{surf}}$, here initially coincident with regularly spaced planes $z =$ constant. Then, the advection of the field c is carried out by solving

$$\frac{\mathrm{d}X_s}{\mathrm{d}t} = u(X_s, Y_s, Z_s) \quad \& \quad \frac{\mathrm{d}Y_s}{\mathrm{d}t} = v(X_s, Y_s, Z_s) , \tag{22}$$

where $Z_s = h_s(X_s, Y_s, t)$, together with

$$\frac{\partial h_s}{\partial t} + u(x,y,h_s)\frac{\partial h_s}{\partial x} + v(x,y,h_s)\frac{\partial h_s}{\partial y} = w(x,y,h_s) . \tag{23}$$

Note that (23) is equivalent to

$$\frac{\mathrm{d}Z_s}{\mathrm{d}t} = w(X_s, Y_s, Z_s)\,. \tag{24}$$

The z coordinate is not treated the same as x and y to allow a direct means of interpolating the field c onto a regular grid from the contours and surfaces (full details are given in Dritschel & Viúdez, 2002). In brief, the present method involves two basic steps: (1) finding the field of c on the surfaces h_s by a fast-fill routine developed originally by Dritschel & Ambaum (2002), and (2) finding the field of c on regularly spaced grid points in z by linearly interpolation between the surfaces which lie on either side of a given grid point in z (for fixed x and y grid points). These operations are done on an xy grid four times finer than the grid used to represent $\boldsymbol{u}$. Likewise, four times as many surfaces are used as grid points in z. This refinement allows for a highly accurate recovery of $\boldsymbol{u}$ from c; in particular, sub-grid-scale features of c contribute to $\boldsymbol{u}$, unlike in conventional numerical approaches. For some quantitative results, see Dritschel & Ambaum (2002).

Once c is found on the regular three-dimensional grid, the velocity field $\boldsymbol{u}$ is found as described in the previous sections. Equal grid lengths are used in all directions. In z, the grid points are given by $z_i = (i-1)\Delta - \frac{1}{2}$, $i = 1, 2, ..., n_z$, where $\Delta = 1/(n_z - 1)$ is the grid length. Note that two of the grid points lie on the plates $z = \pm\frac{1}{2}$ where $\boldsymbol{u}$ is identically zero. (The surfaces h_s are initially spaced at intervals of $\frac{1}{4}\Delta$, and there are $n_{\mathsf{surf}} = 4(n_z - 1) + 1$ of them, two of which lie on $z = \pm\frac{1}{2}$.) Likewise, in y, two of the grid points lie on the boundaries $y = \pm\frac{1}{2}L_y$, while in x, the boundaries are periodic so that $x = \pm\frac{1}{2}L_x$ corresponds to the same grid point.

To solve (22), the velocity components u and v are first found by a tri-linear interpolation from the regular grid to a Lagrangian contour point $\boldsymbol{X}_s$. Each contour in each surface is approximated by a finite, variable number of "nodes", connected together by local cubic splines. The density of nodes is roughly proportional to the square-root of local contour curvature, and nodes are redistributed every other time step to maintain adequate resolution (again, see Dritschel & Ambaum (2002) for details of this and all other contour operations). If two contours of the same level of c or if two parts of the same contour get closer than a prescribed "cutoff" scale δ, they are topologically reconnected by "surgery". This procedure is carried out well below the grid scale, here at $\delta = \Delta/20$, a twentieth of the grid scale. (In practise, little or no surgery occurs in the simulations performed.)

It is assumed that the surfaces $z = h_s$ remain single-valued in time. This assumption is based on the fact that w is typically much smaller in magnitude than u and v, one to two orders of magnitude smaller in the examples considered below, both in r.m.s. and in extreme values. Nevertheless, problems were encountered during the development of the code that suggested either a numerical instability associated with solving (23) directly, or that surfaces truly overturn. These problems however were only encountered at late times in very complex flows, see below. As a result, several other approaches were tried, all of which gave virtually identical results before breakdown. The approach taken in the present version of the code uses the differences in layer thicknesses $\Delta\tau_s \equiv \tau_{s+1} - \tau_s$, where $\tau_s \equiv h_{s+1} - h_s$ is the thickness of layer s. This is a convenient variable since the $n_{\mathsf{surf}} - 2$ independent values of $\Delta\tau_s$ (with

$s = 1, 2 ..., n_{\text{surf}} - 2)$ correspond to the $n_{\text{surf}} - 2$ interior surfaces h_s (with $s = 2, 3 ..., n_{\text{surf}} - 1$). The latter are recovered from the former via a tridiagonal inversion for each xy grid point. The evolution equation satisfied by the layer thicknesses τ_s is obtained by subtracting that for h_s from that for h_{s+1}, giving

$$\begin{aligned}\frac{\partial \tau_s}{\partial t} &+ \left[u(x, y, h_{s+1})\frac{\partial h_{s+1}}{\partial x} - u(x, y, h_s)\frac{\partial h_s}{\partial x}\right] \\ &+ \left[v(x, y, h_{s+1})\frac{\partial h_{s+1}}{\partial y} - v(x, y, h_s)\frac{\partial h_s}{\partial y}\right] \\ &- \left[w(x, y, h_{s+1}) - w(x, y, h_s)\right] = 0 \qquad (25)\end{aligned}$$

then approximating the bracketed terms above by the flux-form expression

$$\frac{\partial \tau_s}{\partial t} + \frac{\partial \bar{u}_s \tau_s}{\partial x} + \frac{\partial \bar{v}_s \tau_s}{\partial y} = 0 \,, \qquad (26)$$

where $\bar{u}_s = \frac{1}{2}[u(x, y, h_s) + u(x, y, h_{s+1})]$ and $\bar{v}_s = \frac{1}{2}[v(x, y, h_s) + v(x, y, h_{s+1})]$ are the layer-averaged horizontal velocity components. This is second-order accurate in the mean layer thickness, $\frac{1}{4}\Delta$, and is preferred numerically because the flux form guarantees conservation of mass in each layer, unlike the original expression (25).

The equation satisfied by $\Delta\tau_s$ is just the difference between that satisfied by τ_{s+1} and by τ_s. It is solved numerically using pseudo-spectral techniques, wherein the flux $(\bar{u}_s\tau_s, \bar{v}_s\tau_s)$ is computed on the xy grid, and then transformed to spectral space before taking x and y derivatives spectrally. To partially de-aliase the nonlinear products, a steep spectral filter of the form $F(k) = \exp(-\theta(k/k_{\max})^{10})$, with θ chosen so that $F(k_{\max}) = 10^{-14}$, is applied when computing derivatives (here k refers to the x or y wavenumber, as the case may be). This filter was first introduced by Broutman *et al* (1997) as a low-cost and relatively benign replacement for de-aliasing, and has been used in analogues of the present code (see Viúdez & Dritschel (2002) and Dritschel & Viúdez (2002)).

For the time integration, an explicit 4th-order Runge-Kutta method is used for updating both the contours (X_s, Y_s) and the thickness differences $\Delta\tau_s$. The time step is limited by the CFL numerical stability condition $C = |\boldsymbol{u}|\Delta t/\Delta < 1$, which is monitored during the course of a simulation. All of the simulations reported below use the same time step $\Delta t = 0.1$, though this is a conservative choice ($C < 0.1$ in all cases). The algorithm is sufficiently efficient to be run in a few minutes to a few hours on a laptop having an Pentium 3 processor.

5. Linear stability analysis

We next consider the stability of a step-function basic state distribution of c, i.e. $c = \frac{1}{2}|y|$, for infinite Ra to complement the finite Ra analysis of Graf *et al* (2002). The x independence of the basic state and of the boundary conditions means that eigenmodes are proportional to $\exp(i\alpha x)$, where $\alpha = 2\pi/L_x$. The shape of the interface $y = \eta(x, z, t)$ dividing the two regions of uniform c determines all other variables, so it is sufficient to seek the interface shape of the eigenmodes. Moreover, we may take $\eta(x, z, t) = -A(z)\cos\alpha x\, e^{\sigma t}$, so the problem reduces to finding the amplitude function $A(z)$ (the "eigenmode") and the growth rate σ (the "eigenvalue") as a function of α.

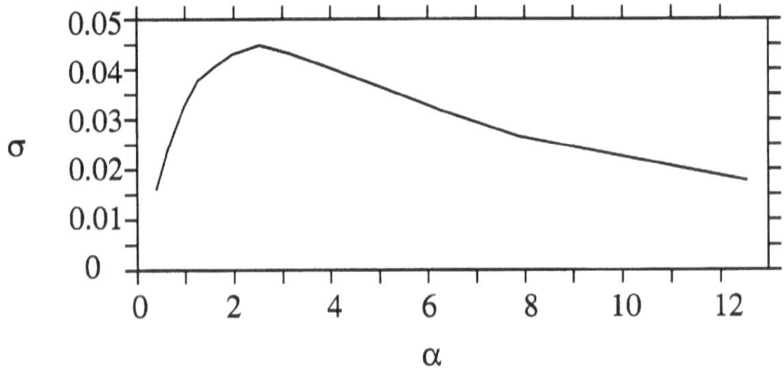

Figure 1: The growth rate σ versus $\alpha = 2\pi/L_x$.

Normally, the equations are linearised in small perturbations to calculate eigenvalues and eigenmodes. Here, instead, the results are obtained directly from the full equations. A guess for the eigenfunction is provided initially, then one time step is taken, the overall r.m.s. value of η is computed and used to renormalise η to its original r.m.s. value, then another time step is taken, and so on until the maximum difference in the r.m.s. value *for each contour* is less that a prescribed tolerance, here 10^{-5}. The initial guess (deduced from the results of Graf *et al* (2002) for finite *Ra*) is given by $\eta = -\epsilon \cos \pi z \cos \alpha x$, with amplitude $\epsilon = L_x/40$, corresponding to a maximum initial slope $|\partial\eta/\partial x| = \epsilon\alpha = \pi/20 = 0.157079...$ (tests indicate that this is small enough to keep nonlinear corrections, which show up as higher harmonics in x, under 1%). This simple procedure takes normally around 1000 iterations to converge. The results are then checked by running simulations starting from the converged solution for η and fitting an exponential $e^{\sigma_{\text{num}} t}$ to the global r.m.s. value of η over the time interval $0 \leq t \leq t_{\text{num}} \sim \sigma^{-1}$ (actually the logarithm of the r.m.s. value is fit by the least-squares method to a linear function of time).

The results of this analysis, together with the simulation parameters used, are listed in Table 1. There is good agreement between σ and σ_{num} with most results overlapping within one standard deviation of the numerical growth rates. The smaller value of t_{num} used for $L_x = 16.0$ was necessary due to the rapid growth of secondary instabilities beyond this time (see next section). In all cases, we have ensured $L_y/L_x \geq 2$ and $L_y/L_z \geq 2$ to keep the effects of the y boundaries insignificant ($\boldsymbol{u}$ appears to decay exponentially in y). For $L_x = 2.5$, doubling L_y/L_x to 64/15 has no effect whatsoever on the value of σ to the number of decimal places reported. Halving L_y/L_x to 16/15 gives $\sigma = 0.044504$, an error of 0.7%. This is consistent with the findings of Graf *et al* (2002).

The growth rates are plotted as a function of $\alpha = 2\pi/L_x$ in figure 1. Maximum growth is observed for $\alpha = 2.5$ (for which coincidentally $L_x = 2.5$). This is again broadly consistent with the high *Ra*, narrow interface results of Graf *et al* (2002), although the wavenumber of peak instability and the growth rates are somewhat higher, as expected for lower density diffusion. The eigenfunction $A(z)$, obtained by projecting each contour on $\cos \alpha x$ then dividing the result by the overall r.m.s. value of η, is shown in figure 2 for selected values of L_x. For large L_x (i.e. $L_z/L_x \ll 1$

L_x	L_y/L_x	n_x	n_y	n_z	σ	σ_{num}	st. dev.	t_{num}
16.0	2	128	257	9	0.01609	0.01621	0.00016	20
12.8	2	128	257	11	0.01940	0.01944	0.00010	52
10.0	32/15	120	257	13	0.02379	0.02367	0.00033	42
8.0	2	64	129	9	0.02793	0.02767	0.00036	36
6.4	2	64	129	11	0.03263	0.03236	0.00032	31
5.0	32/15	60	129	13	0.03774	0.03741	0.00025	26
4.0	2	32	65	9	0.04034	0.04026	0.00029	25
3.2	2	32	65	11	0.04305	0.04329	0.00022	23
2.5	32/15	30	65	13	0.04482	0.04450	0.00020	22
2.0	2	32	65	17	0.04320	0.04380	0.00021	23
1.6	2	16	33	11	0.04053	0.03874	0.00036	25
1.0	2	16	33	17	0.03177	0.03015	0.00037	31
0.8	4	16	65	21	0.02665	0.02557	0.00034	38
0.5	4	16	65	33	0.01769	0.01701	0.00045	56

Table 1: Domain lengths, grid resolutions, and growth rates obtained in the linear analysis. Here n_x, n_y and n_z give the number of grid points used in x, y and in z.

or a narrow gap), the eigenfunction flattens near $z = 0$, while for small L_x, the eigenfunction peaks near $z = 0$, and secondary peaks begin to emerge closer to the plates at $z = \pm\frac{1}{2}$). These results are again broadly similar with the eigenmodes for c obtained by Graf *et al* (2002), though simpler since they reduce here to a single function $A(z)$ describing the entire c field, and hence all fields, for each α.

In summary, the results suggest that a density interface is unstable for all α, but that maximum instability occurs for $\alpha \approx 2.5$, corresponding to a wavelength $L_x \approx 2.5$. There is thus a preferred scale of the instability, which may be expected to emerge in realistically wide domains ($L_x \gg L_z$). We shall examine this situation in the following section.

6. Long-term nonlinear behaviour

We next explore the nonlinear evolution of an unstable interface. Three simulations are described. The first two start with unstable eigenmodes, while the third starts with a random disturbance.

The first, for the most unstable eigenmode having $L_x = 2.5$ (and in a deep domain having $L_y = 32/3$), is displayed in figure 3. The central part of interface at $x = 0$ and $z = 0$ accelerates downward (in y) while the part at the x-periodic edge accelerates symmetrically upward (this symmetric behaviour is a consequence of the Boussinesq approximation used in (1)). By about $t = 60$ when the maximum wave slope $|\partial\eta/\partial x| \sim 1$, the exponential growth of the disturbance gives way to linear growth, as shown in figure 4. A linear least-squares fit of $y_{\max}$ and y_{rms} over the time interval $80 \leq t \leq 280$ gives a slope of $dy_{\max}/dt = 0.01720 \pm 0.00034$ and $dy_{\text{rms}}/dt = 0.01697 \pm 0.00019$ (at times $t \geq 300$, the domain walls in y begin to affect

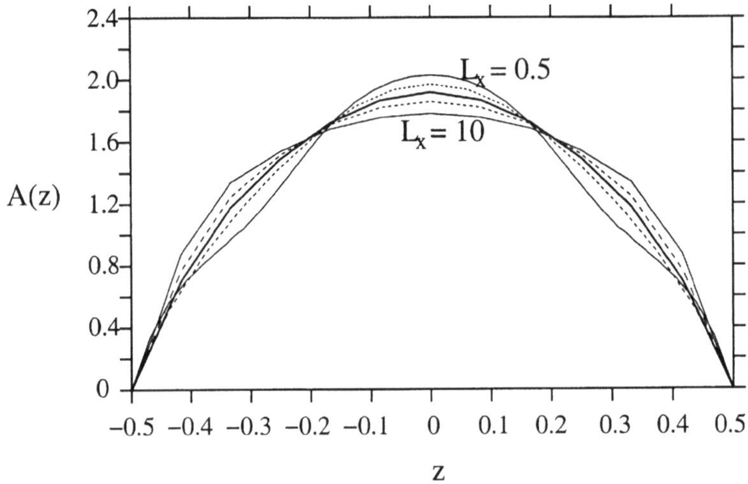

Figure 2: $A(z)$ versus z for selected values of L_x. The line styles are as follows: bold solid for $L_x = 2.5$, dashed for $L_x = 1.0$ and 5.0, and thin solid for $L_x = 0.5$ and 10.0. The curves become progressively flatter around $z = 0$ as L_x increases.

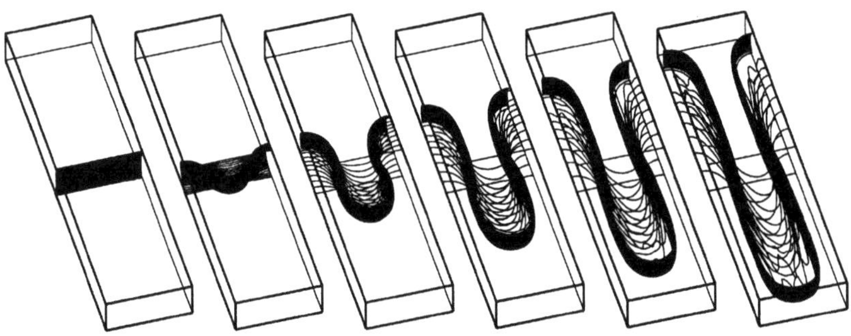

Figure 3: The nonlinear development of the instability for $L_x = 2.5$. The contours of c in each layer are plotted at times $t = 0$, 60, 120, 180, 240 and 300, from left to right. The domain has been rotated first by 15° about the z axis, then by 45° about the original x axis.

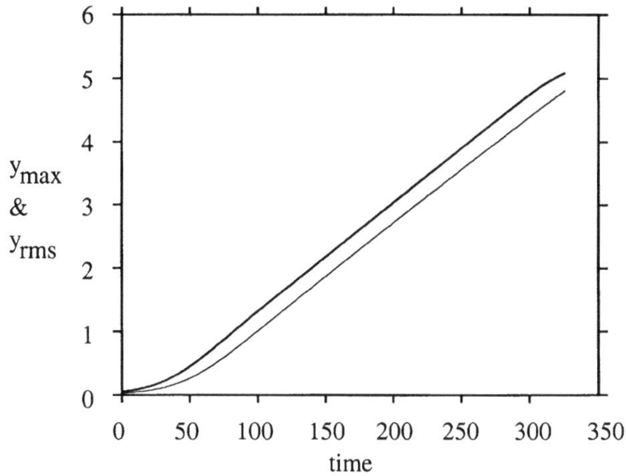

Figure 4: The time evolution of the maximum and r.m.s. y values (shown as bold and solid lines) obtained in the simulation for $L_x = 2.5$ (figure 3).

the behaviour).

A closer look at figure 3 reveals however that there is much more going on than the primary instability. In particular, small-scale "jets" of fluid develop shortly after the interface becomes multi-valued in x. These jets move considerably faster than the primary deformation, with sustained peak y velocities of around 0.049 over the last half of the simulation. This behaviour is found in all simulations conducted, and appears to be characteristic of very high Ra flows — it was not found for instance in the simulations conducted by Graf *et al* (2002)).

We next examine the evolution of the eigenmode found in a much wider domain, $L_x = 12.8$. According to Table 1, this is a much less unstable mode. Figure 5 shows the evolution. Initially, the disturbance grows slowly and with the eigenmode structure, but this soon gives way to a secondary instability having a wavelength 4 times shorter and clearly influenced by the primary instability. The secondary instability has a wavelength comparable to that of the most unstable mode (found for $L_x = 2.5$), and is very likely related. This example shows that long wave instabilities are clearly not preferred in wide domains.

Finally, we examine what happens to an initially random disturbance in a very wide domain having $L_x = 20$ and $L_y = 32/3$. Initially, each contour is represented by 176 nodes, and each of these is displaced in y by a random number uniformly distributed between -0.1 and 0.1. This disturbance is then smoothed by repeatedly applying a 1-2-1 average to the y coordinates (replacing the ith node y_i by $\frac{1}{4}(y_{i-1} + 2y_i + y_{i+1})$). This is done 10 times, and each time the coordinates are scaled to retain the initial r.m.s. value of y. Finally, the disturbance is modulated in z by multiplying by $\cos \pi z$, giving zero displacement at the plates $z = \pm\frac{1}{2}$. Figure 6 shows the flow evolution at 4 selected times after the instability becomes apparent. There are approximately 8 or perhaps 9 disturbance wavelengths visible, consistent with the fastest growing instability wavelength of 2.5 found in the previous section. Note

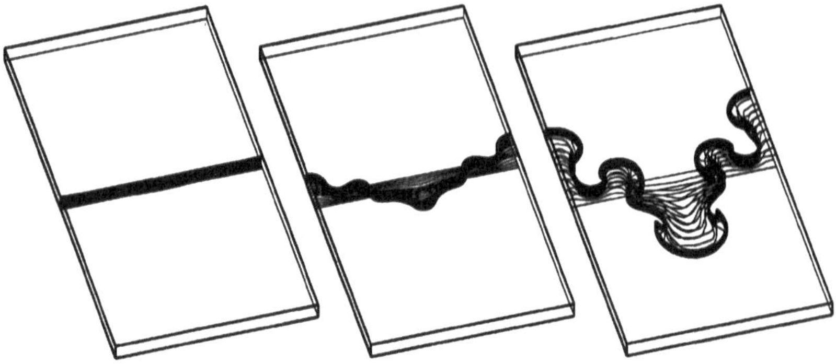

Figure 5: The nonlinear development of the instability for $L_x = 12.8$. The contours of c in each layer are plotted at times $t = 0$, 100 and 200, from left to right. The same perspective is used here as in figure 3.

also that the upward and downward lobes are broadly similar to those found for the evolution of the eigenmode for $L_x = 2.5$ shown in figure 3. Moreover, one also sees here the development of relatively fast moving narrow jets following the overturning of the interface.

7. Conclusions

A new approach has been developed to simulate Stokes flows at infinite *Ra*. In such flows, the density is conserved following fluid particles, and this allows one to represent the density field as material contours. Tracking these contours as they move with the fluid then gives the evolution of the density field.

The new approach employs "contour advection" to track the contours in time, but also uses more conventional grid-based techniques for the "inversion" step, namely recovering the velocity field at any time from the instantaneous distribution of density.

New results have been obtained for the linear stability of a sharp interface (across which the density jumps from one uniform value to another) at infinite *Ra*. Comparing with the finite-*Ra*, smoother-interface results of Graf *et al* (2002), similar instability growth rates, eigenfunctions, and wavenumbers of maximum instability are found. The growth rates and wavenumbers are higher here as it to be expected at higher *Ra*.

The nonlinear development of these instabilities has also been examined. In addition to the primary sinking and rising lobe behaviour found previously by Graf *et al* (2002), a new small-scale secondary instability has been found. This secondary instability takes the form of narrow, relatively high-speed jets and develops soon after the interface overturns.

In future work, it would be interesting to explore this secondary instability in more detail — certainly at higher resolution. Simulations are also needed to examine the very long term consequences of this instability, in particular the net mixing resulting from this instability and any potential long-time saturation.

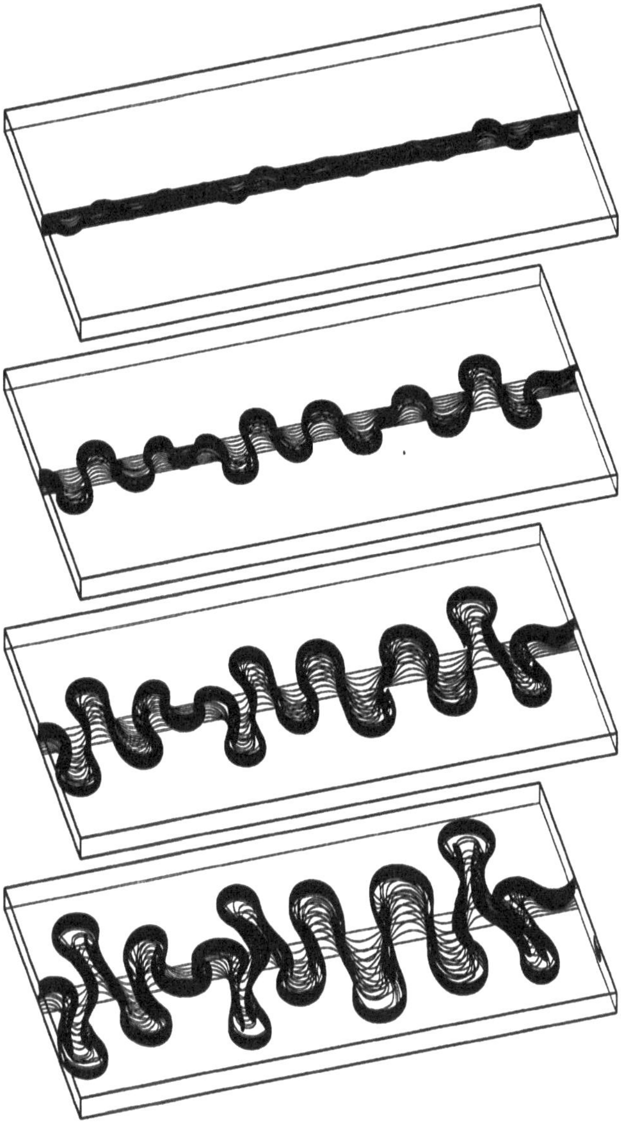

Figure 6: The evolution of a randomly disturbed interface in a domain of dimensions $L_x = 20$, $L_y = 32/3$ and $L_z = 1$. The contours of c in each layer are plotted at times $t = 100$, 150, 200 and 250, from top to bottom. The same perspective is used here as in figure 3. The grid resolution used is $n_x = 240$, $n_y = 129$ and $n_z = 13$, with $n_{\mathsf{surf}} = 49$.

References

BROUTMAN, D., MACASKILL, C., MCINTYRE, M. E., & ROTTMAN, J. 1997 On Doppler-spreading models of internal waves. *Geophys. Res. Lett.* **24**, 2813–2816.

DRITSCHEL, D. G., & AMBAUM, M. H. P. 1997 A contour-advective semi-Lagrangian algorithm for the simulation of fine-scale conservative fields. *Quart. J. Roy. Meteorol. Soc.* **123**, 1097–1130.

DRITSCHEL, D. G., & VIÚDEZ, A. 2002 A balanced approach to modelling rotating stably-stratified geophysical flows. *J. Fluid Mech.* (under review).

GRAF, F., MEIBURG, E., & HARTEL, C. 2002 Density-driven instabilities of miscible fluids in a Hele-Shaw cell: linear stability analysis of the three-dimensional Stokes equations. *J. Fluid Mech.* **451**, 261–282.

VIÚDEZ, A., & DRITSCHEL, D. G. 2002 An explicit potential vorticity conserving approach to modelling nonlinear internal gravity waves. *J. Fluid Mech.* **458**, 75–101.

TURBULENT BOUNDARY LAYERS OVER COMPLIANT WALLS: LOW-DIMENSIONAL MODELS AND DIRECT SIMULATIONS

DIETMAR REMPFER
Illinois Institute of Technology, Chicago, IL, U.S.A,
Rempfer@iit.edu
LOUISE PARSONS, SHENG XU, JOHN LUMLEY
Cornell University, Ithaca, NY, U.S.A

1. Introduction

Since the realization by Reynolds (1883) more than a century ago, that flow turbulence brings with it a very significant increase in the drag forces that a flow exerts on bounding surfaces, reducing this effect has been a desirable goal for many practical applications. Thus, significant research efforts have been invested in trying to determine methods to control turbulence in order to reduce the turbulent drag.

The approaches that were taken can be broadly classified into two groups, the first one attempting to manipulate the flow using active methods, by using appropriate control algorithms to govern the behavior of certain actuators in the flow, and the second one trying to modify the boundary conditions of the flow in some passive manner. While the first of these, active control of turbulence, has lead to some impressive results (Endo & Kasagi 2001; Choi, Moin & Kim 1994), the drag reduction in that case comes at a significant price in terms of the energy that is needed to implement such control schemes. Thus, to our knowledge, there are no experiments that have demonstrated an active control scheme which reduces the total of energy lost through turbulent drag plus energy spent to drive drag-reducing actuators.

Thus the idea of reducing turbulent drag through some passive means is clearly attractive. One option in this category is to try and modify the interaction of flow turbulence with a boundary using a compliant wall with a tailored dynamic response to pressure disturbances from the turbulent wall layer. That idea in fact became very popular for some time, following observations by Kramer (1961) on exceptional swimming capabilities of dolphins. While this approach looked very promising at first, unfortunately, despite extensive research efforts for a number of years as documented by Bushnell *et al.* (1977), it achieved only very limited, if any, success for the case of turbulent flows†. However, with respect to this history, it is interesting to note that the exact reasons for the failure of experiments to verify drag reduction via wall compliance remained unclear, and, in particular, to this day very little can be said definitively about whether or not compliant surfaces may in fact be able to significantly reduce turbulent drag or sound production. We believe that the potential benefits that compliant surfaces may give do warrant taking another, more detailed look at this problem, in our case from a more theoretical perspective and using modern approaches based on low-dimensional modeling of turbulent flows.

In the project we are reporting on below, we want to take a fresh look at the problem of the interaction of near-wall turbulence with a compliant boundary. Below, we will discuss our own approach in more detail, starting with our direct numerical simulations and some of the preliminary results obtained from them, and following up with an introduction into our efforts on low-dimensional modeling of the turbulence-compliant wall interaction.

2. Research Approach

The research reported in this paper rests on two main approaches, which are direct numerical simulation and low-dimensional modeling. In the following, we start by describing our simulations of a turbulent flow in a channel delimited by a compliant and a rigid wall,

† There have been successful demonstrations of compliant-wall technology for the manipulation of transitional flows, however, see, e. g. Gad-el-Hak (1986).

A. Gyr and W. Kinzelbach (eds.), Sedimentation and Sediment Transport, 43–50.

while our low-dimensional modeling effort will be discussed below. The time-varying computational domain we have to deal with can be handled through a time-varying coordinate transformation to eliminate the deformation of the compliant wall as described in the next section. Our choice of the values for the compliant wall parameters is guided by an approximate analysis similar to the one done by Semenov (1991), and it is clear that this approach is not of such a nature as to allow us to directly determine material properties that would guarantee successful turbulent drag reduction. Thus, our DNS must be thought of as an investigative tool which can be used to evaluate the effect of various combinations of materials properties, and which also allows us to obtain some detailed insight into the mechanisms that are at work in the interaction of turbulent flow with a compliant surface. However, DNS by itself will not directly allow us to find material properties that are promising for drag reduction. That role is destined to fall on our low-dimensional models which are described in § 4: Once we are able to construct low-dimensional, i. e. computationally inexpensive models of turbulence in the vicinity of a compliant wall, and once we have established that such models can describe the dynamics of the turbulence-compliant wall system with some appropriate reliability — which is a fact that will have to be established by comparing model predictions with direct numerical simulations — it is straightforward to examine the influence of characteristic parameters of the compliant coating — like stiffness and internal damping — on the dynamical behavior of the wall layer, and it will be feasible to optimize these parameters in order to maximize a desired effect, like reducing the bursting frequency of the near-wall layer which may result in a proportional reduction of the turbulent drag.

3. Direct Numerical Simulation

The problem we are considering is basically a fluid-solid problem. On the fluid side, we have the incompressible NS equation as the model for fluid motion. On the solid side, we use a spring-supported elastic plate to model the compliant wall. The equation governing the wall-normal motion nondimensionalized by the average half channel height h and the laminar centerline velocity u_0 reads

$$C_0\frac{\partial^2\eta}{\partial t^2}+C_1\frac{\partial\eta}{\partial t}+C_2\left(\frac{\partial^4\eta}{\partial x_1^4}+\frac{\partial^4\eta}{\partial x_3^4}+2\frac{\partial^4\eta}{\partial x_1^2x_3^2}\right)$$
$$+C_3\eta-C_x\frac{\partial^2\eta}{\partial x_1^2}-C_z\frac{\partial^2\eta}{\partial x_3^2}=-p_w, \tag{1}$$

where η is the wall-normal displacement, p_w is the pressure disturbance at the wall and C_0, C_1, C_2, C_3, and $C_x(C_z)$ are the nondimensional wall properties defined as

$$C_0=\frac{\rho_m b}{\rho h},\quad C_1=\frac{1}{Re}\frac{dh}{\rho\nu},\quad C_2=\frac{1}{Re^2}\frac{B}{\rho h\nu^2},$$
$$C_3=\frac{1}{Re^2}\frac{K_E h^3}{\rho\nu^2},\quad C_x=\frac{1}{Re^2}\frac{T_x h}{\rho\nu^2},\quad C_z=\frac{1}{Re^2}\frac{T_z h}{\rho\nu^2}. \tag{2}$$

The wall parameters are: the plate density ρ_m and thickness b, the wall damping coefficient d, the flexural rigidity of the plate B, the streamwise (spanwise) tension per unit width $T_x(T_z)$ and the spring stiffness K_E; ρ is the density of the fluid and ν is its kinematic viscosity. The turbulent channel flow we simulate is assumed periodic in the streamwise(x_1 or x) and spanwise(x_3 or z) directions and has a rigid upper and a compliant lower wall that are separated by an average distance of $2h$ in the wall-normal (x_2 or y) direction. Re is its Reynolds number based on h and u_0.

The main difficulty of the simulation is to handle the boundary conditions at the moving compliant wall. We use a time-varying coordinate transformation to eliminate the deformation of the compliant boundary in the computational domain. The continuity equation in the

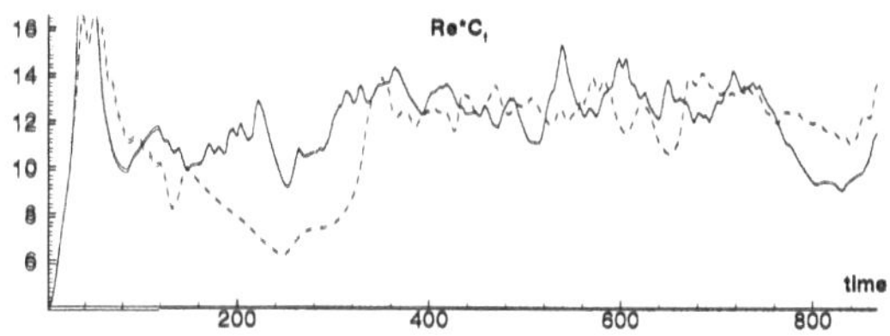

FIGURE 1. Time traces of skin friction coefficients on the compliant wall and the rigid wall. solid line: on the compliant wall; dashed line: on the rigid wall.

computational coordinate system becomes

$$u^i_{;i} = 0, \tag{3}$$

and the momentum equation in the rotational form becomes

$$\frac{\partial u^i}{\partial t} = -\frac{\partial T^{-1}}{\partial t}\frac{\partial u^i}{\partial x_2} + u^j g^{ik}(u_{j;k} - u_{k;j}) - g^{ij}\frac{\partial p}{\partial x_j} + \frac{1}{Re}(g^{jk}u^i_{;k})_{;j}, \tag{4}$$

where u^i is a contravariant velocity component, the semicolon denotes covariant differentiation, T^{-1} is the inverse coordinate transformation function in the wall-normal direction, and g^{ij} is the contravariant metric tensor of the transformation. The equation for pressure p is obtained by taking a covariant derivative with respect to x_i of the momentum equation and can be written as

$$\left(g^{ij}\frac{\partial p}{\partial x_j}\right)_{;i} = \left[-\frac{\partial T^{-1}}{\partial t}\frac{\partial u^i}{\partial x_2} + u^j g^{ik}(u_{j;k} - u_{k;j})\right]_{;i}. \tag{5}$$

The boundary conditions for velocity must also be transformed into the new coordinate system. The boundary conditions for pressure can then be determined indirectly from the incompressibility condition via an influence matrix technique. A Fourier-Galerkin and Chebyshev-Tau pseudospectral method for spatial discretization and a three-sub-step Runge-Kutta method for time advancement are used for solving the equations.

3.1. NUMERICAL RESULTS

We have carried out a direct numerical simulation of a turbulent flow with Reynolds number equal to 3000 in a "half-compliant" channel (lower wall compliant, upper wall rigid) with the values of the compliant wall properties given as $C_0 = 2$, $C_1 = 3000$, $C_2 = 360$, $C_3 = 1.8 \times 10^6$, $C_x = 1.8 \times 10^6$, $C_z = 1.8 \times 10^6$. The computational domain sizes in the streamwise and spanwise directions are chosen to be $\frac{4\pi}{5}h$ and $32 \times 96 \times 65$ grid points are used in these two directions and the wall-normal direction.

Figure 1 shows the history of the skin friction coefficients on the compliant wall and the rigid wall. Based on the samples we have so far, the averaged skin friction coefficient on the compliant wall is 12.45, which is a little higher than the value of 12.27 on the rigid wall. Instantaneous compliant wall shapes and the wall-normal velocity at the wall at one instant are shown in Figure 2.

At this point, the question is whether we just did not choose the right parameters for successful drag reduction, or whether there is something more generic to the negative result above. We performed a more detailed analysis of the dynamics of our wall to get more insight into this issue. In brief, the result of this analysis is that it is not possible to choose mechanical properties of the wall in such a way that the wall moves in an analogous manner to the active wall investigated by Kang & Choi (2000), which resulted in reduced drag. From this result it seems unlikely that the simple compliant wall we have chosen can lead to successful drag reduction. However, it is important to emphasize that our study so far has only explored

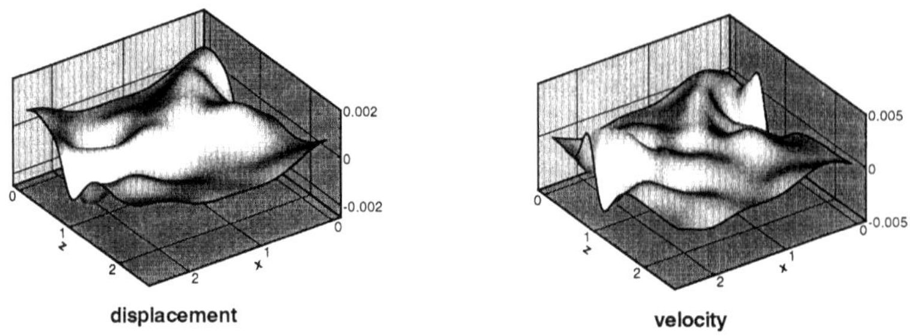

FIGURE 2. Wall-normal displacement η and velocity v_w of the compliant wall nondimensionalized by the average half channel height h and the laminar centerline velocity u_0.

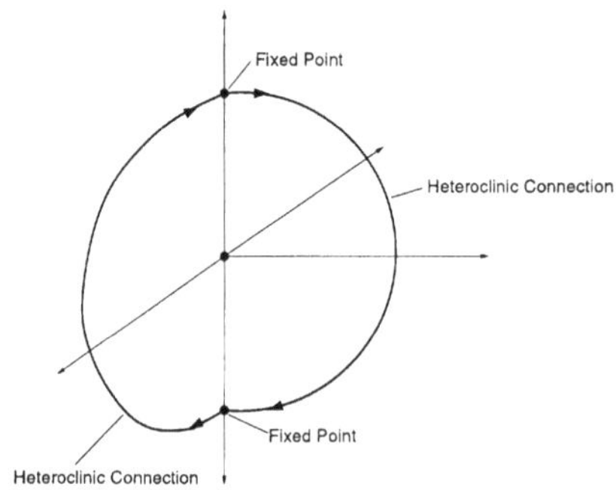

FIGURE 3. Schematic diagram of a heteroclinic cycle.

one way to design a compliant wall, with a homogeneous, isotropic wall material. While our results suggest that this choice may not be promising, we cannot make any predictions as to the probability of success of more sophisticated wall coatings, that may be inhomogeneous and/or anisotropic. In particular, there are theoretical reasons to speculate that the finite compliant panels that have been used with success in the case of transitional flows by Carpenter and coworkers Carpenter, Davies & Lucey (2000) might also hold promise for turbulent flow.

4. Low-Dimensional Model: Dynamical Systems Approach

4.1. INTRODUCTION

In this section we will discuss the low-dimensional models we have used to describe the interaction of a compliant surface with a turbulent wall-layer. For these models, we have restricted ourselves to consider a wall which may deform in the wall-normal direction only. This was done to simplify the equations (at least for this preliminary work) but was also motivated by the assumption that this deformation mode will be the most important one.

The model equations governing the fluid motion are derived from a hybrid POD (Proper Orthogonal Decomposition) - Galerkin Projection technique in which the Navier-Stokes equations are projected onto a set of basis functions for the fluid motion. These basis functions are usually derived from a POD, although in our early work we also used additional basis functions, which we will describe in more detail in section 4.2. It is important to note that these basis functions typically capture a large percentage of the energy in the flow. Hence, by modeling the evolution with time of the coefficients of an expansion of the flow variables in terms of these functions we can draw some significant conclusions about the time-evolution of the flow. In the system we are investigating, since the flow is homogeneous in the spanwise and streamwise directions, the POD functions are simply Fourier modes in those directions. Further, we restrict our attention to functions in which the streamwise Fourier mode is zero since the flow structures are very long (almost unchanging) in this direction.

Thus, in order to arrive at a model that is as simple as possible, and since this assumption has proven successful in the low-dimensional models of near-wall turbulence by Aubry *et al.* (1988), we will only consider the streamwise invariant component of the flow, resulting in equations that have a structure similar to the ones of Aubry *et al.* (1988) (see below). The wall-motion has been modeled as a damped mass-spring system in a space of spanwise Fourier modes,

$$M^k \ddot{\hat{\xi}}^k + D^k \dot{\hat{\xi}}^k + K^k \hat{\xi}^k = -\hat{p}(0,k), \tag{6}$$

where $\hat{\xi}_k$ is the wall displacement in Fourier space.

The wall-displacement is driven by the pressure of the fluid at the wall. The wall-pressure term in this equation is coupled to the wall-pressure term in the flow equations, hence we do not actually need to solve for the pressure at any time. The wall-displacement variables, ξ_k are related to the flow variables, $a_k^{(n)}$, by the boundary conditions which will be discussed in section 4.2.

4.2. MODEL DEVELOPMENT

4.2.1. *Basis Functions*

In order to use POD basis functions in a low-dimensional model, these functions must first be extracted from a data set (either experimental or computational) for the flow of interest. When we began our early work we did not have such a data set available for a turbulent flow over a compliant wall. In order to overcome this problem we used POD basis functions from a rigid-wall flow along with *"ad-hoc"* basis functions from a Stokes flow with oscillating boundary conditions. These additional basis functions will model the flow induced by wall oscillations and allow the no-slip boundary conditions to be satisfied.

4.2.2. *Model Equations*

We can use the boundary conditions of the problem to eliminate $\hat{\xi}_k$ and its derivatives from the model equations. Then the equations are coupled using the pressure terms. The boundary conditions may also be combined to give a third equation which is essentially a relationship between $\dot{\hat{\xi}}$ and $\hat{\xi}$. The resulting model equations are

$$\begin{aligned}
\dot{a}_k^{(1)} &= \sum_p b_{kp}^{(1)} a_k^{(p)} + \sum_{k',p,q} c_{(k',k-k')_{pq}}^{(1)} a_{k'}^{(p)} a_{k-k'}^{(q)} \\
&\quad + \sum_{r,k',p,q} d_{rk'pq}^{(1)} a_k^{(r)} Re\left(a_{k'}^{(p)} a_{k'}^{(q)*}\right) \\
\dot{a}_k^{(2)} &= \frac{\tilde{p}}{\phi_{2_k}^{(2)}\Big|_w} \left[\sum_p b_{kp}^{(2)} a_k^{(p)} + \sum_{k',p,q} c_{(k',k-k')_{pq}}^{(2)} a_{k'}^{(p)} a_{k-k'}^{(q)}\right] \\
&\quad - \tilde{d}\, a_k^{(2)} + \tilde{k} \left.\frac{\phi_{1_k}^{(3)}}{\phi_{2_k}^{(2)} \frac{\partial U}{\partial y}}\right|_w a_k^{(3)} \\
\dot{a}_k^{(3)} &= - \left.\frac{\phi_{2_k}^{(2)} \frac{\partial U}{\partial y}}{\phi_{1_k}^{(3)}}\right|_w a_k^{(2)}.
\end{aligned} \tag{7}$$

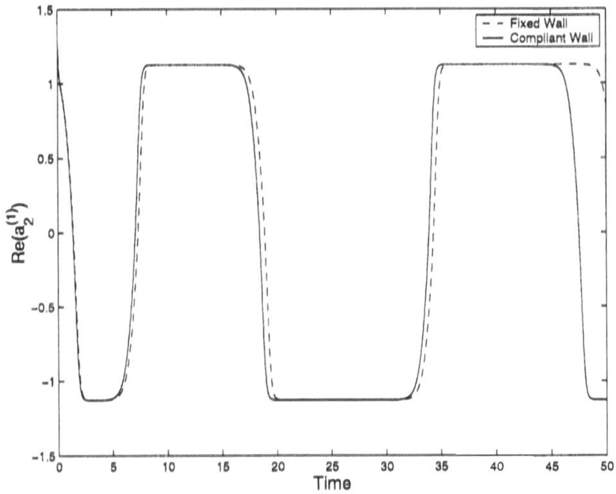

FIGURE 4. Time series of the dynamics of both rigid and compliant wall models with $\tilde{k} = 5000$, $\tilde{d} = 0.1$ and $\tilde{\rho} = 0.1$.

The variables, $a^{(j,k)}$, represent the amplitude of the basis functions $\varphi^{(j,k)}$, where j is the index (or "quantum number") of the POD function and k represents the Fourier wavenumber in the spanwise direction. $\tilde{d} = d/\tilde{m}$, $\tilde{k} = k/\tilde{m}$, $\tilde{\rho} = \rho/(\tilde{m}\phi_{2_k}^{(2)^*}|_w)$, $\tilde{m} = m + \rho/(\phi_{2_k}^{(2)}\phi_{2_k}^{(2)^*})|_w$. Subscript w implies evaluation at the equilibrium position of the wall ($y = 0$).

Three equations are needed for each Fourier number, k, in order to satisfy the boundary conditions. We will consider a system of two Fourier wavenumbers so our system will have a dimension of 6 complex variables (12 real variables).

4.3. RESULTS

Similar POD/Galerkin Projection models have been studied by Aubry *et al.* (1988) and ? for rigid wall flows. They found that heteroclinic cycles exist in the phase space of variables. A schematic diagram of such a cycle is shown in Figure 3.

The fixed points of the cycle correspond to a situation when the amplitudes of the basis functions are not changing with time. These fixed points are joined by heteroclinic connections which form a cycle. The flow "bursts" as it moves away from a fixed point along a heteroclinic trajectory and then reforms as it approaches another fixed point. This behavior is repeated, and since the heteroclinic cycle is attracting, the period of motion tends to infinity as time tends to infinity.

In our compliant wall models we find similar heteroclinic cycles in the phase space. In order to compare the compliant wall model with the rigid wall model we would like to compare the period of motion or the "bursting frequency" for the two. The hope is that the compliant wall model will exhibit a longer bursting period which would indicate a drag reduction.

We systematically varied the compliant wall parameters $\tilde{d}$ and $\tilde{k}$ to find optimal values for drag reduction. Since the important dynamics take place in the linear region of the fixed points, we also completed a linear analysis of the system in order to determine optimal wall parameters. The overall conclusion was that, for this model, and taking into account only the linear analysis (near the fixed points) of the corresponding dynamical system, no such optimal parameters exist. In many cases the compliant wall model performed worse than the rigid wall case. The best case that was found was when the compliant wall and rigid wall dynamics were the same.

A time series plot comparing the dynamics of the rigid wall and compliant wall models (using the same parameters as before) is given in Figure 4.

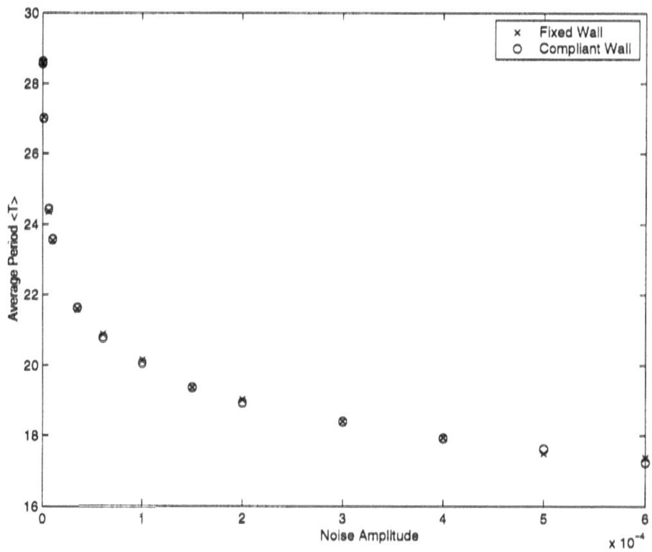

FIGURE 5. Average bursting period as a function of noise amplitude for both rigid and compliant wall models with $\tilde{k} = 5000$, $\tilde{d} = 0.1$ and $\tilde{p} = 0.1$.

4.4. STOCHASTIC MODELING

At this point we would like to re-introduce the pressure term at the outer boundary which has previously been neglected from our equations. This term has been modeled as white noise in (Aubry *et al.* 1988) and (Stone & Holmes 1989). Although the pressure signal at the outer boundary differs from white noise, it may be approximated in this way without any ill effects.

The stochastic differential equations are given in the following form,

$$da^{(j,k)} = f(a^{(j,k)})dt + \epsilon dW,$$

where dW is a random process, f is the function of the eigenfunction coefficients derived from the Galerkin projection and ϵ is its amplitude. After introducing this stochastic term to the equations the dynamics take on a different behavior. Although the phase portraits still show evidence of heteroclinic cycles, the trajectories no longer continue to approach this cycle as time tends to infinity (resulting in a period of motion that tends to infinity as well). Rather, the trajectories are perturbed by the stochastic terms resulting in a period of motion that is random, with a mean value. In effect, the pressure signal at the outer boundary triggers bursts in the flow and modifies the time scale of the process.

By integrating sample paths we may find an average period. Results for both the rigid and compliant wall (with $\tilde{k} = 5000$, $\tilde{d} = 0.1$, $\tilde{p} = 0.1$) are shown in Figure 5.

As can be seen, for a wide range of noise amplitudes, ϵ, the compliant wall model appears to do no better than the rigid wall case, as we encountered in the previous section.

5. Conclusions

The material described above clearly represents work in progress, and has gone through its ups and downs as they are associated with any original work. We had found early indications of reduced drag both in low-dimensional models and direct simulations, only to later dismiss them as being caused by modeling flaws, or by numerical inaccuracies. However, through this process we have constantly improved both the numerical methods for our direct simulation and the basis for our low-dimensional models, and we are confident that both our approaches, taken together, will give us a much clearer picture of the interaction of turbulence with a compliant wall.

Will we be able to ultimately find compliant walls with properties such that turbulent drag can be reduced significantly? At this point, it may be wise for us to be cautious, and simply answer with a "remains to be seen". Ultimately, the question we are asking is the following: If we look at the whole range of parameters for the elastic properties of a compliant wall, are there any parameters that would result in average wall shear stresses that are lower than the ones we get for the limiting case of an infinitely rigid wall? Intuitively, to us at least, it seems unlikely that the answer to that question should be negative. We might also note that if indeed the worst-case scenario did turn out to be true, and if indeed for fundamental reasons that we do not yet understand a rigid wall turns out to be the one that produces the lowest possible amount of turbulent drag, this would constitute a very surprising and important new result on the physics of turbulence.

Until we do have definite answers to these questions, we do believe that our research in this field is worthwhile, and we remain confident that we will be able to give such answers by following the path we have taken. Thus, although our results so far have not been positive for drag reduction, it is our hope that with improved simulations and low-dimensional models which include more of the relevant physics, we will ultimately arrive at a firm understanding of the interaction of compliant surfaces with near-wall turbulence.

Acknowledgements

This work has been supported by the AFOSR under contracts F49620-96-1-0329, F49620-98-1-0397, F49620-99-1-0012, and F49620-00-1-0029.

References

REYNOLDS, O. (1883) On the Experimental Investigation of the Curcumstances which Determine Whether the Motion of Water Shall Be Direct or Sinuous. *Phil. Trans. Royal. Soc.* **174**, pp. 953–982.

ENDO, T.; KASAGI, N. (2001) Active control of wall turbulence with wall deformation. *Fluids and Thermal Engineering* **44** (2), pp. 195-203.

CHOI, H.; MOIN, P.; KIM, J. (1994) Active turbulence control for drag reduction in wall-bounded flows. *J. Fluid Mech.* **262**, pp. 75–110.

BECHERT, D. W.; BRUSE, M.; HAGE, W.; VAN DER HOEVEN, J. G. T.; HOPPE, G. (1997) Experiments on drag-reducing surfaces and their optimization with an adjustable geometry. *J. Fluid Mech.* **338**, pp. 59–87.

CHOI, H.; MOIN, P.; KIM, J. (1993) Direct numerical simulation of turbulent flow over riblets. *J. Fluid Mech.* **255**, pp. 503–539.

KRAMER, M. O. (1961) The Dolphin's Secret. *J. Am. Soc. Nav. Engr.* **73**, pp. 103–107.

BUSHNELL, D. M.; HEFNER, J. N.; ASH, R. L. (1977) Effect of Compliant Wall Motion on Turbulent Boundary Layers. *Phys. Fluids A* **20**, S31.

GAD-EL-HAK, M. 1986 Boundary Layer Interactions with Compliant Coatings: An Overview. *Appl. Mech. Rev.* **3**, pp. 511–524.

CHOI, K. S.; YANG, X.; CLAYTON, B. R.; GLOVER, E. J.; ATLAR, M.; SEMENOV, B. N.; KULIK, V. N. (1997) Turbulent Drag Reduction Using Compliant Surfaces. *Proc. R. Soc. Lond. A* **453**, pp. 2229–2240.

KANG, S.; CHOI, H. 2000 Active wall motions for skin-friction drag reduction. *Phys. Fluids* **12**, 3301.

CARPENTER, P. W.; DAVIES, LUCEY, A. D. 2000 Hydrodynamics and compliant walls: Does the dolphin have a secret? *Current Science* **79** (6), pp. 758–765.

SEMENOV, B. N. 1991 On conditions of modelling and choice of viscoelastic coatings for drag reduction. In: *Recent Developments in Turbulence Management*, (ed. K.-S. Choi), pp. 241-262.

HOLMES, P.; LUMLEY, J. L.; BERKOOZ, G. *Turbulence, Coherent Structures, Dynamical Systems and Symmetry.* Cambridge University Press, 1996.

AUBRY, N.; HOLMES, P.; LUMLEY, J. L.; STONE, E. (1988) The dynamics of coherent structures in the wall region of a turbulent boundary layer. *J. Fluid Mech.* **192**, pp. 115–173.

STONE, E,; HOLMES, P. (1989). Noise-induced intermittency in a model of a turbulent boundary layer. *Physica D* **37**, pp. 20–32.

THE ENERGETICS AND THE HEAT AND MASS TRANSFER IN UNSTEADY STRATIFIED TURBULENCE

HIDESHI HANAZAKI
Institute of Fluid Science, Tohoku University, 2-1-1 Katahira, Aoba-ku, Sendai, 980-8577, Japan, e-mail: hanazaki@ifs.tohoku.ac.jp

1. Introduction

The effects of stable stratification on the fluid motion must be fully investigated since even a weak stratification often gives significant effects on the heat and mass transport. Previous studies have clarified some of the most special characteristics of strongly stratified turbulence including the counter-gradient heat/density flux which makes the eddy diffusivity coefficient negative.

An important remaining question would be the passive scalar diffusion in this system. However, there have been little studies which investigated the difference between the passive and active (density or heat) scalars. It is true that if the initial conditions and the molecular diffusion coefficients are the same for the passive and active scalars, subsequent time development is the same and the eddy diffusion coefficients also agrees. However, if the initial condition is different, subsequent time development will differ from each other.

In this study we consider the passive scalar diffusion in stably stratified turbulence for the first time as an initial value problem. We use the rapid distortion theory (RDT)[1] which has given good agreement with DNS[2] for strongly stratified turbulence without a passive scalar. The theory is based on the linearized governing equations and therefore has limitations in its applicability conditions, but it is the only theory which can take into consideration the effects of initial conditions.

2. RDT Equations

We consider an inviscid Boussinesq fluid with uniform vertical density (active scalar) and passive scalar gradient. We define $\boldsymbol{u} = (u_1, u_2, u_3)$ as the perturbation velocity, $\overline{\rho}(x_3)$ and $\overline{c}(x_3)$ as the undisturbed density and passive scalar distribution, $\rho_0(= \overline{\rho}(x_3 = 0))$ as the mean density, ρ and c as the density and passive scalar perturbation from $\overline{\rho}(x_3)$ and $\overline{c}(x_3)$, and x_3 as the unit vector in the vertical direction. Then, we substitute the spectral decompositions

$$u_i(\boldsymbol{x}, t) = \sum_{\boldsymbol{k}} \hat{u}_i(\boldsymbol{k}, t) e^{i\boldsymbol{k}\cdot\boldsymbol{x}} \quad (i = 1, 2, 3), \tag{1}$$

$$\frac{g}{\rho_0} \rho(\boldsymbol{x}, t) = \sum_{\boldsymbol{k}} \hat{\rho}(\boldsymbol{k}, t) e^{i\boldsymbol{k}\cdot\boldsymbol{x}}. \tag{2}$$

$$c(\boldsymbol{x}, t) = \sum_{\boldsymbol{k}} \hat{c}(\boldsymbol{k}, t) e^{i\boldsymbol{k}\cdot\boldsymbol{x}}. \tag{3}$$

A. Gyr and W. Kinzelbach (eds.), Sedimentation and Sediment Transport, 51–54.

into the governing equations, and obtain equations for the spectral components $\hat{u}_i$, $\hat{\rho}$ and $\hat{c}$.

If all the nonlinear advection terms can be neglected in the governing equations, we can use the linearised equations. The condition for which the linearisation is valid is given by $Fr_l(\equiv u_l/Nl) \ll 1$, where l is the eddy length scale and $u(l)$ is the corresponding eddy velocity. Noting that $u_0/l_0 = O(u/l)$ (l_0: integral scale, u_0: corresponding eddy velocity), and hence $Fr(\equiv u_0/Nl_0) \sim Fr_l$ holds at moderate and low Reynolds numbers, $Fr \ll 1$ is the condition for which we can use the linearlised (i.e., rapid distortion theory: RDT) equations:

$$\frac{d\hat{u}_i}{dt} = \left(\frac{k_i k_3}{k^2} - \delta_{i3}\right)\hat{\rho}, \tag{4}$$

$$\frac{d\hat{\rho}}{dt} = N^2 \hat{u}_3, \tag{5}$$

$$\frac{d\hat{c}}{dt} = -\gamma \hat{u}_3, \tag{6}$$

where $N^2 = -(g/\rho_0)(d\overline{\rho}/dx_3)$ is the Brunt-Väisälä frequency and $\gamma = d\overline{c}/dx_3$ is the mean passive scalar gradient.

3. Solutions of RDT Equations

In this study we assume that the initial fluxes of density and passive scalar are zero as in the usual experiments, i.e., $\Phi_{\rho i}(\boldsymbol{k}, t = 0) = \overline{\hat{\rho}_0^* \hat{u}_{i0} + \hat{\rho}_0 \hat{u}_{i0}^*}/2 = 0$ and $\Phi_{ci}(\boldsymbol{k}, t = 0) = \overline{\hat{c}_0^* \hat{u}_{i0} + \hat{c}_0 \hat{u}_{i0}^*}/2 = 0$ $(i = 1, 2, 3)$, where overline denotes the ensemble average and subscript 0 denotes initial values. If we further assume that the initial fluctuations are isotropic as in the usual experiments, we obtain the variances and the covariances, e.g. as

$$\begin{aligned} \overline{cu_3}(t) &= -\frac{1}{2N}(\overline{c\rho}(0) + 2\gamma PE_0)\int_0^\pi d\theta \sin^2\theta \sin(Nt\sin\theta) \\ &- \frac{\gamma}{4N}(KE_0 - 2PE_0)\int_0^\pi d\theta \sin^2\theta \sin(2Nt\sin\theta), \end{aligned} \tag{7}$$

$$\overline{\rho u_3}(t) = \frac{N}{4}(KE_0 - 2PE_0)\int_0^\pi d\theta \sin^2\theta \sin(2Nt\sin\theta), \tag{8}$$

where KE_0 and PE_0 are the initial kinetic energy and potential energy repectively.

Since the turbulent diffusion coefficients for the density and passive scalar fluxes are given by $K_\rho(t) = \overline{\rho u_3}(t)/N^2$ and $K_c(t) = -\overline{cu_3}(t)/(d\overline{c}/dx_3) = -\overline{cu_3}(t)/\gamma$ in the notation of this paper, $K_\rho(t) = K_c(t)$ is equivalent to the condition $\overline{cu_3}(t)/\overline{\rho u_3}(t) = -\gamma/N^2$. Equations (7) and (8) show that this is equivalent to $\overline{c\rho}(0) + 2\gamma PE_0 = 0$ in the initial condition. Then, we can conclude that for the difference between K_ρ and K_c, initial density fluctuation ($PE_0 \neq 0$) is necessary, since if $PE_0 = 0$, $\overline{c\rho}(0) = 0$ also holds. This suggests that, in the usual grid-generated turbulence where $PE_0 \sim 0$ and the initial correlation $\overline{c\rho}(0) \sim 0$ is small, $\overline{cu_3} \sim (\gamma/N^2)\overline{\rho u_3}$ (i.e., $K_\rho \sim K_c$) holds.

This conclusion apparently disagrees with the previous arguments[3,4] that "When $\overline{c\rho}(0) = 0$, there will be a large difference between $\overline{cu_3}$ and $\overline{\rho u_3}$ in the subsequent time development". However, in the previous studies the discussions have been made based on the energy-flux budget equations for "stationary" turbulence and there have been no considerations on the unsteadiness or the initial conditions. Differences in the results show that the identification of steadiness or unsteadiness of turbulence is important in modeling the turbulent fluxes.

We also note that the effects of PE_0 and $\overline{c\rho}(0)$ appear as the slowly oscillating components with frequency N, and they are superimposed on the rapidly oscillating components with frequency $2N$.

To see the actual development, we show in figure 1 the time development of $R_{c\rho}$, R_{c3} and $R_{\rho 3}$. Here we have used the same initial conditions as used by Kaltenbach et al.[4], i.e., $PE_0 = \overline{c\rho}(0) = 0$ and $N^2\overline{c^2}(0)/(\gamma^2 KE_0) = 3.67$.

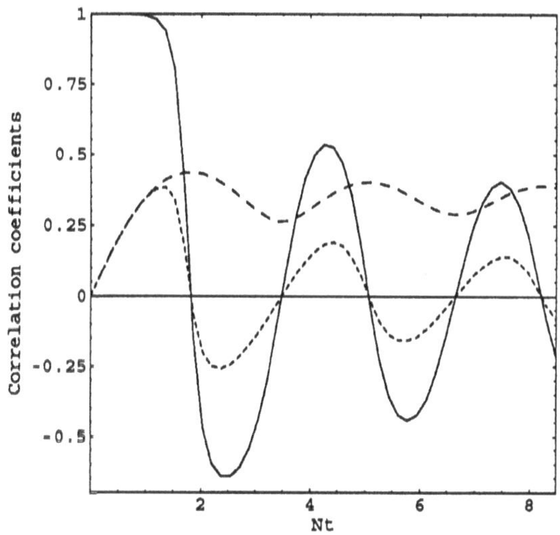

Figure 1: Time development of the correlation coefficients obtained by RDT for $PE_0 = \overline{c\rho}(0) = 0$ and $N^2\overline{c^2}(0)/(\gamma^2 KE_0) = 3.67$. ————, $R_{\rho 3}$; — — — —, $R_{c\rho}$; - - - - - - - - -, R_{c3}.

If $PE_0 = \overline{c\rho}(0) = 0$, all the time developments are determined by $N^2\overline{c^2}(0)/(\gamma^2 KE_0)$ in RDT. For example, R_{c3} at arbitrary time is given by

$$R_{c3}(t) = \frac{-\frac{1}{4} sign(\gamma) \int_0^\pi d\theta \sin^2\theta \sin(2Nt\sin\theta)}{\left(\frac{N^2\overline{c^2}(0)}{\gamma^2 KE_0} + \frac{1}{2} - \frac{1}{4}\int_0^\pi d\theta \sin\theta\cos(2Nt\sin\theta)\right)^{1/2}} \times \left(\frac{1}{3} + \frac{1}{4}\int_0^\pi d\theta \sin^3\theta \cos(2Nt\sin\theta)\right)^{-1/2}. \quad (9)$$

We should note that due to the mean shear ($Ri = 0.5$) in DNS by Kaltenbach et al.[4], long-time development does not agree with the RDT results obtained for

$Ri = \infty$ (no shear). However, short time development ($Nt < 2$) agrees well. Indeed, the DNS by Kaltenbach et al. also showed $-R_{c\theta}(= R_{c\rho}) = R_{c3}$ and $-R_{\theta 3}(= R_{\rho 3}) \sim 1$ (θ: temperature fluctuation) for small Nt.

Long-time asymptotics of the correlation coefficients under the conditions of $PE_0 = \overline{c\rho}(0) = 0$ are

$$R_{c\rho}(Nt \gg 1) \sim -\frac{sign(\gamma)}{\left(1 + \frac{2N^2\overline{c^2}(0)}{\gamma^2 KE_0}\right)^{1/2}}, \qquad (10)$$

$$R_{\rho 3}(Nt \gg 1) \sim \frac{6^{1/2}}{4}\sqrt{\frac{\pi}{Nt}} \sin\left(2Nt - \frac{\pi}{4}\right), \qquad (11)$$

and

$$R_{c3}(Nt \gg 1) \sim -\frac{\frac{6^{1/2}}{4} sign(\gamma)\sqrt{\frac{\pi}{Nt}} \sin\left(2Nt - \frac{\pi}{4}\right)}{\left(1 + \frac{2N^2\overline{c^2}(0)}{\gamma^2 KE_0}\right)^{1/2}}. \qquad (12)$$

These results show that if $PE_0 = \overline{c\rho}(0) = 0$ and $N^2\overline{c^2}(0)/(\gamma^2 KE_0) \ll 1$ holds, the correlation coefficients satisfy $|R_{c3}| \sim |R_{\rho 3}|$ and at the same time $|R_{c\rho}| \sim 1 (Nt \gg 1)$, although if $N^2\overline{c^2}(0)/(\gamma^2 KE_0) \gg 1$, the correlation coefficient becomes small ($|R_{c\rho}| \ll 1$ for $Nt \gg 1$). Therefore, initial uncorrelation ($\overline{c\rho}(0) = 0$) does not assure the subsequent uncorelation between the passive and active scalars.

On the other hand, in the turbulent diffusion coefficients, $K_\rho(t) = K_c(t)$ holds independent of the value of $\overline{c^2(0)}$ only if $PE_0 = \overline{c\rho}(0) = 0$. This shows that $K_\rho = K_c$ ($Nt \gg 1$) and $|R_{c\rho}| = 1$ ($Nt \gg 1$) do not have direct correspondence.

4. Conclusions

Analytical solutions of the RDT equations for the initially isotropic turbulence show that some of the previous arguments based on the 'stationality' of turbulence is not correct in unsteady turbulence. This shows that the initial conditions and the unsteadiness are important to estimate the turbulent diffusion coefficient in 'unsteady' turbulence. The results will be useful for the parameter determinations in future DNS, experiments and modeling of the turbulent diffusion.

References

[1] Hanazaki, H. and Hunt, J. C. R., Linear processes in unsteady stably stratified turbulence, J. Fluid Mech. **318**, 1996, 303–337.
[2] Gerz, T. and Yamazaki, H., Direct numerical simulation of buoyancy-driven turbulence in stably stratified fluid. J. Fluid Mech. **249** , 1993, 415–440.
[3] Warhaft, Z., Heat and moisture flux in the stratified boundary layer, Q. J. Roy. Met. Soc. **102**, 1976, 703–707.
[4] Kaltenbach, H. -J., Gerz, T. and Schumann, U., Large eddy simulation of homogeneous turbulence and diffusion in stably stratified shear flow, J. Fluid Mech. **280**, 1994, 1–40.

II

Turbulence and Sediment Transport

Turbulence and Sedimentation

THE PROBABILITY OF SEDIMENT MOVEMENT AT THE THRESHOLD OF MOTION AND TIME DEPENDENT FLUID PROCESSES

CLINTON L. DANCEY (*Virginia Tech, Department of Mechanical Engineering, Blacksburg, VA, USA, 24061-0238, cld@vt.edu*)
PANAYIOTIS DIPLAS (*Virginia Tech, Department of Civil and Environmental Engineering, Blacksburg, VA, USA, 24061, pdiplas@vt.edu*)

1. Introduction

The characterization and prediction of the motion of sediment near the threshold are long-standing problems in the study of sediment transport. In this paper a new criterion for the threshold condition, recently introduced by the authors, is reviewed and relevant physical mechanisms that may be necessary for the accurate prediction of sediment motion near the threshold are elaborated. The new criterion is based upon an operational definition of the probability of grain movement. The relevant physical mechanisms to be discussed relate to inherently time-dependent, transient flow effects - inertia forces and impulse (force/time history) effects.

2. Review: The Probability of Movement

The threshold of motion for sediment in open channel turbulent flow has been characterized by a variety of experimental approaches over the last 60 – 70 years (see for example review article by Buffington and Montgomery, 1997). Recently the authors (Dancey et al, 2002) have proposed a criterion that is based upon similarity arguments (similar to Neill and Yalin, 1969) with an emphasis on the measured fractional rate of grain entrainment (similar to Shvidchenko and Pender, 2000). The criterion employs an operationally defined "probability of grain movement" that includes the rate of grain dislodgment over an area of the sediment bed, the number of mobile available grains, and the time scale characteristic of the turbulent channel flow. The threshold condition is specified by a particular (experimentally assigned) value of this probability of grain movement. The approach is quantitative, objective, and permits a consistent characterization of the threshold condition.

This criterion assumes that individual sediment grains are dislodged by randomly occurring turbulent structures (yet to be experimentally identified). The probability of individual grain movement, given that the grain has been exposed to such a structure, is used to assign the threshold condition. That is, the probability of individual grain movement is assigned a low but non-zero value at the threshold. This low specified value of the probability is the threshold condition. The expression for the probability of individual grain movement is given by (Dancey et al, 2002)

$$\Gamma = \frac{\dot{n}\Im}{n}$$

where $\dot{n}$ is the rate of grain dislodgment over an area of the bed (number per unit time); n, is the number of grains available to be dislodged over the area; and $\Im$ is the relevant time scale of the flow. In Dancey et al (2002) it is assumed that $\Im = H/U$, where H is

A. Gyr and W. Kinzelbach (eds.), Sedimentation and Sediment Transport, 57–60.

the channel flow depth, and U is the flow cross sectional area averaged channel velocity. It is clear that this definition is operational, based purely upon the measured rates of grain movement and the time scale of the flow, without regard to the precise processes or mechanisms actually responsible for grain dislodgment. However, individual grain movements are initiated by local flow/sediment interactions, which, in the mean, are responsible for the measured value of $\dot{n}$. In the remainder of this paper, hydraulic processes that may be associated with individual grain dislodgment are discussed. The role of inertia forces and impulse effects are emphasized. Because of space limitations, these concepts will only be introduced here, with illustrative examples to demonstrate their relevance to grain movement.

3. Discussion

As an example, the forces exerted by an unsteady but uniform incompressible flow on the isolated stationary sphere are considered. For simplicity only the component aligned with the uniform flow, U, is considered here. The hydraulic force acting on the sphere is commonly modeled by a sum of forces:

$$F = C_D \frac{1}{2} \rho_f U^2 A + C_m \rho_f \forall \frac{DU}{Dt} + F_{History}$$

The first term on the right-hand-side is the quasi-steady viscous drag force (C_D is the drag coefficient); the second term on the right is the "inertia" force (representing both the added mass and the local fluid acceleration). The last term on the right hand side is the history integral. A is the frontal area of the exposed sphere and $\forall$ is the volume. $C_m = 1 + C_{AM}$, where C_{AM} is the added mass coefficient. For the purposes of the present illustration, the Basset history integral is neglected and C_D and C_m are assumed constant.

The relevance of the inertia force is easily demonstrated by assuming U(t) is periodic, e.g. $U = U_o + U' \sin(\omega t)$. U_o is the average flow velocity, U' is the fluctuation amplitude and ω is the frequency of the oscillation. By substituting U(t) into the previous equation the instantaneous, quasi-steady viscous drag and inertia force contributions can be obtained. The instantaneous "inertia" force is given by

$$F_{inertia} = C_m \rho \forall U' \omega \cos \omega t$$

The amplitude of the instantaneous inertia force is directly proportional to the oscillation frequency, ω, as well as the fluctuation amplitude, U'. However, for sinusoidal U(t), the average force, which is the integral of the total instantaneous force over an integral number of cycles is exclusively associated with the viscous drag term. The average inertia force is zero since cos ωt integrates to zero over an integral number of oscillations. That is, the time average force, which traditionally acts as a surrogate for the critical stress, does not account for instantaneous "inertia" forces that arise due to the unsteadiness of the flow. In fact, depending upon the magnitudes of C_D, C_m, U_o, U', and ω, the instantaneous magnitude of the total force may significantly exceed the average. For large U' and ω the "inertia" force may in fact be larger than either the instantaneous or average viscous drag force, and yet it does not contribute at all to the time average force. In such circumstances the mean bed stress does not reflect the forces responsible for sediment dislodgment. In light of this fact, the probability distribution of DU/Dt in the flow is likely to be relevant for the proper modeling of grain movement; yet, to the authors' knowledge, it has been ignored in all characterizations of the threshold condition and in models for the bedload transport rate.

The time dependent nature of the hydraulic forces is also relevant since the physical movement of a particle is associated not only with the instantaneous magnitude of the force applied to the particle, but also with the duration of the applied force. That is to say, the hydraulic force must exceed some minimum value to initiate movement, but the force must remain in effect over some time interval to actually dislodge a particle from the sediment bed. This can be illustrated with another simple example. In this instance the motion of a spherical particle submerged in water is analyzed under the application of a vertical, transient lift force that is opposed to gravity. For simplicity viscous drag and inertia forces are neglected in this illustration. A constant lift force, F, is applied at t = 0 for a time period T. A sufficient but brief impulse can kick the particle upwards. For this illustration, an upward vertical displacement of one particle diameter is regarded as representative of significant particle dislocation, and results will be summarized here for the case of an 8 mm diameter glass sphere. Figure 1 is a plot of the solution: the required force to weight ratio, F/W, as a function of force duration, T, to achieve a vertical dislocation of 8 mm. For a specified value of the force to weight ratio, the duration must equal or exceed the value given in the figure to achieve the stated displacement of 8 mm. Clearly no unique value of the applied force to cause a particle dislocation of one diameter exists; both F and T are relevant.

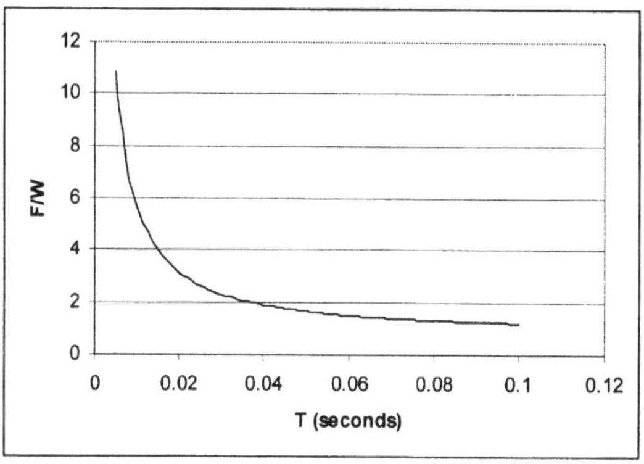

Fig. 1. F/W, vs T, to cause vertical displacement of one particle diameter.

In the context of the earlier model with a sinusoidal oscillating flow, a qualitative time history of the hydraulic force is shown in Figure 2. The average force is also indicated. In this illustration it is assumed that the critical magnitude of the applied force (necessary to cause an unbalanced moment on the sediment particle and initiate motion) is greater than the average but less that the instantaneous maximum, as indicated in the figure. The impulse associated with particle movement is given by

$I = \int (F - F_{crit})dt$ and it is represented by the shaded area in the figure. If the impulse is too small, the particle may move within its "pocket" but will not be dislodged to a new pocket. The unbalanced force must be applied over a sufficiently long time to transfer enough momentum to the particle to cause complete dislodgement of the particle from its pocket. Extremely large unbalanced forces require only short periods of time to impart sufficient impulse (Figure 1), while small

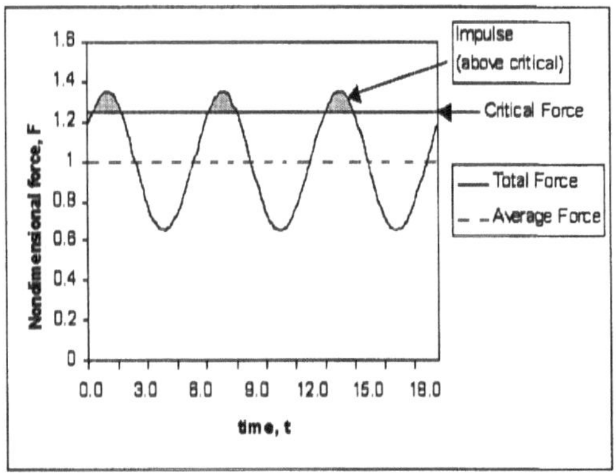

Fig. 2. Periodic hydraulic force, average and threshold force, and impulse.

unbalanced forces above the critical require a longer duration to impart the same impulse. That is, in any unsteady flow the probability distribution of the impulse, I, is likely to be very relevant to grain entrainment, but it has never been even indirectly measured, to the authors' knowledge; and it has not been considered in any published model for sediment transport.

4. References

Buffington, J. and Montgomery, D. (1997). "A systematic analysis of eight decades of incipient motion studies, with special reference to gravel-bedded rivers." Water Resources Research, 33(8), pp. 1993-2029.

Dancey, C.L., Diplas, P., Papanicoloau, A., and Bala, M. (2002). "Probability of individual grain movement and the threshold condition." To appear in the Journal of Hydraulic Engineering.

Neill, C.R. and Yalin, M.S. (1969). "Quantitative definition of beginning of bed movement." Journal of the Hydraulic Division, 95 (1), pp. 585-588.

Shvidchenko, A. B., and Pender, G. (2000). "Flume study of the effect of relative depth on the incipient motion of course uniform sediment." Water Resources Research, 36 (2), pp. 619-628.

TURBULENCE DYNAMICS IN A TWO-PHASE FLOW

ARMIDA GAION
Padua University – Engineering Faculty - Department IMAGE
Via Loredan, 20 – I 35131 Padova, Italy – e-mail: gaion@idra.unipd.it

1. Introduction

Under proper conditions the turbulent flow over a sand bed picks up solid particles. The initial pick up, the suspended transport and the interaction between the two phases depend on the parameters:

$$Re = \frac{Uh}{\nu} \qquad Fr = \frac{U}{\sqrt{gh}} \qquad Re_s = \frac{|U_f - U_s|d}{\nu} \qquad R = \frac{w}{ku^*} \qquad St = \frac{t^*U}{L}$$

and on the time scales: t^* of the particles, t_e of the energetic eddies and t_η of the dissipative eddies, that is on:

$$t^* = \frac{d^2 \rho_s}{18 \nu \rho_f} \qquad t_e = \frac{k\nu}{u^{*2}} y^+ \qquad t_\eta = \frac{k^{1/4}\nu}{u^{*2}} (y^+)^{1/4} \quad .$$

For $t^* < t_e$ the particles follow the energetic eddies with a tendency to decrease the fluid turbulence intensity. Even at low values of the solid concentration a strong interaction between the two phases in the wall region, and mostly with regard to the turbulent part of the flow field, has been experimentally detected [7]. Relevant consequences, like drag reduction or, in different flow conditions, drag enhancement can follow. Thus the possibility of the wall shear stress control by solid transport is of interest [8].
The wall pressure induced by the wall stream-wise turbulence is responsible for the incipient movement of the solid particles [5]. In the wall region the solid concentration is the highest, the transport of the suspended solid is strongly intermittent, as well as the burst, and the correlation between the fluctuating concentration and the fluctuating fluid velocity is strong. The burst period, which in wall variables is T^+=100, is correlated to the period of the span-wise coherent structures in the outer region of the boundary layer, which carry most of the Reynolds stress [4] and which, when the condition $t^* < t_e$ is satisfied, transport the solid particles.
For particle density ρ_s=2.6 g/cm^3, diameters d=100μm and d=200μm and for a volume concentration $\overline{c} \cong 10^{-3}$ the mean and turbulent velocities have been measured by PDA in an open channel, in the boundary layer region, both for the fluid and for the solid, down to y^+=5 (Fig. 2) [6].The comparison between the different time scales with reference to the measured data is shown in Fig. 1.

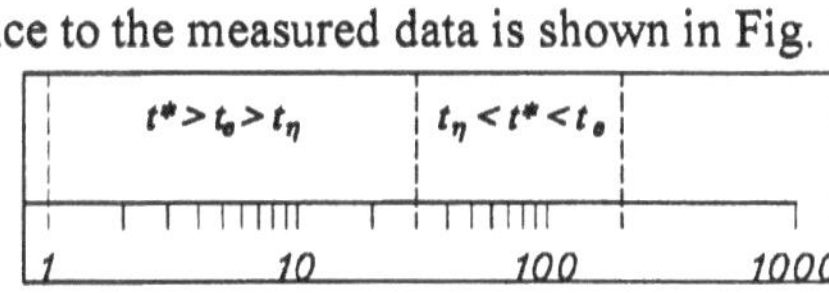

Figure 1 Comparison between time scales, energetic and dissipative eddies with reference to the data of Fig.2

A. Gyr and W. Kinzelbach (eds.), Sedimentation and Sediment Transport, 61–65.

The results show, due to the solid transport, a fluctuating velocity and a Reynolds-stress increase for $y^+ \lesssim 30$ and, in the logarithmic region, a decrease of the Reynolds-stress and of the turbulent fluctuations, stronger for v', and a change of the value of the ratio u'/v', connected to the shape of the main vortices. The particle velocity is shifted with respect to the fluid velocity.

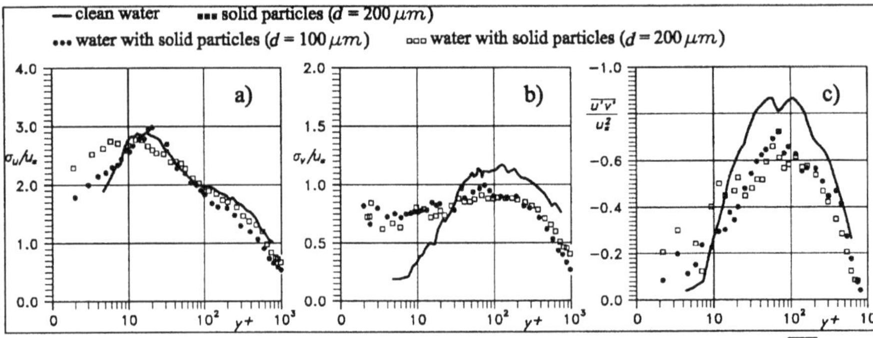

Figure 2 – Adimensionalized mean square values a): u', b): v', c): Reynolds-stress $\overline{u'v'}$.(from: Gaion, Righetti, Romano [4]). Re_f=1.17*10^4, Fr_f=1.7, u*=4.05 cm/s. $u^*=\sqrt{gR_H i_b}$, i_b: channel slope.

h cm	R_H cm	ν m²/s	U_f cm/s	u* cm/s	$Re_f 10^3$	Fr_f	w cm/s	$\bar{c}$?10^3	t* s	$t_\eta/(y^+)^{1/4}$ s	t_e/y^+ s
2	1.43	$939 \cdot 10^{-8}$	76	4.05	11.75	1.70	3.83	1.25	$63 \cdot 10^{-4}$	$45 \cdot 10^{-5}$	$229 \cdot 10^{-6}$

Table 1 – Variables and parameters values referred to the data of Fig. 1.

2. The theoretical model

The momentum and continuity equations for the two phases are obtained under the hypothesis: a) c << 1, b) negligible viscous stress, c) small value of Re_s.
In their non-dimensional formulation they can be written:

$$(1-c)\frac{D\vec{u}_f}{Dt} = (1-c)p + \frac{1}{Fr^2}(1-c)h - \vec{F} \quad (1)$$

$$\frac{f}{ft}(1-c) + \left[(1-c)\vec{u}_f\right] = 0 \quad (2)$$

$$c\frac{Du_s}{Dt} = -cp + \frac{1}{Fr^2}h + \vec{F} \quad (3)$$

$$\frac{fc}{ft} + (c\vec{u}_s) = 0 \quad (4)$$

$\vec{F}$ is the interaction between the two phases, which is supposed to be the only drag force

$$F = 18(d/L)^{-2}\,\mathrm{Re}_f^{-1}\,c\left|U_f - U_s\right| = \beta\,\mathrm{Re}_f^{-1}\,c\left|U_f - U_s\right|$$

After introducing the mean and fluctuating velocity $\boldsymbol{U}$ and $\boldsymbol{u}'$ and the mean and fluctuating concentration $\bar{c}$ and c', following Drew [2] in the averaging procedure, we obtain from equation 1:

$$(1-\bar{c})\frac{DU_f}{Dt}=-(1-\bar{c})p+\overline{c'p'}+\frac{1}{Fr_f^2}(1-\bar{c})h-\frac{\beta}{\mathrm{Re}_f}\left[\bar{c}\left|U_f-U_s\right|-\overline{c'u'}\right]+$$

$$\frac{1}{\mathrm{Re}_f}\left[(1-\bar{c})\overline{u_f'u_f'}\right] \tag{5}$$

The particle mean velocity can be expressed as a function of the fluid velocity through the parameter $St/Fr<1$, following Druhizhin [3]. The perturbation momentum equation follows from equations 1 and 5. Considering the boundary layer region $y^+>50$, periodical cross-wise vortices are superimposed to the mean flow, which is in the x direction:

$$u_f(x,y,t)=U_f(y,t)+u_f'(x,y,t), \qquad v_f(x,y,t)=v_f'(x,y,t).$$

From equation 5 it follows that the effect of the disturbance on the mean flow comes from the y behaviour of the Reynolds-stress and from the correlation terms $\beta\,\overline{c'u'}$ and $\overline{c'\ p'}$.

The problem is then formulated in terms of the perturbation stream function Ψ and of the vorticity ω:

$$\Psi(x,y,t)=a_1(y,t)\cos(\lambda x)+a_2(y,t)\sin(\lambda x)=A(y,t)\cos(\lambda x-\alpha(y,t))\ ;$$

$$\omega(x,y,t)=b_1(y,t)\cos(\lambda x)+b_2(y,t)\sin(\lambda x), \quad \text{with } A^2=a_1^{\,2}+a_2^{\,2};$$

Initial conditions refer to a trivial perturbation symmetrical with respect to the y axis and to a uniform concentration inside vortices decaying at the end of the burst cycle. Boundary conditions refer to the burst region.

3. Observations on the Reynolds-stress

From the stream function Ψ, one obtains:

$$-\overline{u'v'}=-\frac{\lambda}{2}A^2\frac{f}{fy}\alpha(y,t),$$

showing the dependence of the Reynolds-stress on the y behaviour of the phase α of the vortex, that is on its orientation with respect to the mean flow. Fig. 3a) shows a stable vortex, getting energy from the mean flow and giving a positive contribution

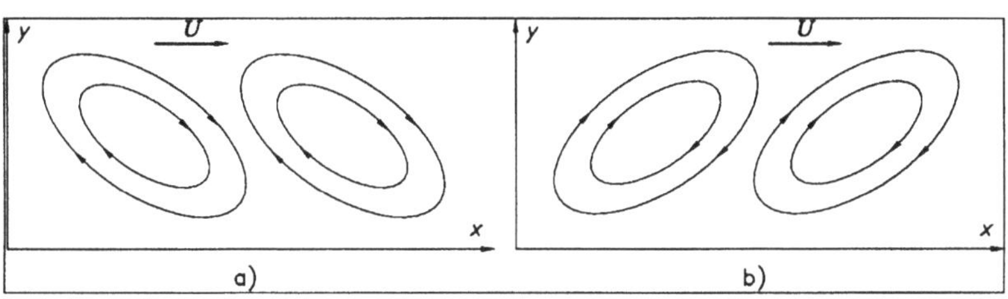

Figure 3 – Stream function of the perturbation. a): the phase is decreasing with increasing y and the Reynolds-stress in positive; b): the phase is increasing with y and the Reynolds-stress is negative.

to the Reynolds-stress and Fig. 3b) shows a vortex giving back energy to the mean flow and giving a negative contribution to the Reynolds-stress.
Integration of equation 5 through the boundary layer depth shows, in comparison with clean water, a Reynolds-stress decrease due to the contribution of the terms $\beta\,\overline{c'u'}$ and $\overline{c'\ p'}$; their relevance on the balance comes from their physical meaning: the mean values are strongly dependent on the intermittency in the burst region.

4. Solution procedure and results

Equation 1 for mean flow, the perturbation vorticity equations and the continuity equation are the starting point for the solution. As a first approximation non linear terms are forgotten, which means no interaction between different wave-lengths is considered.
A Runge–Kutta fourth order step-by-step integration is the numerical method followed, which will be exposed in detail in a different paper. The parameters values follow from the experimental data we refer to and are reported in table 1. The results show that the concentration moves towards the external region of the vortex; during the burst period the vortex, when the solid is transported, becomes less stable and for about 60% of the period it shows negative viscosity conditions, so contributing to a Reynolds stress decrease, and to a decrease of v'. The apparent Re_f is so decreased and the dissipation mechanism is affected. Small vortices become stable and the total dissipation is decreased.

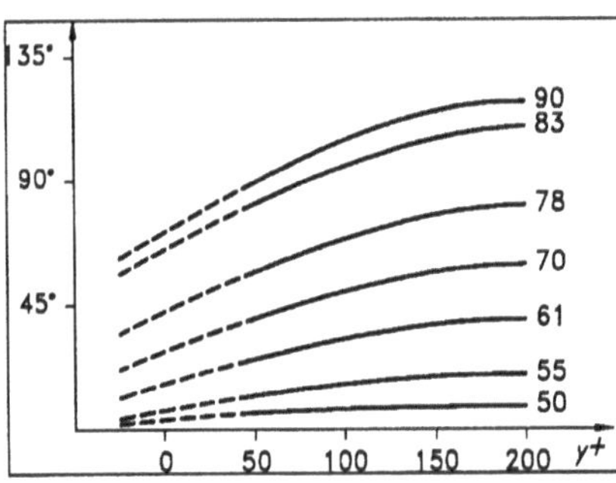

Figure 4 – Phases of the function of the trial perturbation at different times t^+ as function of y^+.

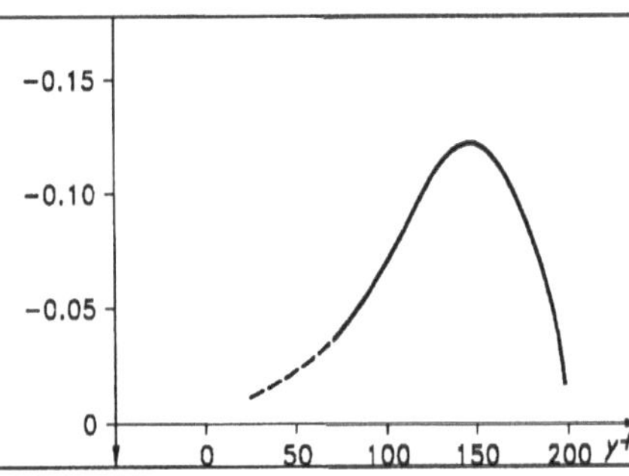

Figure 5 – Reynolds-stress value of the trial perturbation at the time t^+=90.

Main Notations

u	non dim. instantaneous velocity	k	von Kàrmàn constant
U	non dim. mean velocity	$(.)'$	fluctuating value
h	water depth	$(.)_f$	referred to the fluid
d	particle diameter	$(.)_s$	referred to the solid
w	particle fall velocity	$(.)^+$	non-dimensional wall variable
R_H	hydraulic radius	ρ	mass density
ν	kinematic fluid viscosity		

References

[1] Bellomo E. *Finite disturbances in plane parallel laminar flows. A numerical program.* Proc. of the NATO Advanced Study Institute on "Surface Hydrodynamics". 95-108. 1966.

[2] Drew D.A. *Mathematical modelling of two phase flow.* Ann. Rev. Fluid Mech. *15*, 261-291. 1983.

[3] Druzhinin O.A. *On the two way interaction in 2D particle laden flows. J.F.M. 297*, 49-76. 1995.

[4] Elgobashi S. E. *On predicting particle-laden turbulent flows.* J. Appl. Scientific Research, 52, *309-329*. 1994.

[5] Gaion A. *Moto incipiente di una particella sul fondo di un canale.* X Conv. AIMETA. 731-734. 1990.

[6] Gaion A., Righetti M., Romano G.P. *Influenza del trasporto solido sulle caratteristiche del moto medio e della turbolenza del fluido.* Conv. Turbolenza. Capri. 21-27. 1992.

[7] Gyr. A *Structure of turbulence and drag reduction.* Proc. IUTAM Symp. 1989. Springer. 1990.

[8] Lumley J., Blossey P. *Control of turbulence.* Ann. Rev. Fluid Mech. *30*, 311-328. 1998.

[9] Righetti M. Interazione tra fluido e solido in sospensione in prossimità del fondo. Ph.D. Thesis, Padua University, 1994.

[10] Robinson S.K. *Coherent motions in the turbulent boundary layer.* Ann. Rev. Fluid Mech. 23, 601-639. 1991.

[11] Sumer M.B. *Particle motions near the bottom in turbulent flow in an open channel.* J. Fluid Mech. 86, 109-127. 1981.

INVESTIGATION OF PARTICLE SIZE DISTRIBUTION IN TURBULENT OPEN CHANNEL FLOW

THOMAS DREHER and BERNHARD WESTRICH
Inst. für Wasserbau, Universität Stuttgart, Pfaffenwaldring 61, 70550 Stuttgart, E-mail, Dreher@iws.uni-stuttgart.de, Westrich@iws.uni-stuttgart.de

1. Introduction

Deposition of suspended sediments in turbulent open channel flows is dominated by complex interaction of hydrodynamic processes in turbulent boundary layer and the self-organized macro and micro scale roughness at the bed. The present knowledge about physical processes controling sedimentation is still poor. Particularly with regard to the effects of near bed turbulence charateristics on the onset of fractional sedimentation and the influence of bed roughness, depositional features and bed forms, such as low speed streaks, ripples and dunes. The experimental investigation is focussed on the fractional sedimentation of suspended particles with a size spectrum ranging from 2 to 150 μm and the interaction with a fixed smooth and rough bottom. The experimental results provide a more detailed insight into sedimentation processes, based on detailed measurements of flow, particle size and partial concentrations. The very near wall behaviour of particles is measured with a special non-intrusive particle sensor.

2. Experimental set up

All experiments were conducted in a laboratory flume with with a recirculating flow. It has a length of 8m, a width of 0.25m and a constant slope of $I_0=10^{-3}$ and is equipped in a way, that deposition of suspended sediments in the piping and the backflow channel can be excluded. Hydraulically smooth conditions were provided by a channel bed of steel, whereas for hydraulically rough conditions a single layer of closely packed sand particles (average diameter 1mm) was fixed on the channel bed. All flow measurements were non intrusive and performed with a two-component laser doppler anemometer in clear water.

2.1 PARTICLE MEASUREMENTS (GRAIN SIZE)

The particle size distributions and the partial concentrations were determined by using Aucoteams particle sizer Partmaster L, based on the principle of light attenuation [1]. It measures the size of individual particles. The instrument is calibrated for spherical particles and allows to measure 128 grain size intervals between 1.6 and 293 μm. Measurements can be made up to approx. 6000 particles per milliliter sampling volume.The particle volume $\pi d_a^3/6$, determined from the measured diameter d_a, does not necessarily correspond to the volume of the not spherical sand particle.

(1)
$$\psi_i = \frac{m_{zu,i}}{\sum_i^N \frac{n_j}{6} \pi \, d_{a_j}^3 \, \varrho_s}$$

So a material-specific calibration is required. The grain size-dependent calibration factor

A. Gyr and W. Kinzelbach (eds.), Sedimentation and Sediment Transport, 67–74.

ψ_i is defined in equation 1, where j denotes the particle intervals from 1 to 128, n_j and d_j the number of particles and the diameter for interval j, m_{zui} the mass of the admitted sand mixture of the species i with a density ϱ_s. If enough sand species i, with narrow particle size distributions, are available, a calibration for the entire measuring range can be provided. The sand for the calibration was obtained by wet sieving and by separation in settling columns. ψ_i refers thereby to the average grain size of the species i. The grain size d_s is defined as the size of a spheric particle with the same volume as that of the arbitrarily formed sand particle, i.e. $d_s = \psi^{1/3} d_a$.

2.2 USED SAND MATERIAL

For runs with smooth and rough bed, sand of the type SP10 and SP6 are used, respectively (Figure 1). The error bars are taken from the manufacturers data sheet.

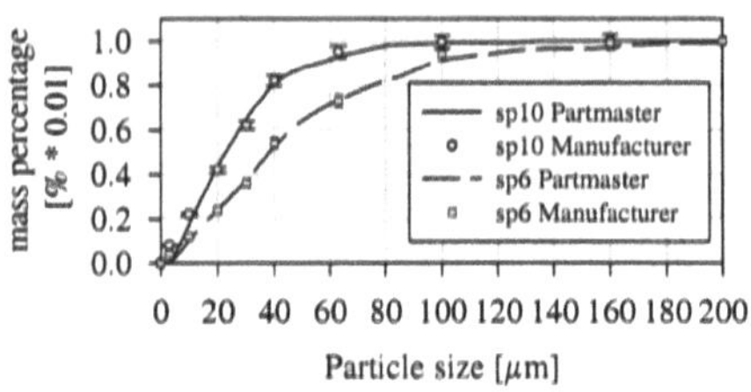

Fig. 1

2.2 PARTICLE SENSOR

The movement of particles with direct contact to the wall, is observed with a self-developed particle sensor, schematically shown in Figure 2. The toothed structure acts as a capacitor, which is part of a bridge circuit that is adjusted to a minimum output for clear water. If a particle moves along the sensor, the electrical field, which has a maximum intensity at the gaps, is disturbed. In Figure 2 the output voltage as function of time is schematically shown for a solitary particle passing the surface. The time between the peaks is proportional to the particle velocity. With presence of a sediment layer, the output signal is proportional to the thickness of the layer. Figure 3 shows an example of such a sensor signal. A signal voltage of approximately 1,62 V corresponds to the particle free sensor. A total uncovered sensor surface occurs intermittently for short time periods, as a consequence of turbulent bursting. This so called ejection event is well known in

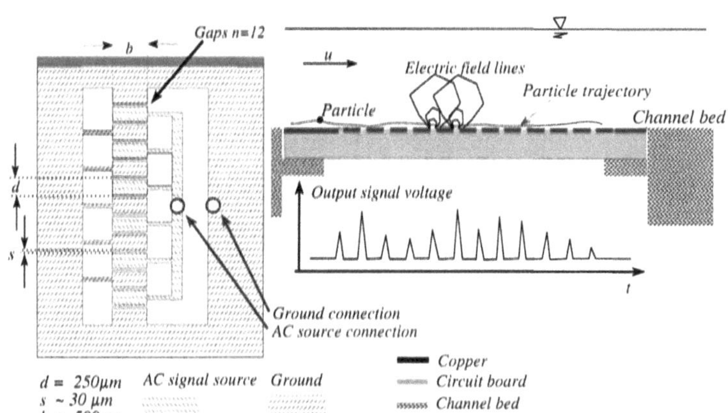

Fig. 2 structure and functional principle of the particle sensor

open channel hydraulics [2][3]. Most time the signal voltage varies around mean values clearly above the output signal for a particle-free sensor, i.e. the bed is then covered with a mobile sediment layer. If an ejection event is identified for a signal voltage below a certain threshold value, it is possible to quantify the average time between bursts t_b, or the burst frequency f_b. This threshold value was defined as $0.15*(V_{max}-V_{cw})+V_{cw}$, where V_{cw} is the value for clear water and V_{max} is the maximum signal voltage of about 2.16 volts. The signal voltage breaks down if the sediment layer, covering the sensor surface, is much greater compared to the width of the gaps. In this case the output is below 1.62 volts without any fluctuations. In Figure 3 a ripple starts passing the sensor at t~115 s.

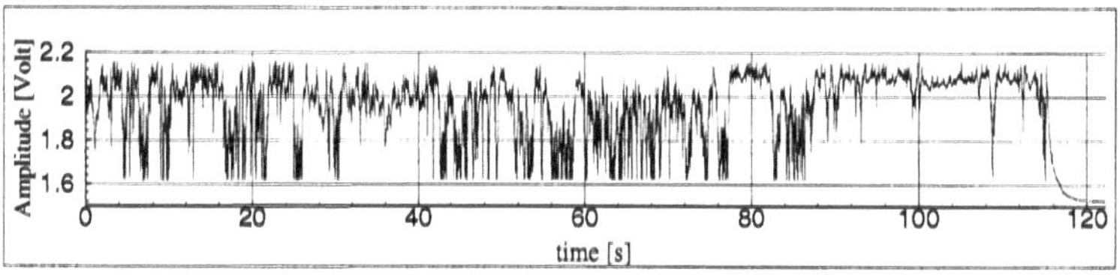

Fig. 3

3. Hydraulically smooth conditions

Experiments were carried out for three different discharges Q. The hydrodynamic parameters, as well as particle characteristics are specified in table 1 for clear water conditions. The bottom shear velocity u_* was determined by extrapolating the Reynolds stress $<-\varrho u'v'>$, obtained by LDA measurements, to the channel bed. The integral parameters $I_e=u_*^2/(g\ r_{hy})$ und $\epsilon_{average} = 1/(\varrho_w V)\ dE/dt$, well known in open channel hydraulics [2], were used to calculate the dissipation rate to $\epsilon_{average} = u_*^2 u/r_{hy}$.

Discharge	Q	5	10	15	l/s
Depth	h	4.59	6.80	8.79	cm
Velocity	u=Q/A	.436	.588	.683	m/s
Sheare velocity	u_*	0.0242	0.0266	0.0294	m/s
Reynolds number	Re	20010	39980	60000	-
Froude number	Fr	.65	.72	.74	-
Dissipation rate	$\epsilon_{average}$	.00762	.00946	.0114	m^2/s^3
Kolmogoroff scale	$\eta=(\nu^3/\epsilon)^{1/4}$	107	101.4	96.7	μm
Viscous sublayer	$5\text{-}10\ \nu/u_*$	206-413	188-377	170-340	μm

Table 1

I_e is the energy gradient, $r_{hy} = bh/(b+2h)$ the hydraulic radius, g the acceleration due to gravity, ϱ_w the density of water, E the dissipated energy, h the water depth and V the unit volume. The dissipation rates correspond to the values obtained for y~0.1h by Eq.2 [2].

(2)
$$\epsilon = \frac{u_*^3}{h}\ 9.8\ \frac{1}{\sqrt{y/h}}\ e^{-3y/h}$$

The maximum Rouse parameter $R=v_s/(\kappa u_*)$ of all runs is 1.2 for $d_s = 125\mu m$ and Q=5l/s. v_s and $\kappa=0.4$ denotes the fall velocity adapted from Olson [4] and the von Karman constant, respectively. The values from table 1 show that the particles are smaller than the viscous sublayer and smaller or in same order of magnitude as Kolmogoroffs micro scale $\eta=(\nu^3/\epsilon)^{1/4}$, where ν is the kinematic viscosity. For moderate concentration such particles do not suppress turbulence [5] The Shields parameter [2]for the three discharges, $\tau_* = \varrho_w\ u_*^2/[(\varrho_s-\varrho_w)gd_s]$, varies for $2<d_s<163\mu m$ between 0.3 and 30 for

0.05<Re*<5, where $Re_* = d_s u_* / \nu$ is the particle Reynolds number. Thus, τ_* is above the classical Shields curve and in the range of ripples and low speed streaks for diverse modified Shields diagrams after, for example [2][6].

3.1 INVESTIGATIONAL PROCEDURE AND RESULTS

Starting with clear water conditions, the concentration during an experiment was increased stepwise by feeding sediment material. The feeding took place in a time interval larger than the circulation time of the system $t_u = V_{tot}/Q$, so a quasi-steady increase of concentration is ensured. V_{tot} is the total volume of circulating water. The samples were taken approx. 20 minutes after each loading at y=h/2, y=h/4 and y=5mm. The contents of the of the micro sediment trap, consisting of a 1mm hole in the bed, was withdrawn for different concentrations $C_R = m_{in}/V_{tot}$, where m_{in} denotes the entire mass of sand already added to the flow. The sensor signals were recorded after each loading step over a time interval of 90-120s with a sampling rate of 1.6 kHz. Figure 4 shows for ξ=y/h=0.5, the total suspended concentration $C_{tot} = \Sigma_j C_{pj}$ as the sum over all partial concentrations C_{pj}, depending on the discharge Q. The ratio C_{tot}/C_R varies between almost 100% for Q=15l/s, where d_{50} with approx. 23μm corresponds to the SP10 sand, and approx. 50% for Q=5l/s with d_{50}~15μm. The ratio C_{tot}/C_R is almost constant for Q=15 and Q=10l/s. In case of Q=5l/s the relative concentration, and the average particle size, increases clearly with

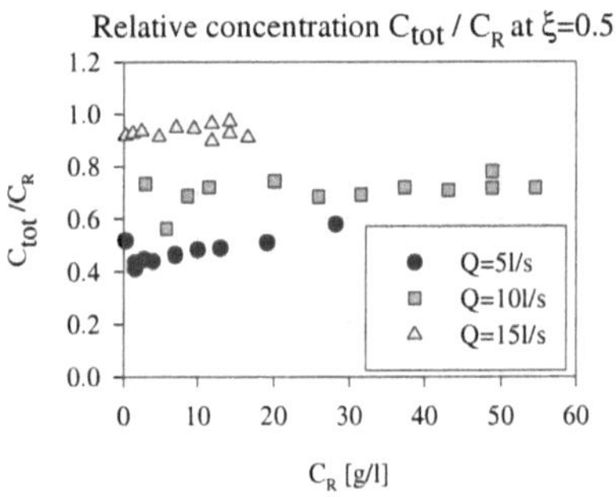

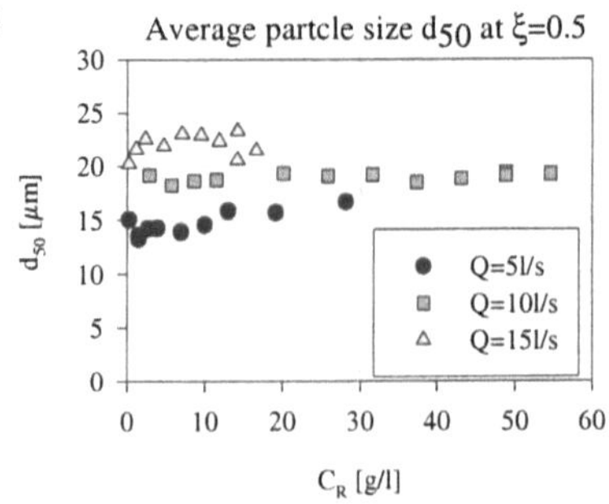

Fig. 4a

Fig. 4b

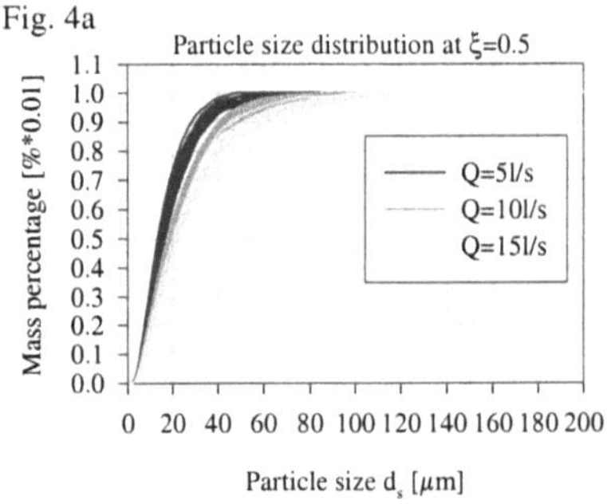

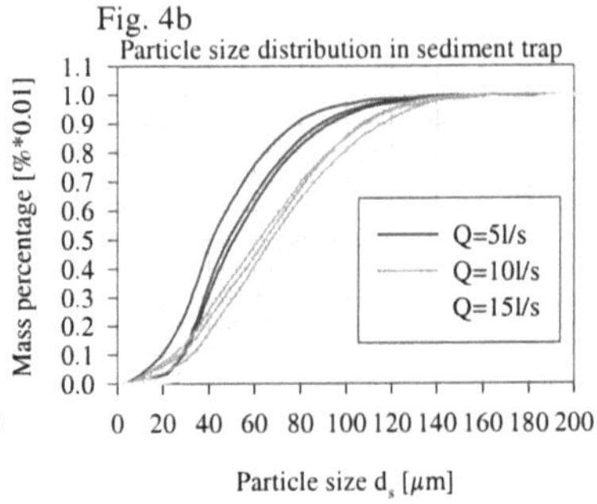

Fig. 5a

Fig. 5b

increasing sediment supply. This is caused by the formation of sickle shaped bed forms, which develop with increasing availability of near bed material to ripple-like formed mobile structures. These bed forms occur for C_R>1.5g/l, but only at Q=5l/s. They were observed by the sensor signal and visually through a window pane in the channel bed. These ripple structures lead to an increased vertical momentum diffusion [7], which results in an increase of C_{tot}/C_R. In Figure 5 the particle size distribution of suspended particles at ξ=0.5 (5.a) and the particles in the sediment traps (5.b) is shown for all concentrations C_R and for 3 discharges. A overlapping range between the fractions transported in suspension at ξ=0.5, and those creeping along the wall, is indicated clearly. The presence of these exchange fractions is caused by the fact that the particles, which are smaller than the viscose sublayer, move along low speed streaks and were entrained in suspension by randomly occuring ejections events (u'<0, w'>0). A detailed description of these phenomenon is given in Nino and Garcia [3].

3.2 SENSOR SIGNALS

Using the above mentioned 15% threshold, the average time between the ejections and

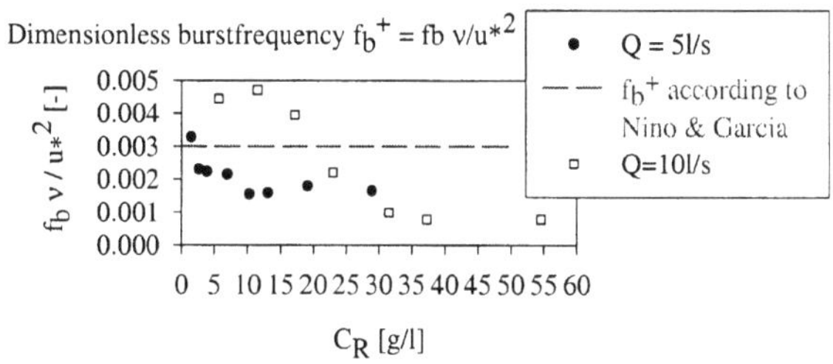

Fig. 6

their duration can be estimated. The dimensionless burst frequency $f_{b+}=f_b\nu/u_*^2$ is indicated by Nino and Garcia as $f_b+=3*10^{-3}$, the characteristic expansion of these coherent structures as approx. 100-200 wall units $y_w=\nu/u_*$, and the life time of the structures with approx. $70\nu/u_*^2$. y_w for the given discharges is between 34 and 41 μm, so that the sensor with a length of 2.75mm is smaller than the characteristic length of the turbulent structures. Figure 6 shows the dimensionless burst frequency as function of C_R. The burst frequency decreases with increasing concentration, and is for small concentrations close to $3*10^{-3}$. Since in this evaluation a ejection event is only detected if the sensor is free of particles, then the decreasing of the burst frequency is to be explained by the fact that for high concentrations only high intensity events were detected, in case of low C_R all of them. With increasing concentration C_R, a immobile layer is formed and can be interpreted as a net sediment deposition. The duration of ejections is estimated by the average time intervall, for which the signal is below the threshold, and is about 0.1 seconds, consistent with the data of Nino and Garcia.

3.3 PARTIAL CONCENTRATIONS

The grain size spectrum was divided into 9 not-equidistant intervals (see table 2).

Interval i	1	2	3	4	5	6	7	8	9
d_s [μm]	0-5	5-15	15-30	30-50	50-80	80-120	120-170	170-230	230-300

Table 2

The partial concentrations C_{pi}, belonging to the interval i, are normalized by the concentration $C_{pRi}=m_{in,i}/V_{tot}$, where $m_{in,i}$ is the added mass of particles in interval i. (we omit the index i for the following). In Figure 7 the normalized partial concentrations at ξ=0.5 are plotted for Q=5 and 15l/s. For the higher discharge, the partial concentrations are quite evenly distributed around 1. The symbols for interval 5 are somewhat above 1. Considering, that only 7% of the added material is within this range, a small uncertainity in determining the particle size distribution of the SP10 sand can be responsible for this. Generally, the scatter for the measured values is increasing with increasing grain size.The limited total number of the particles for the analysis tends to decrease the number of coarse particles in a sample. For Q=5l/s, a selective deposition is shown in Figure 7b. For particles smaller than 15μm the concentration is quite constant, and about 80% of the added sediments is transported in suspension. For interval 3, 4 and 5 the portion transported in suspension is reduced significantly. The increase of the total concentration C_{tot} in Figure 4a is identified as an increase of the partial concentrations of particles greater than about 15μm. Most of these particles are transported near the wall and provide the material for the formation of bed forms.

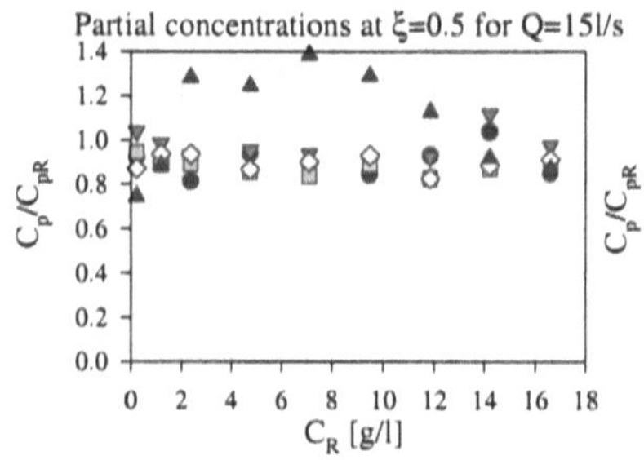

Fig. 7a

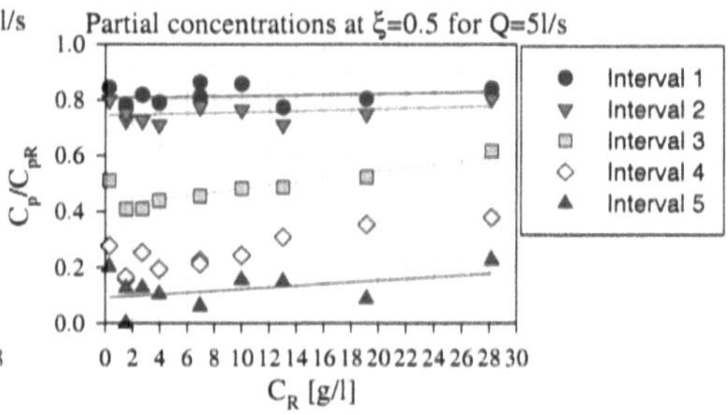

Fig. 7b

The vertical profiles for Q=5 and 10l/s, were normalized with Cp(ξ=0.5) and compared with the vertical distribution of the concentration after Rouse [8]. The Rouse profile has thus the following form:$C(\xi)/C(\xi=0.5)=(\xi^{-1}-1)^{Vs/(\kappa\beta u^*)}$,with β=1. In Figure 8a and 8b the averaged measured values for all concentrations C_R (symbols) and the values computed according to the Rouse formula (lines) are shown. The agreement is quite good, although only 2 and 4 points in the vertical direction are present. A clear deviations is only shown in case of Q=5l/s for interval 4 in the near wall region (triangle down symbol at ξ=0.25). The presence of bed forms directly upstream the sampling tube may cause that the values near the wall are biased to lower concentraions, due to the intensified vertical mixing. However, turbidity does not allow to observe the bed while taking the samples.

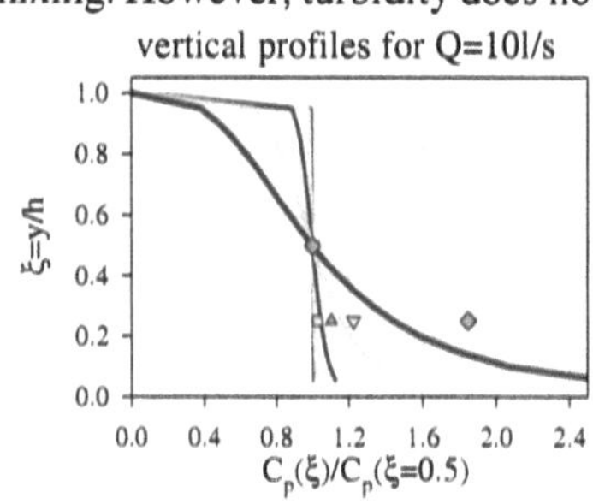

Fig. 8a

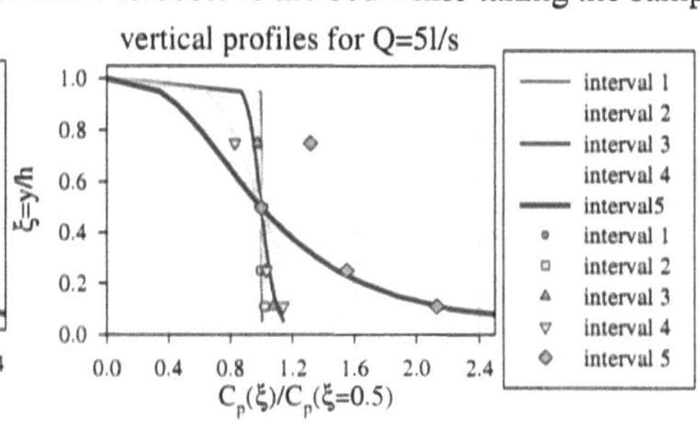

Fig. 8b

4. Transition to rough conditions

The feeding of sediments of type SP6 took place only once per experiment, for different runs, however, in different quantity. Samples were taken at ξ=0.5, at y=5mm and with the sampling tube attached at the channel bed, so that the center of the tube is 6mm above the wall, defined as $\xi=0$ for the following. After the sediment input, concentration and average size of suspended particles deacreases with time. This is affected by the roughness, acting like a sediment trap. Particles are dispersed between the roughness elements, which is known as colmation effect [9]. After each experiment the particles catched in a certain area by the rough wall were removed by suction, so the mass per unit area of trapped sand can be estimated. From table 3 follows that the dimensionless roughness $k_s^+=k_s\, u_*/\nu$ is less than 70, so the flow is in the transient domain between hydraulically smooth and rough regime [2].

Discharge	Q	5	10	15	l/s
Depth	h	4.78	7.17	9.07	cm
Velocity	u=Q/A	.418	.558	.662	m/s
Shear velocity	u_*	0.0235	0.0316	0.0361	m/s
Reynolds number	Re	19980	40008	60430	-
Froude number	Fr	.611	.665	.702	-
Dissipation	ϵ	.00664	.01224	.0164	m^2/s^3
Kolmogg. scale	η	110.8	95.1	88.4	μm
Dimensionless roughness	k_s^+	23.5	31.6	36.1	-

Table 3

4.1 EXPERIMENTAL RESULTS

In Figure 9a and 9b the total concentration C_{tot} and the average grain size d_{50} at $\xi=0.5$ is shown as a function of time for $C_R=1$ and 0.1g/l and for Q=5, 10 and 15l/s. The reduction of the concentration is time-dependent and a function of the amount of sediment input. Thus the concentration of the suspended particles decreases after 200 minutes in the case of Q=5l/s to approx. 5% of the original value for $C_R=0.1$g/l, and to approx. 20% for $C_R=1$g/l.The average suspended particle size is reduced from its original value of $d_{50SP6}\sim 38\mu m$, to 6 and 16μm respectively, so the porosity of the roughness elements significantly affect sediment transport. For a higher initial concentration the limited storage volume of the sand bed, consiting of a single layer of grains, leads to an equilibrium state after about 3.5 h. The partial concentrations show a strongly grain size-dependent behavior. For the different size intervals, the temporal evolution of the normalized concentration C_p/C_{pR} for two initial concentrations C_R and for Q=5,10 and 15l/s, is plotted in Figure 10. Lines are marking $\xi=0.5$ for each case. The temporal reduction of the concentration is proportionally to the particle size and inversely proportional to the discharge. The probability that a particle vanishes between the roughness elements, from where it cannot be resuspended any more and thus is withdrawn from the circulating flow, seems to be a function of particle size and shear velocity. The curves for

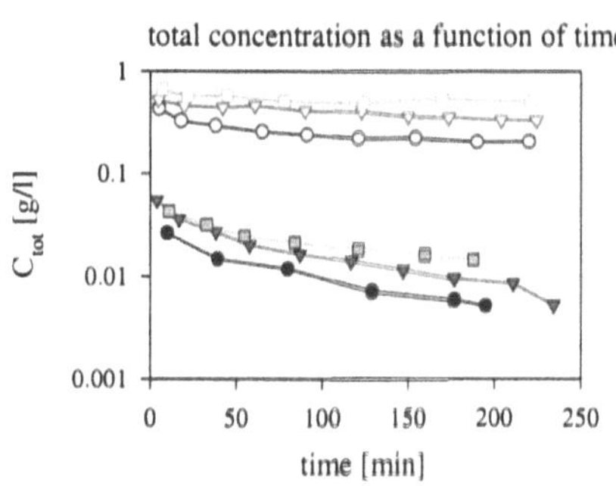

Fig. 9a

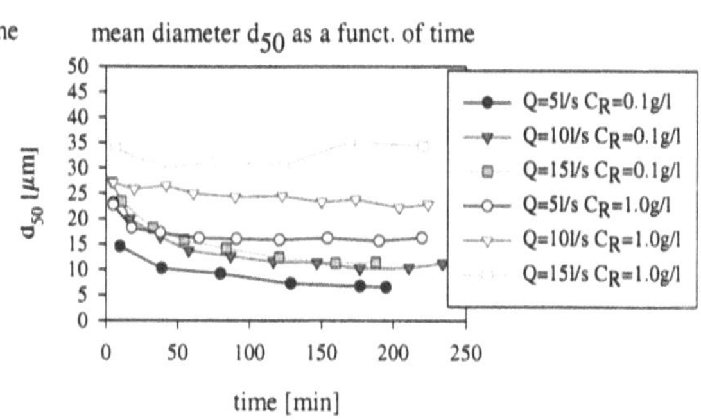

Fig. 9b

particles smaller than 30 μm are more or less straight, which suggests, due to the log scale of the graph, a first order decay, i.e. $dC_p/dt \propto C_p$. With increasing amount of deposited sediments the interstitial volume becomes smaller, so that larger particles can no longer be trapped and the concentration remains constant. For C_R=1g/l the concentration decreases rapidly during the first few minutes, particularly for the coarser particles which move along the bed and can fill up the storage volume very quickly.

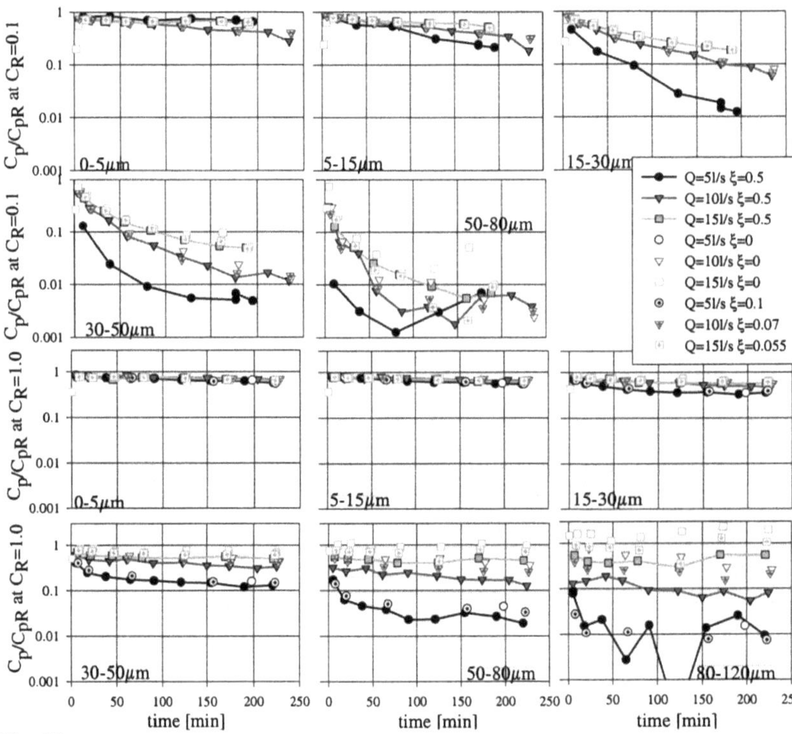

Fig. 10

References

[1]Rudolph, A. ; Heidenreich, E. (1991)."Flexibel anpassbare optische Sensoren zur Partikelmesstechnik". Tagungsband MessComp, p474-479

[2]Nezu, I ; Nakagawa, H. (1993)."Turbulence in open channel flows". A.A. Balkema, Rotterdam

[3]Nino, Y; Garcia,M.H.(1996)"Experiments on particle-turbulence interaction in the near wall region of an open channel flow:implications for sediment transport". J. Fluid. Mech, vol 326 p285-319

[4]Zanke, U.(1982)."Grundlagen der Sedimentbewegung". Springer Velag, Berlin

[5]Nikora, V.I. (2002)." Fluctuations of suspended sediment concentration and turbulent sediment fluxes in an open channel flow".J. of Hydr. Eng., Feb. 2002, p214-224

[6]Sumer, B.M. (1986)."Recent developments on the mechanics of sediment suspension". Transport of suspended solids in open channels(ed. W. Bechteler), Rotterdam:Balkema,p 3-22

[7]Cellino, M; Graf, W.H. (2000)"Experiments on suspension flow in open channels with bed forms". J. Hydr. Res., Vol. 38, N°4, p 289-298.

[8]Rouse, H. (1937). "Modern conceptions of the mechanics of turbulence".Transactions of the ASCE, Vol. 102.

[9]Schälchli, U. (1993). "Die Kolmation von Fliessgewässersohlen: Prozesse und Berechnungsgrundlage". Mitteilung der Versuchsanstalt für Wasserbau, Hydrologie und Glaziologie der ETH, 124, Zürich, 273 p.

ENTRAINMENT OF LARGE PARTICLES FROM GRANULAR BED PROTECTIONS UNDER LOW-MOBILITY TRANSPORT CONDITIONS

BAS HOFLAND, ROB BOOIJ AND HENRI L. FONTIJN
Fluid Mechanics Section, Faculty of Civil Engineering and Geosciences, Delft University of Technology. PO BOX 5048, 2600 GA Delft, The Netherlands. e-mail: B.Hofland@citg.tudelft.nl

1. Introduction: Bed protections and sediment transport

In this paper the stability of bed protections is treated as a special form of sediment transport. The influence of turbulence fluctuations on the stability of bed protections is discussed and examined with the aid of measurements.

Layers of stone are often used to protect the soil around hydraulic structures against scour. The protective layers are constructed in order not to move, although some movement of rocks is allowed. The design value for the Shields factor, $u_*^2/((\rho_s/\rho-1)gd)$, is low, around 0.035 (Pilarczyk, 2001). The stones can be very large (decimeters to meters), giving very high particle Reynolds numbers (typically: $Re_* \equiv u_*d/\nu \approx 10^3$ to 10^5). The rocks are usually obtained directly from a quarry, making the shape very angular, and the size distribution relatively narrow. Furthermore, as the protections have a finite size and are always applied near structures obstructing the flow, non-equilibrium flow and transport are of importance. However, the influence of non-equilibrium turbulence profiles on transport is not clear. This is therefore the main focus of study.

Damage to a bed protection already occurs under low mobility transport conditions (Shields factor under the "critical" value 0.056). In this regime no bed forms are present. Stones move occasionally during extreme events in the flow in the form of sliding and rolling. Only some stones are small, unstable, and/or exposed enough to move. This implies that the flow can be regarded as a flow over a fixed bed, and there will be a limited influence of the moving stones on each other. The characteristics mentioned above (narrow size distribution, flow as over a fixed bed, high Re_*), imply that transport of stones from bed protections can be regarded as a simplified form of sediment transport. It becomes feasible to describe the probability of entrainment by regarding the probability density function (p.d.f.) of the positions of the most exposed stones (varying per stone) and of the forces on the stones (varying in time) as introduced by Grass (1971). The micro bed topography can be regarded as a stochastically distributed boundary value for the strength of a single particle, which may only vary slowly in time, because of repositioning of the top particles.

2. Measurements

Simultaneous pressure and velocity measurements were executed in a 15 m long and 0.5 m wide flume in order to see which turbulence events cause the extreme forces on a particle.

Miniature, low range, piezometric pressure transducers (HONEYWELL) were installed in three sides of a bed-mounted cube (model stone, see Figure 1). The fluctuating pressures could be measured accurately (error < 10 N/m^2, sampling frequency 500 Hz). The cube was part of a one layer thick granular bed consisting of stones with a diameter, d, of about 3 cm. The flow type was open channel flow. A two component laser doppler set-up was used to measure flow velocities. The effect of the protrusion

A. Gyr and W. Kinzelbach (eds.), Sedimentation and Sediment Transport, 75–78.

of the cube, Π, on the character of the fluctuating pressures was investigated by changing Π from 0 to 1 cm. The measurements show that the fluctuating pressures on the cube change significantly when altering Π. The ratios of lift to drag, and of fluctuating pressures to mean pressure are not constant, for instance. Nevertheless, only the cube with $\Pi = 1$ cm ($\approx 0.3d$) is discussed next, as this protrusion is representative for the most unstable particles in a uniformly sized bed. In the following we will focus on the extreme velocity events that cause the maximum forces.

3. Velocity indicators

In order to see which velocity fluctuations induce large forces on a stone, a few mechanisms that possibly generate forces will be discussed, and subsequently linked to the measured pressures on and velocities near the cube.

Indicators for the fluctuating drag and lift forces are obtained from the measured pressures by $\Delta p'_D = p'_1 - p'_3$ and $\Delta p'_L = -p'_2$, respectively (see Figure 1).

The average drag force is mainly caused by the stagnation pressure and the fact that (because of flow separation) the pressure does not build up at the lee side of the stone. The lift force is due to the asymmetric flow over the cube, which causes differences in pressure on either side (Bernoulli's law). Both relations have the form $F = \frac{1}{2}\rho C A u^2$. If u slowly changes, then this "quasi-steady" flow leads to fluctuating lift and drag forces on the stone proportional to $Uu' + u'^2$ ($\propto u'$ for $u' << U$).

Booij (1998) suggested that the pressure fluctuations that are inherent to a turbulent flow field can also cause net forces on a stone, which might aid in the entrainment of particles. These fluctuating pressures are linked to the inertia of the accelerating and decelerating water parcels above the bed. The horizontal and vertical components of the resulting forces on a stone caused by this mechanism will be addressed as the potential drag and lift force. The adjective "potential" is used, as the velocity-pressure relation is described by potential flow theory, contrary to the quasi-steady drag described before which is caused by viscosity related flow separation at the downstream side of the stone (Lighthill, 1986). In order to see which velocity fluctuations might be linked to these forces we start with the Euler equation (no viscous terms, as Re_* is high):

$$-\frac{1}{\rho}\nabla p = \frac{D\vec{u}}{Dt} \qquad (1)$$

and combine this with the buoyancy force on a body in a constant pressure gradient:

$$F_x = -d^3 \frac{\partial p}{\partial x} \qquad (2)$$

Subsequently we take: V=0, W=0, $\partial U/\partial x$=0, $\partial U/\partial z$=0, and $u_i' \partial u'/\partial x_i << U \partial u'/\partial x$ (the last simplification is a reasonable assumption, but not generally true). Now we obtain:

$$F_x' \propto v' \frac{\partial U}{\partial y} + \frac{Du'}{Dt} \qquad (3)$$

The material derivative of u' in eq. (3) cannot be measured by a one-point measurement, but v' can.

It is also reasonable to assume that, when the momentum transfer towards the bed is large, the resulting forces on the stones are increased. Although the average momentum transfer is equal to $-\overline{u'v'}$, the fluctuating momentum transfer is: $-(uv)' \approx -Uv' - u'v'$, for extreme values of u.

Finally, the presence of certain coherent structures is often linked to the position of the velocity vector on the u',v' plane, which is divided into four quadrants, Q_i. During sweeps (Q_4: $u'>0$, $v'<0$) and ejections (Q_2: $u'<0$, $v'>0$) the value of $-u'v'$ is positive. This means that this term enables us to link the presence of coherent structures to the extreme forces on particles, although only a point measurement is available. This term is not equal to the momentum transfer.

Summarizing, the following velocity indicators can be obtained. Extreme values of u' should correspond to extreme drag and lift forces caused by quasi-steady forces. The largest quasi-steady forces are expected to have a longer duration, as the low frequency u-fluctuations contain more energy. Extreme values of $+v'$ could correspond to the extreme potential drag force, see eq. (3). This force can be expected to have a short duration, as the pressure gradients increase with decreasing eddy size. The fluctuating momentum transfer is given by $-Uv'-u'v'$. The term $-u'v'$ is related to certain coherent structures. The last two indicators describe indirect causes of increased forces on a stone, and the corresponding velocity indicators can therefore be expected to have a correlation to the forces, when measured at higher positions in the flow.

Now we examine whether the maximum values of these indicators correspond to increased values of the forces on the cube in our present measurements. The length scale of the fluctuations is of importance, as this indicates the size of the eddies responsible for the extreme forces. Therefore the following procedure is followed. First of all, a band-pass filter is applied to the velocity-indicator signals. The central frequency of the pass-band is chosen to correspond to a certain length scale, λ, (obtained from Taylor's hypothesis, using $\lambda = 0.8U(y)/f$, where f is the frequency and $U(y)$ the time averaged velocity at height y). This is done for two λ's, representing small scale (stone size) and large scale (water depth) eddies. Next we see whether the instants of maximum velocity for a certain length scale coincide with those of extreme values of the pressures on the cube. For every filtered velocity indicator, $\Delta p'_D$ is averaged over all instants at which this indicator (at a certain height, and with a certain length scale) has a maximum that exceeds a threshold value (three standard deviations). The results for $\Delta p'_D$ are depicted in Figure 2. The measured p.d.f. of $\Delta p'_D$ is plotted for reference. The skewed log-normal shape resembles the p.d.f. of shear stresses on a smooth bed. This could be expected, as the combined drag forces on the particles form the bed shear stress. The markers on the bars on top represent the conditionally averaged values of $\Delta p'_D$ (for the maxima of the velocity indicators, at a certain height and for a given length scale). So, a marker placed to the right of the mean value of $\Delta p'_D$ (vertical line), shows that extreme values of the considered velocity indicator coincide with large $\Delta p'_D$. This implies that this velocity indicator indeed indicates increased, potentially displacing, drag forces.

We can see that low frequency u'-maxima near the cube correspond to the largest $\Delta p'_D$ (indicated by the circular marker on the lowest bar). Further away from the cube ($y = 54$ mm $\approx 1.5d$) the sweeps coincide with the highest values of $\Delta p'_D$. Sweeps have previously been seen to correspond to transport at higher Re_* (Williams *et al.* 1986; Nelson *et al.*, 1995). This would mean that the drag force is the force responsible for displacing the stones, as the conditional averages of $\Delta p'_L$ show that $\Delta p'_L$ is negative during the sweep events. Conversely, ejections cause a decreased $\Delta p'_D$ (see diamond-shaped markers in Figure 2), and an increased $\Delta p'_L$. It is interesting to see that the small scale extreme values of $+v'$ near the cube correspond to slightly increased $\Delta p'_D$ (Figure 2, square marker), which might indicate a small influence of the potential forces as well.

4. Conclusions

We consider the protruding cube ($\Pi/d = 0.3$), as it is a realistic model for an unstable stone. The horizontal pressure gradient (both mean and fluctuating) was much larger than the vertical gradient. This means that drag is probably the main cause of entrainment, although the angle of repose and the exposed areas of the stone still have to be incorporated in the analysis. Fluctuating lift might be necessary for the entrainment if the angle of repose is high (smaller particles). Near the cube extreme values of u' coincides with the largest $\Delta p'_D$. At higher elevations in the flow (y = 54 mm) sweeps coincide with larger $\Delta p'_D$ than extreme values of u'. It must be mentioned that the type of flow structure leading to a "sweep" (Q_4-event) near a rough bed is not clearly defined at present, contrary to the smooth bed case.
Flow configurations with an altered turbulence structure will be studied as well. A PIV installation is being set up for measurements of entire flow fields.

5. Literature

Booij, R. (1998) *Erosion under a geometrically open filter*. Tech. Report. 2-98, Delft University of Technology. In Dutch.

Grass, A.J. (1971) *Initial instability of fine bed sands*. Journal of the Hydraulics Division, Proc. of ASCE, **96**(HY3), 619–632

HONEYWELL, http://catalog.sensing.honeywell.com/datasheet.asp?PN=24PCEFA6D. Data sheet of pressure sensors, type 24PCEFA6D.

Lighthill, J. (1986). *Fundamentals concerning wave loading on offshore structures*. Journal of Fluid Mechanics **173**. 667–681.

Nelson, J.M. Shreve, R.L. McLean, R. and Drake, T.G. (1995) *Role of near-bed turbulence structure in bed load transport and bed form mechanics*. Water Resources Research **31**(8).

Pilarczyk, K.W. (2001) *Unification of stability formulae for revetments*. Proc. XXIX IAHR Congress, Beijing, theme E.

Williams, J.J. Thorne, P.D. and Heathershaw, A.D. (1986). *Measurements of turbulence in the benthic boundary layer over a gravel bed*. Sedimentology **36**. 959–970.

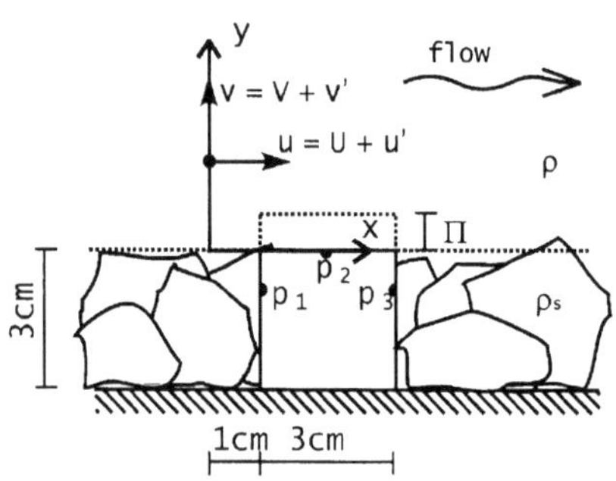

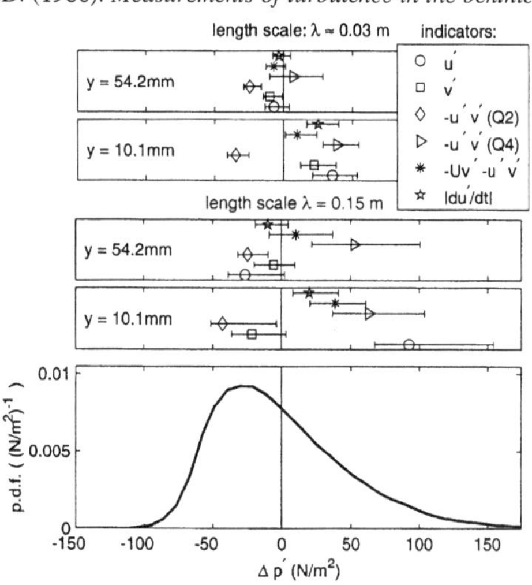

Fig 1. (left) Pressure transducers in granular bed with definition of symbols. Fig 2. (right) Markers: conditionally averaged $\Delta p'_D$ on the basis of band-pass filtered signals of various velocity indicators (see legend). Below: p.d.f. of $\Delta p'_D$. All horizontal axes have the same scale.

INTERACTION BETWEEN TURBULENT OPEN-CHANNEL FLOW AND PRESSURE FLUCTUATIONS IN SUBSURFACE GRAVEL LAYER

Martin Detert[1)], Gregor Kühn[1)], Michael Klar[2)]

[1)] *Institute for Hydromechanics (IfH), University of Karlsruhe, Kaiserstrasse 12, D-76128 Karlsruhe, Germany, detert@ifh.uka.de resp. kuehn@ifh.uka.de*

[2)] *Interdisciplinary Center for Scientific Computing (IWR), University of Heidelberg, Im Neuenheimer Feld 368, D-69120 Heidelberg, Germany, michael.klar@iwr.uni-heidelberg.de*

1. Approach

The stability of a gravel bed in an open-channel flow is usually described by the Shields (1936) parameter, τ_*. It is roughly defined as the ratio of the frictional load on the grain ($\sim \rho_w \times u_*^2 \times d^2$) to the gravitational force on the grain that resists movement ($\sim g \times (\rho_s - \rho_w) \times d^3$). The value of $\tau_* = 0.03$ is often used as the upper limit for bed protections. This and other correlations, e.g. Meyer-Peter/Müller (1949) yield a good estimation for practical engineering problems. However, these equations do not fully account for the physical processes that cause the instability of gravel bed.

For a more detailed consideration of stability problems, the analysis of the influence of fluctuating forces is important. Fluctuating forces are caused by turbulence of the open-channel flow, including the occurrence of coherent structures, as well as the interactions among the whole hydrodynamic system, which consists of free surface, subsurface gravel layer and eventually groundwater flow.

The subject of this work is on the investigation of turbulent pressure fluctuations in open-channel flow at the top of the gravel layer and on the hydrodynamic reaction in a porous subsurface gravel layer. The long-term goal of this project is to quantify the influence of turbulent fluctuations on the bed stability of waterways.

2. Experimental Facility and Measurement Setup

The measurements were performed in an experimental flume ($L = 40$ m, $B = 0.9$ m, $Q_{max} = 0.15$ m³/s) located at the Federal Waterways Engineering and Research Institute (BAW), Karlsruhe. A sand layer of 0.5 m thickness covered by a gravel layer of 0.1 m thickness was placed in the flume on a length of 20 m. The grain diameter of the gravel was $d_{50} = 10$ mm, the grain diameter of the sand was about 1 mm. The measurement section was located 12 m downstream of the water inlet. Glass windows on a length of 4 m allowed optical access. The water level in the flume could be adjusted by a movable weir at the channel outlet. The maximum depth h was 0.4 m. Variation in discharge and water depth provided flow conditions with Reynolds number $Re = v \times h/\nu = 50{,}000$ to 140,000. The values of the Shields parameter ranged from $\tau_* = 0.18 \times 10^{-3}$ to 3.06×10^{-3} and the grain size Reynolds Number Re_* varied in the range from 150 to 660.

The following parameters were measured in the free surface flow:

- water level fluctuations with ultrasonic probe and pressure gauge;
- fluid velocity with 3D-PTV (3D-Particle-Tracking-Velocimetry, developed by the IWR, University of Heidelberg, see below), also LDA (Laser Doppler Anemometer) and ADCP (Acoustic Doppler Current Profiler)

A. Gyr and W. Kinzelbach (eds.), Sedimentation and Sediment Transport, 79–82.

Measured parameters for the gravel layer were:

- pressure fluctuations with pressure gauges at measuring points located at the surface of the gravel layer and at z = -1,5 to -7 cm (downward direction);
- flow velocity in an adapted artificial gravel pore with a specially designed flexible endoscopic 3D-PTV setup (IWR) and additional pressure gauge, located in vertical positions varying from z = -1 to -10 cm

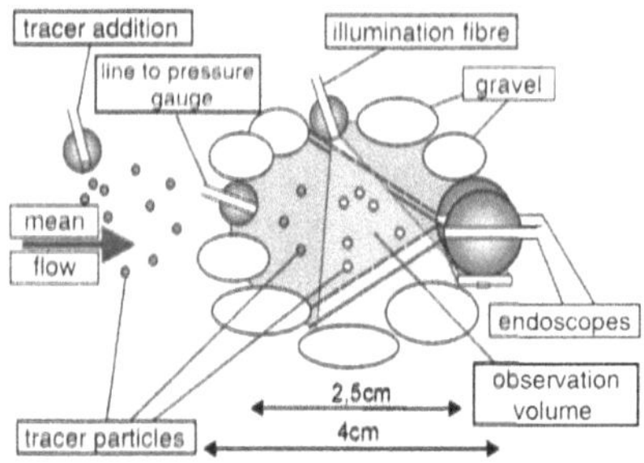

Figure 1: Artificial pore (Klar, 2001)

The 3D-PTV-System yields the Langrangian representation of a flow field, i.e. it measures trajectories of seeding particles added to the fluid. Towards this end, the seeding particles have to be identified in the images and traced from one image to the next (Stybalkowski, 2001, Klar, 2001). Fig. 1 shows the miniaturized stereoscopic setup for measurements of flow and pressure in an artificial pore in the gravel layer.

3. Results

The aim of the experiments was to analyse the interaction between the turbulence in open-channel flow and the pressure fluctuations in the subsurface gravel layer.

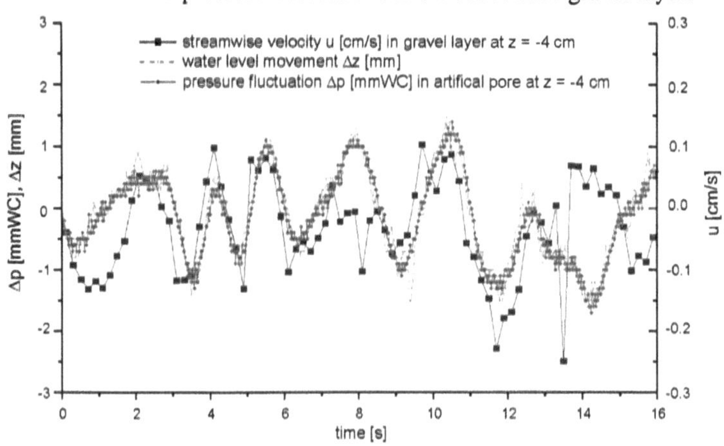

Figure 2: pressure fluctuation and streamwise velocity in the gravel layer including water level movement. Mean velocity u = 0.3 m/s and water depth h = 0.40 m

However, the actual experimental conditions could not avoid the development of long surface waves with a typical period of 1.5 to 3 s. Thus the measuring instruments gauged both influences of turbulence and of water level movement. Instantaneous values of simultaneous measurements of pressure and velocity in all vertical positions of the gravel layer showed a temporal behaviour coincident with the long wave fluctuation of water level. Superposed fluctuations of pressure and velocity due to large eddies in the free surface flow could be detected only in the upper layer of the granular filter. For instance, Fig. 2 shows a time series with the interaction between the turbulent velocity of the free surface flow and the flow in the gravel layer. The data set shows a clear correlation between the fluctuation of the wave induced dynamic pressure and the velocity in the gravel layer (Lang et al., 2001).

To consider only the turbulent pressure fluctuations, the wave induced movement of the water level was eliminated from the measurement signal. Fig. 3 shows only the turbulent pressure fluctuations denoted by the standard deviation $\sigma(p') = rms(p')$. The impact on the pore flow in the upper grain layer becomes clearly apparent. Moreover, all series of experiments show an exponential damping of these fluctuations with an increasing gravel depth. At a relative filter depth of $z/h = -0.2$ the fluctuations are barely above the detection limit.

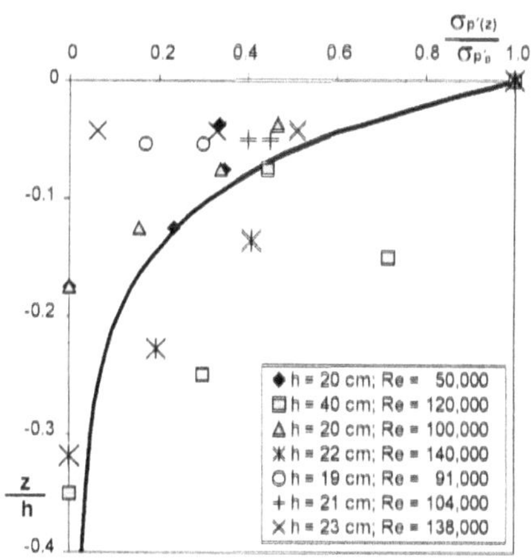

Figure 3: Decrease of turbulent pressure fluctuations as a function of depth z/h in the gravel layer (Daebel, 2001)

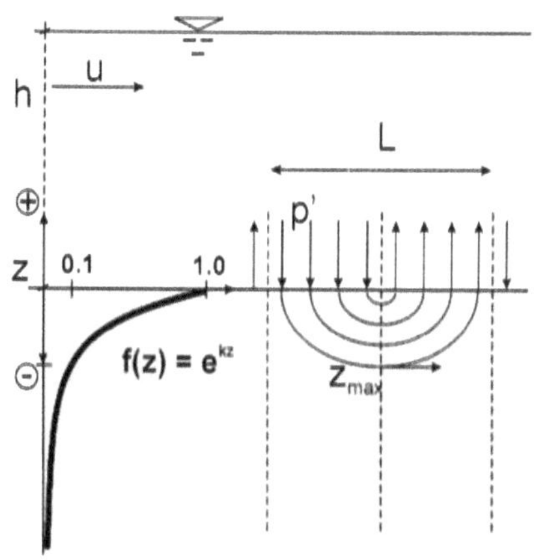

Figure 4: Sketch for deduction of spatial damping

Assuming Darcy´s law, the flow in the gravel layer can be described by Laplace´s equation $\nabla^2\Phi = 0$. This differential equation likewise depicts the pressure field. Laplace´s equation can be solved by the exponential function e^{kz}. According to Fig. 4 this damping of pressure due to harmonic wave fluctuation can be approximated by

$$\sigma p`(z) = e^{kz} \times \sigma p`_0$$

with $k = 2\pi/L$ and L = scale of a fluctuating phenomena. Supposing that the turbulent pressure fluctuations at the surface of the gravel layer are damped to 10 % in deeper layers the equation can be solved to $z = -0.37\times L$. With $L \approx 0.5\times h$ this confirms the experimental results as mentioned before. Thus the filter layer with 10 cm height protects the sand layer below just due to its extension. Local erosion of the sand as the result of turbulent fluctuations cannot be expected.

Further results showed, that the time averaged turbulent pressure fluctuations at the surface of the gravel layer were about three times the bed shear stress τ_0. The maximum fluctuation values were about ten times τ_0. This corresponds to known results from Yalin (1992) and Dittrich (1996).
The time-averaged velocity profiles as well as the turbulent velocity fluctuation profiles of the free surface flow agreed with the logarithmic law behaviour known from Nezu & Nakagawa (1993). Moreover, the results of measurements in the free surface flow also showed that the instantaneous velocity is dominated by the long wave fluctuating water level.

4. Further Investigations

The variation of the thickness of the gravel bed with and without hydraulic contact to the sand layer below will be studied in the forthcoming experiments. Therein an improved measurement equipment (high resolution dynamic pressure gauge and endoscopic setup) and improved flume conditions (upstream shift of the inlet, increased flow rate) will be employed.

In addition to these experiments an immobile porous channel bed will be investigated by numerical simulation. Therefore, the flow field both above and within the gravel layer will be calculated by numerical large-eddy simulations (LES) with temporary and spatial highly resolved Cartesian grids. In this manner a detailed analysis of eddy structures and instantaneous forces on grains may be given. In a follow-on project it is proposed to calculate the lift-up of single grains.

5. References

Daebel, H., Einfluss einer turbulenten Gerinneströmung auf den momentanen Porenwasserdruck in einer Kiessohle, Vertieferarbeit, Institut for Hydromechanics, University of Karlsruhe, Germany, 2001

Dittrich, A., Nestmann, F. & Ergenzinger, P., Ratio of lift and shear force over rough surfaces, in: Coherent flow structures in open channels, edited by Ashworth, P.J. et al., pp. 125-146, John Wiley & Sons Ltd., New York, USA, 1996

Klar, M., 3-D Particle-Tracking Velocimetry applied to Turbulent Open-Channel Flow over a Gravel Layer, Interdisciplinary Center for Scientific Computing, University of Heidelberg, Germany, 2001

Lang, C., de los Santos Ramos, F. & Kühn, G., Untersuchungen zur Stabilität poröser Gerinnesohlen, report no. 782, Institut for Hydromechanics, University of Karlsruhe, Germany, 2001

Meyer-Peter, E. & Müller. R., Eine Formel zur Berechnung des Geschiebetriebs, Schweizerische Bauzeitung, Vol. 67, No. 3, pp. 29-32, Suisse, 1949

Nezu, I. & Nakagawa, H., Turbulence in Open-Channel Flows, IAHR/AIRH Monograph Series, Balkema Publishers, Rotterdam, The Netherlands, 1993

Shields, A., Anwendung der Ähnlichkeitsmechanik und der Turbulenzforschung, Mitteilungen der Preußischen Versuchsanstalt für Wasserbau und Schiffbau, Heft 26, Berlin, Germany, 1936

Stybalkowski P., Strömungsmessung in Sedimentporen mittels 3-D Particle Tracking Velocimetry, diploma thesis, Interdisciplinary Center for Scientific Computing, University of Heidelberg, Germany, 2001

Yalin, M. S., River Mechanics, 1st edition, Pergamon Press Ltd., Oxford, England, 1992

TWO-PHASE FLOW ANALYSIS OF SEDIMENT VELOCITY

BLAIR P. GREIMANN
United States Bureau of Reclamation, Denver Federal Center, Bldg. 67, P.O. Box 25007 (D-8540), Denver, CO 80225-0007, bgreimann@do.usbr.gov

1. Introduction

Computing sediment velocity is critical to predicting sediment transport rates and the effect of sediment on the environment. There has been a large amount of research looking at the velocity of bed-load particles. The research has suggested that the shear velocity governs the bed-load sediment velocity, whereas the flow velocity governs the suspended sediment velocity. However, a single method to compute the velocity of bed-load and suspended load has remained elusive. In this paper, two-phase flow analysis is used to compute the velocity of bed-load and suspended load. The two-phase approach has an advantage in that it can account for the inertia of particles and the effect of turbulence on their transport.

2. Governing Equations

The general two-phase flow equations have been previously derived in Greimann et al. (1999) and Greimann and Holly (2001). Here they are applied to two-dimensional uniform flow. It is assumed that the bed is sediment-starved and that the sediment is of a single size and density. It is also assumed that the concentration is small ($\Phi_s << 0.1$). Therefore, the derivation does not directly apply to natural rivers where bed forms are present, to well-graded sediment, or to high concentrations of sediment. However, this simplified formulation will be used here to assess the ability of the two-phase formulation to calculate the velocity of sediment.

The two-phase flow equations are used to derive the velocity and concentration profiles of the sediment only. The fluid eddy viscosity profiles are obtained by fitting log-profiles to the fluid velocity data and empirical relations are used to evaluate the turbulence intensities. Two-phase flow equations could also be derived for the fluid velocity profiles and turbulence intensities, but the author's opinion is that currently there is not sufficient experimental data or analytical understanding of particle-turbulence interaction to develop a reliable two-phase flow model for the flow considered here.

The x-momentum equation for the sediment is:

$$\begin{aligned}0 = &-\frac{d}{dy}\left(\Phi_s \rho_s \overline{u_s' v_s'}\right) + \Phi_s \rho_s \tau_p^{-1}\left(U_f - U_s + U_d\right) + \\ &\rho_f C_A\left[\frac{d}{dy}\Phi_s\left(\overline{u_f' v_f' - u_s' v_s'}\right) + \frac{d}{dy}\Phi_s V_d\left(U_f + U_d\right)\right]\end{aligned} \tag{1}$$

A. Gyr and W. Kinzelbach (eds.), Sedimentation and Sediment Transport, 83–86.

$$-\overline{u'_s v'_s} = \nu_s^t \frac{dU_s}{dy}, \qquad \nu_s^t = D_{yy} + \tfrac{1}{2}\tau_p \overline{v_s'^2} \tag{2}$$

where ν_s^t is turbulent eddy viscosity of the sediment phase. The variable U indicates a velocity in the horizontal direction and V a velocity in the vertical direction. The subscript s signifies the sediment phase and the subscript f signifies the fluid phase. The variable U_d is the drift velocity. The y-momentum equation for the sediment can be rearranged as:

$$\frac{d\Phi_s}{dy}\left\{D_{yy} + \tau_p\left[\overline{v_s'^2} + \frac{\rho_f C_A}{\rho_s}\left(\overline{v_s'^2} - \overline{v_f'^2}\right)\right]\right\} = \Phi_s \tau_p\left[\left(\frac{\rho_f}{\rho_s} - 1\right)g - \frac{d\overline{v_s'^2}}{dy} + \frac{\rho_f}{\rho_s}\frac{d\overline{v_f'^2}}{dy} + \frac{C_A \rho_f}{\rho_s}\frac{d\left(\overline{v_f'^2} - \overline{v_s'^2}\right)}{dy}\right] \tag{3}$$

The above equation is similar to an equation derived in Greimann and Holly (2001), but here it is assumed that the concentration is small.

Equations (1) and (3) are used to solve for the horizontal velocity and concentration of sediment. The fluid velocity is assumed using an empirical fit to the log-law of Spalding (1961). The fluid turbulence intensities are computed using the correlations of Nezu and Rodi (1986). The sediment turbulence intensities are assumed to be related to the fluid turbulence intensities by a constant of 1.2, which is the value calibrated to match concentration profile data in Greimann and Holly (2001).

The implementation of the boundary conditions is very critical. A variety of methods have been used. Johnson et al. (1990) and Jenkins and Hanes (1998) apply a stress balance at the wall with the stress due to wall collisions computed as:

$$\tau_w = \frac{\sqrt{3}\pi\varphi'\Phi_s\rho_s|U_s|\sqrt{T_s}}{6\Phi_{s,\max}} \tag{4}$$

where $T = \frac{1}{3}\overline{u'_i u'_i}$ and their original expression has been simplified assuming small concentrations. The parameter φ' is a coefficient to account for the amount of momentum loss a particle experiences when it contacts the bed. Johnson et al. used 0.25 for simulations of smooth aluminum at the bottom boundary and 0.6 for 160 grit sand paper. For all the results shown in this paper, the boundary condition is applied at a distance of 0.5 d_p from the wall. One difficulty in this analysis is that it was found that the coefficient, φ', was dependent upon the Shields parameter. Using the experimental data the following relation was found: $\varphi' = \frac{1}{2}\sqrt{\theta_c/(\theta - \theta_c)}$, where θ_c is a critical Shields number. The critical Shields number is dependent upon the particle shape and bed characteristics when the particle Reynolds number is large.

3. Comparison with Experiments

The sediment velocity model was compared against the experimental data of Francis (1973) and Hu and Hui (1996). The three modes of sediment transport were observed in these experiments: rolling, saltating and suspension. The critical shields number was calibrated for each particle type simulated (Table 1 and results are shown in Figure 1.

Particle Type	Critical Shields Number (θ_c)
Smooth spherical (Hu and Hui)	0.001
Natural Gravel (Francis, Series K and C)	0.015
Angular (Francis, Series A)	0.08
Angular (Francis, Series L, LL)	0.25

Table 1. Parameters used in the simulations.

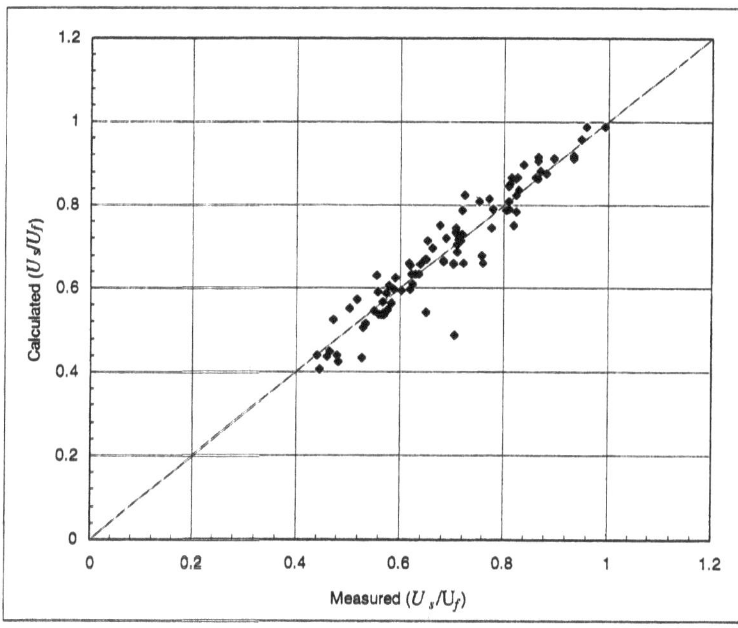

Figure 1. Comparison of predicted velocity and measured.

It should be noted that the formulation given here has the correct behavior as the particles become purely suspended. Because diffusion is taken into account, fine particles will become uniformly distributed throughout the water column. As shown in Greimann et al. (1999) fine particles will lag the fluid velocity on the order of the sediment fall velocity. As the fall velocity decreases the sediment horizontal velocity approaches that of the fluid.

4. Summary

The two-phase flow approach is used to compute the average velocity of rolling, saltating and suspended sediment in a laboratory flume. The two-phase approach simultaneously accounts for particle inertia and the affects of fluid turbulence. Therefore, one set of equations can describe the velocity of bed-load and suspended load sediment.

It was necessary to calibrate the critical Shields number for each different particle shape. However, all other parameters could be kept constant. Further work needs to be done to determine if the boundary condition is generally applicable.

5. References

Francis, J.R.D. (1973) "Experiments on the Motion of Solitary Grains Along the Bed of a Water-Stream," Proc. R. Soc. London A, 332:443-471.

Greimann, B.P., Muste, M. and Holly, F.M. (1999). "Two-Phase Formulation of Suspended Sediment Transport," *J. Hydr. Res.*, 37(4):479-500.

Greimann, B.P. Holly, F.M. (2001). "Two-Phase Flow Analysis of Concentration Profiles," *J. Hydr. Engr.*, ASCE, 127(9):753-762.

Hu, C., Hui, Y., "Bed load Transport. I:Mechanical Characteristics", *J Hydr. Engr.*, ASCE, 122(5):245-261, 1996.

Jenkins, J.T., and Hanes, D.M.. (1998). "Collisional Sheet Flows of Sediment Driven by a Turbulent Fluid," *J. Fluid Mech*, 370:29-52.

Johnson, P.C., Nott, P. Jackson, R. (1990). "Frictional-Collisional Equations of Motion for Pariculate Flows and Their Application to Chutes," J. Fluid Mech. 210:501-535.

Nezu, I., and Rodi, W. (1986). "Open-Channel Flow Maeasurements with a Laser-Doppler Anemometer," *J. of Hydr. Engr.*, ASCE, 112, 335-355.

Spalding, P.R. (1961). "A Single Formula for the Law of the Wall," J. Appl. Mech, 28:455-457.

EXPERIMENTAL MEASUREMENT OF SEDIMENT SUSPENSION AND PARTICLE KINETIC STRESS TRANSPORT WITHIN A HORIZONTAL CHANNEL FLOW

Ken Kiger, Chunhui Pan, and Angel Rivero

Department of Mechanical Engineering, University of Maryland, College Park, MD 20742, kkiger@eng.umd.edu

One of the important objectives in the science of sediment transport is to be able to predict the evolution of the sediment flux suspended within a turbulent carrier fluid. Unfortunately, this task is complicated by a multitude of difficulties that stem from how the particles are organized within the flow structure, and how the dynamics of the particle feedback alters the turbulence characteristics of the carrier flow. The goal of the current work is to utilize novel optical methods to experimentally characterize the instantaneous interaction between the particles and the fluid within a prototypical, fully developed, turbulent channel flow. This information can then be used to understand how the particle/turbulence interaction is manifested within such flows and to act as a database for comparison with model development and simulation within a well-posed and controlled environment.

Toward this end, experiments have been conducted within a 488 x 36 x 4 cm smooth-walled, closed channel, under both single-phase and two-phase conditions (while maintaining a constant flow rate). The flow was fully developed, with a friction velocity of approximately 2.8 cm/s for the single-phase conditions, giving a Reynolds number of 570 when based on the channel half-height and friction velocity. The dispersed phase consisted of nominally spherical glass particles 200 μm in diameter, with bulk particle volume fractions of $\alpha = 2\times10^{-4}$. Simultaneous measurements of both the instantaneous carrier phase and the dispersed phase were made using a single-camera two-phase PIV method developed within our laboratory [1], as applied to approximately 500 image pairs of the two-phase flow. From these measurements,

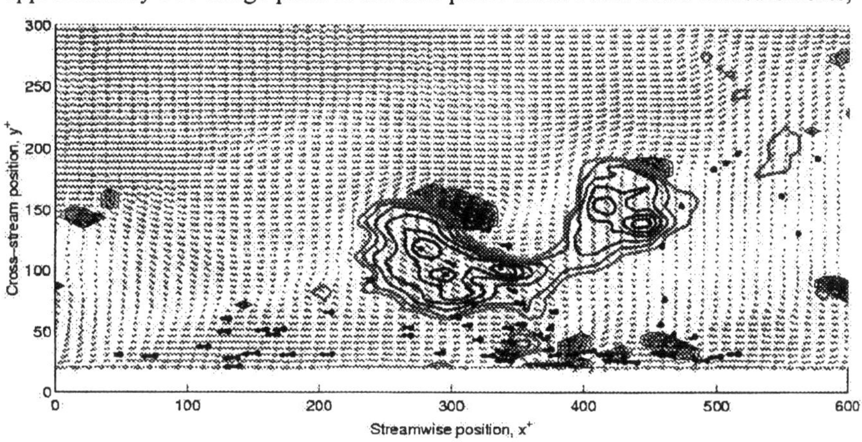

Figure 1: Instantaneous realization of particle suspension event. Fluid velocity vectors are shown in blue, sediment locations and velocity are shown in black. The line contours represent positive contributions to the Reynolds stress (quadrant 2 and 4 events with $-u'v' > 30$ cm^2/s^2), while the filled contours identify regions of strong swirl.

A. Gyr and W. Kinzelbach (eds.), Sedimentation and Sediment Transport, 87–90.

mean and second-order moments of the fluid, the dispersed phase, and their interaction can be calculated, along with instantaneous reconstructions of the particle/fluid flow field. An example of a single realization from this data set is shown in Figure 1, which depicts a burst event as characterized by an ejection of slower-moving wall fluid, entraining with it particles away from the wall. In this case, the flow structure appears to be similar to the vortex packet structures identified by Adrian, et al [2], as indicated by a repeating array of strong swirl events followed by an upstream ejection of slower moving wall-fluid entrained by the hairpin structure. Recent analysis of this data [3], indicates that the organization of the particles relative to these structures is important to the development of the mean concentration and velocity profiles of the dispersed phase.

In addition to the physical insight provided by the above data set, it can also be used to help further the development of transport models for the dispersed phase. Classical tools used to predict the suspended load of sediment has typically relied upon low-order turbulence closures, such as mixing length models. While these methods can be made to agree with observed data under restricted conditions, the agreement is not universal, and higher-order methods that account for the particle-fluid interaction are needed to achieve further progress. One such method that has been proposed in the context of gas/solid two-phase systems [4], is to use a two-fluid equation set that utilizes a particle kinetic stress transport equation to track the energetic evolution of the particulate field. For the restricted set of conditions where the particle-fluid interaction is dominated by a quasi-steady viscous drag force, the turbulent kinetic stress of the particulate phase can be found by multiplying the equation of motion for an isolated particle by its fluctuation velocity, $u''_{2,j}$, performing the ensemble averaging procedure (indicated by the angle brackets), and then adding the result to the same equation with the subscripts reversed (for a detailed description of the procedure, the reader is referred to Simonin, *et al*, 1993 [5]). The results give:

$$\frac{f}{ft} + U_{2,m}\frac{f}{fx_m}\left\langle u''_{2,i}u''_{2,j}\right\rangle_2 = P_{2,ij} + D_{2,ij} + \Pi^d_{2,ij} + \Pi^p_{2,ij}, \qquad (1)$$

$$D_{2,ij} = -\frac{1}{\alpha_2}\frac{f}{fx_m}\left[\alpha_2\left\langle u''_{2,i}u''_{2,j}u''_{2,m}\right\rangle_2\right] \qquad (2)$$

$$P_{2,ij} = -\left\langle u''_{2,i}u''_{2,m}\right\rangle\frac{fU_{2,j}}{fx_m} - \left\langle u''_{2,j}u''_{2,m}\right\rangle\frac{fU_{2,i}}{fx_m} \qquad (3)$$

$$\Pi^p_{2,ij} = \left\langle \frac{\rho_1}{\rho_2}\frac{3}{4}\frac{C_d}{d}|v_r|\left[u''_{1,i}u''_{2,j} + u''_{1,j}u''_{2,i}\right]\right\rangle_2, \qquad (4)$$

$$\Pi^d_{2,ij} = -\left\langle \frac{\rho_1}{\rho_2}\frac{3}{2}\frac{C_d}{d}|v_r|u''_{2,i}u''_{2,j}\right\rangle_2, \qquad (5)$$

where the numerical subscript indicates the phase (1=carrier, 2=dispersed), the alphanumeric subscript indicates the coordinate direction (1=streamwise, 2=wall-normal, 3=spanwise), the first terms on the right-hand side, $P_{2,ij}$, represent the mean-shear production of particle kinetic stress, $D_{2,ij}$ is the transport of particle kinetic stress by particle fluctuations, $\Pi^d_{2,ij}$ and $\Pi^p_{2,ij}$ are the dissipation and production of particle kinetic stress by interaction with the carrier flow turbulence, and C_D is the particle drag coefficient, which is approximated with the empirical expression $C_D=24(1+0.15Re_p^{0.687})/Re_p$. In order for this equation (1) to be useful, closure models appropriate for sediment transport need to be developed for the last three terms, as

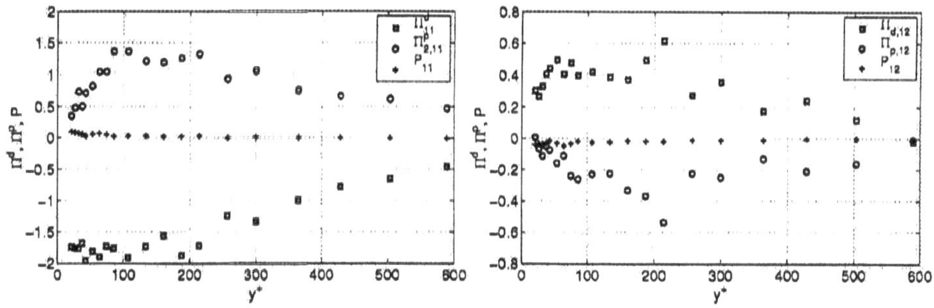

Figure 2: Particle/fluid interaction term of the particle kinetic stress equation. a) streamwise direction, b) Reynolds stress contribution. Quantities are non-dimensionalized using wall units.

they depend explicitly on third-order particle velocity moments, and the particle-fluid covariance. In the context of the present experiments, the statistical sample is not yet sufficient to allow for reliable calculation of third-order moments of the particles, but the important particle-fluid momentum transfer terms can be calculated, and are shown in Figure 2 for two components. It can be observed that for the current conditions, the dissipative contribution of the particle/fluid interaction term is generally larger than the production term, but tends to equilibrate in the outer region of the flow. The production by mean shear, $P_{2,ij}$, is also found to be quite small in comparison to the particle-fluid interaction terms. In comparing the above results to those of Wang, *et al* [4], the qualitative shape of the curves are similar, but with significant differences in the magnitudes and the locations of the peak values (note that the results in [4] must be divided by the magnitude of their Reynolds number, $Re_\tau = 180$, to scale the values in wall units). Specifically, both cases show a dominance of Π_d over Π_p (indicating a net dissipation due to the particle-fluid interaction), particularly in the streamwise direction, while there seems to be a greater symmetry in the outer flow region for the i=1, j=2 component.

Given the differences in the flow conditions between the experiments and the simulations of Wang, *et al.* (one-way coupling, no body force, uniform mean concentration, large density ratio, etc.), even the qualitative similarities may be considered somewhat surprising. Some of the differences, particularly the magnitude, can likely be explained by the large difference in the effective response time of the particles in the two cases. Examining equation (4) and (5), the factors in front of the velocity correlations correspond to the inverse of the effective particle response time:

$$\tau_{12}^F = \frac{\rho_2}{\rho_1}\frac{4}{3}\frac{d}{\langle C_D\rangle\langle|\mathbf{v}_r|\rangle} \qquad (6)$$

When scaled in wall units, this gives a Stokes number of around 4 for the current experiments, and 120 for the simulations, which indicates that particles in the current experiments should be very responsive to the majority of the fluid timescales within the channel, and hence the transport of kinetic stress due to this interaction term would be correspondingly higher. This is also born out by the similarity in magnitude of the production term, P_{11}, which in wall units both have non-dimensional peak magnitudes around 0.1 to 0.2. Wang, *et al* [4], also tested a simplified model (henceforth referred to as model 1) for the correlation terms given by:

$$\Pi^p_{2,ij} = \frac{1}{\tau^F_{12}}\left\langle\left[u''_{1,i}u''_{2,j} + u''_{1,j}u''_{2,i}\right]\right\rangle_2 \quad \text{and} \quad \Pi^d_{2,ij} = -\frac{1}{\tau^F_{12}}\left\langle u''_{2,i}u''_{2,j}\right\rangle_2 \qquad (7)$$

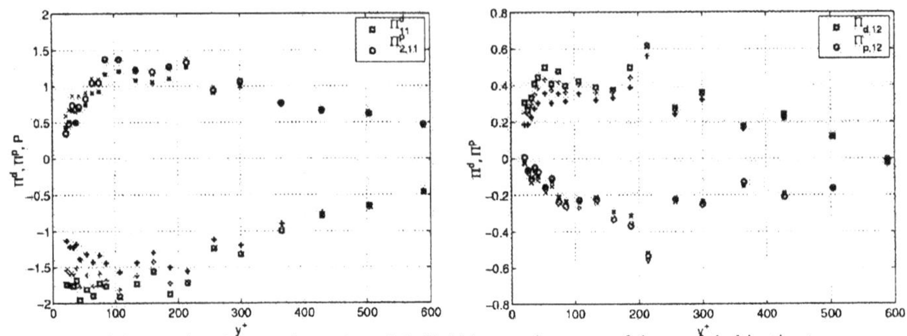

Figure 3: Measured and approximated particle/fluid interaction term of the particle kinetic stress equation. a) streamwise direction, b) Reynolds stress contribution. Model 1 results are given by the grey symbols (+,x), while Model 2 results are given by the black symbols (+,x)

Equation (7) reduces to an exact equation for the case of small particle Reynolds number, where the drag force is given by Stokes drag. For finite Reynolds numbers, the approximation should work well where the correlation between slip velocity and the particle-fluid stresses are small. Another approximation that could be done would be to use the particle viscous response time ($\tau_p=\rho_2 d^2/18\rho_1 \nu$) instead of τ_{12}^F, which will be referred to as model 2. The results of these models are both presented in Figure 3. From these results, it can be seen that both models give reasonable results (typically within 5% over most of the flow, in some cases varying around 20 to 30% approaching the wall). Model 1 typically performs slightly better close to the wall, as it accounts for the mean variation in the particle response time as a function of the wall-normal coordinate. This is consistent with the fact that the mean Reynolds number varies from approximately 2 in the outer flow, up to 15 near the wall where the corrections to the Stokes drag coefficient will be most appreciable.

The above results represent only a partial set of observations to study the two-fluid coupling and how it can be modeled, and we conclude with a list of items that should be given attention in future work. As mentioned above, one also needs the third-order moments in the transport term (equation 2) to complete the particle kinetic stress transport. This can be provided from the above data with an increased statistical sample. The particle-fluid cross-correlation in equation (4) also requires a separate evolution equation to complete the closure (see references [4] and [5] for details on how this can be addressed). Finally, in deriving equation (1), the effect of other particle force terms such as added mass, lift, and far-field pressure were neglected. For sediment particles, these forces may play a measurable role in the dynamics, and their effects warrant further investigation.

[1] Kiger, K. T. and C. Pan (2000) "PIV technique for the simultaneous measurement of dilute two-phase flows," *J. Fluids Eng.*, **122**, pp. 811-8.

[2] Adrian, R. J., Meinhart, C. D., and C. D. Tomkins (2000) "Vortex organization in the outer region of the turbulent boundary layer," *J. Fluid Mech.*, **422**, pp. 1-54.

[3] Kiger, K. T. and C. Pan, (2002) "Suspension and turbulence modification effects of solid particulates on a horizontal turbulent channel flow,"*Journal of Turbulence*, 3(19), p. 1-21.

[4] Wang, Q., Squires, K. D., and O. Simonin (1998) "Large eddy simulation of turbulent gas-solid flows in a vertical channel and evaluation of second-order models," *Int. J. Heat and Fluid Flow*, **19**, pp. 505-11.

[5] Simonin, O, Deutsch, E. and Boivin, M. (1993) "Large eddy simulation and second-moment closure model of particle fluctuating motion in two-phase turbulent shear flows," *Turbulent Shear Flows 9: selected papers from the International Symposium on Turbulent Shear Flows, Kyoto, Japan, August 16-18, 1993*, Springer-Verlag, Berlin, p. 85-115.

III

Turbulent Structures and Sediment Transport

TURBULENT STRUCTURE OF EXCITED AXISYMMETRIC IMPINGING JET

S. ALEKSEENKO, A. BILSKY, O. HEINZ, B. ILYUSHIN, D. MARKOVICH
Institute of Thermophysics Siberian Branch of RAS, Lavrentyev Ave., 1, Novosibirsk, 630090, Russia, e-mail: dmark@itp.nsc.ru

Abstract

The turbulent structure of axisymmetric submerged impinging jet has been experimentally studied for the conditions of external low-amplitude periodical forcing. The conditional sampling technique is applied along with 2D PIV and electrodiffusion method. The triple decomposition of velocity pulsations is applied in order to distinguish the coherent and stochastic parts of turbulent kinetic energy and statistical characteristics. The major contribution of coherent component in the near field of the jet is ascertained. With the purpose of determining the fine structure of hydrodinamical characteristics the high-resolution PIV experiments (up to 0.17 mm/vector) have been carried out. This approach allowed detecting the travelling waves of pulsating characteristics in the core of impinging jet including close vicinity of stagnation point.

1. Introduction

The intensity of sediment transport near the walls can be influenced effectively by proper organisation of turbulent flow structure. Thus, strong dynamic impact may be provided by different types of impinging flows (impinging jets, for instance). Impinging jets are the hydrodynamical objects with wide spectrum of physical effects: in different flow regions one can observe both free and wall shear layers, large-scale vortex structures (LSVS) developing in the free jet, interacting with the impingement surface and in high degree affecting the sediment transport. Additional effects appear to be caused by sharp curvature of stream lines in the vicinity of jet turn and, as a consequence, appearance of Gortler instability. PDF of turbulent fluctuations in impinging jet is far from the Gauss distribution due to the presence of LSVS and anisotropy of pulsations in the near-wall region is quite different from one at usual boundary layer conditions.

There are only few experimental studies of impinging jet hydrodynamics with detailed consideration of turbulent flow structure. Thus, enough complete information on the main turbulent characteristics is presented in the paper by Cooper et al. (1993) where authors performed hot-wire measurements of average velocity components, RMS pulsations and Reynolds stresses for the jet issuing from the round tube and impinging onto the normal obstacle. Landreth and Adrian (1990) presented data on the PIV measurements of impinging jet characteristics. Their results contained instant distributions of velocity and vorticity in the central cross-section of the jet and also ensemble averaged velocity field in the neighborhood of stagnation point. Triple correlations were measured with the aid of PTV in the work of Nishino et al. (1996). In

A. Gyr and W. Kinzelbach (eds.), Sedimentation and Sediment Transport, 93–100.

all these studies impinging jet was considered at its natural conditions. Practically only work by Didden and Ho (1985) contains the results on conditionally averaged velocity components in a boundary layer of wall region of forced impinging jet. Local unsteady separations of boundary layer have been registered which were induced by large vortices penetrating into wall region. However, the presented information is restricted by mean characteristics.

In order to create the complete pattern of the large-scale coherent structures development in the impinging jet shear layer the information is necessary about the "frozen structure" of these vortex formations. For this aim the conditions should be realized when the large-scale structures are excited by external forcing and conditional sampling technique needs to be used for data acquisition and processing. At the same time for the validation of mathematical models the comparison on statistical moments of different order is necessary for adequate description of mixing and transfer processes. Along with these reasons an accurate experimental information will allow more understanding of the physical mechanisms of instabilities development and energy transfer between different turbulent scales.

This work is devoted to the experimental study of turbulent structure of submerged axisymmetric impinging jet under the conditions of external periodical forcing. Major attention is made on the quantitative description of the flow and accurate measurements of statistical characteristics. Obtained results include observations of new hydrodynamic effects that obviously should be comprehensively studied in the future works.

2. Description of experiment

The experiments were carried out at the hydrodynamical loop with exchangeable working sections (plexiglas channels) for using different experimental techniques. The set up was equipped by the system of pumps and flow meters, reservoir, connecting tubes and apparatus for measurements. Well-profiled round nozzles with diameters of 10 and 15 mm were inserted through the side (or bottom) wall of a channel. The submerged round jet issuing from the nozzle impinged normally on the opposite wall. To measure wall shear stress and local liquid velocities the electrodiffusion method (EDM) was applied. The details of this technique are described in the author's work (Alekseenko and Markovich, 1994).

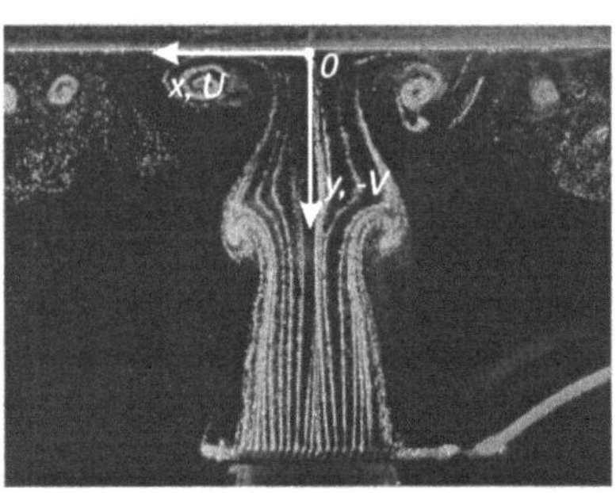

Fig. 1. The photograph of impinging jet flow. Re = 1400, H/d = 3, Sh = 0.5.

Measurements of instant velocity fields were performed using 2D PIV system (Dantec FlowMap PIV System based on PIV1100 processor, ES 1.0 Kodak camera 1K x 1K and 50 mJ NdYAG laser). The software developed allows to calculate the statistical characteristics of turbulent flow of impinging jet including high-order statistical moments. To obtain more precise estimation of high statistical moments the large number of instant frames of velocity distributions were processed (up to 9,000

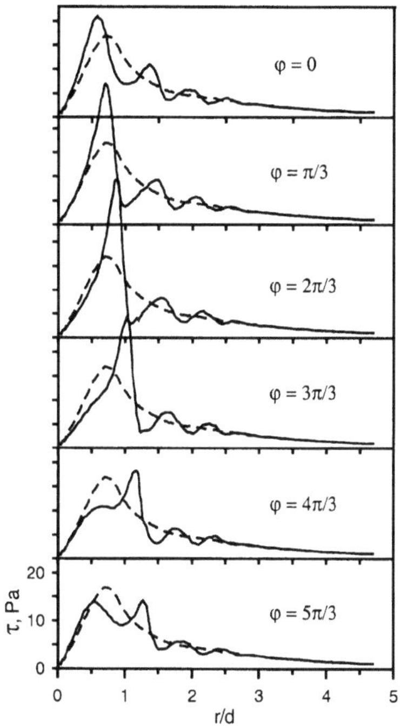

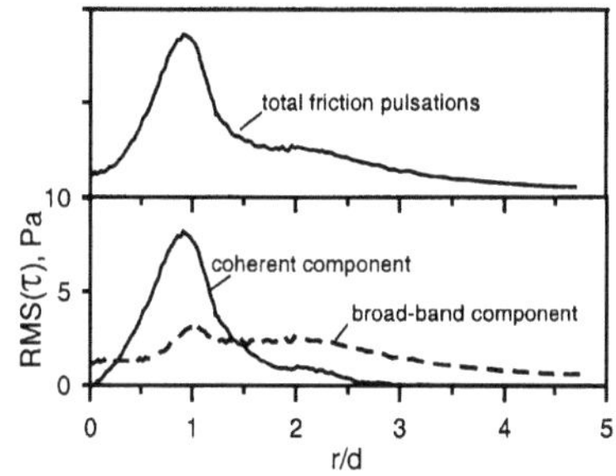

Fig. 2. Conditionally averaged distributions of wall shear stress (left) and decomposition of its RMS pulsations onto coherent and broad-band components (right). EDM measurements. Re = = 12700, $H/d = 2$, $d = 10$ mm, Sh = 0.53.

for separate phase and 20,000 for whole statistics).

The excitation of the jet was provided by a standard electrodynamic vibration exciter connected by the instrumentality of the silphone with the plenum chamber. The sinusoidal excitant signals conveyed from the generator through the power amplifier to the exciter. The initial oscillations of flow embodied the axisymmetric mode (m = 0) and their RMS value changed from $\tilde{u}/U_0 = \sqrt{\tilde{u}^2}/U_0$ = 0.0001 to 0.03 depending on the experimental conditions. The forcing frequency f_f , was characterized by the Strouhal number, $Sh_d = f_f \times d / U_0$.

The velocity measurements near the nozzle exit, performed for high values of Reynolds number, have shown that the imposed oscillations of lowest level do not influence the initial flow characteristics. The level of natural turbulence measured in the vicinity of nozzle exit was for different nozzles in the range of $u'/U_0 = \sqrt{u'^2}/U_0 = 0.005 \div 0.06$ at the nozzle axis and $0.05 \div 0.1$ at the centre of shear layer.

During the experiments the range of Reynolds numbers was tested: Re = 7600 ÷ 25200. Here $\mathrm{Re} = U_0 \times d / n$, U_0 is the mean flow rate velocity at the nozzle exit, d is the nozzle diameter and ν is the kinematic viscosity of electrochemical solution equal to $1.04 \times 10^{-6} m^2/s$. The non-dimension distance H/d between the edge of a nozzle and the plate was varied from 2 to 4.

3. Results and discussion

The impinging jet flow under the conditions of periodic forcing with the frequencies from the range of most jet's sensitivity (Sh = 0.4 ÷ 0.7) is strictly periodic in the

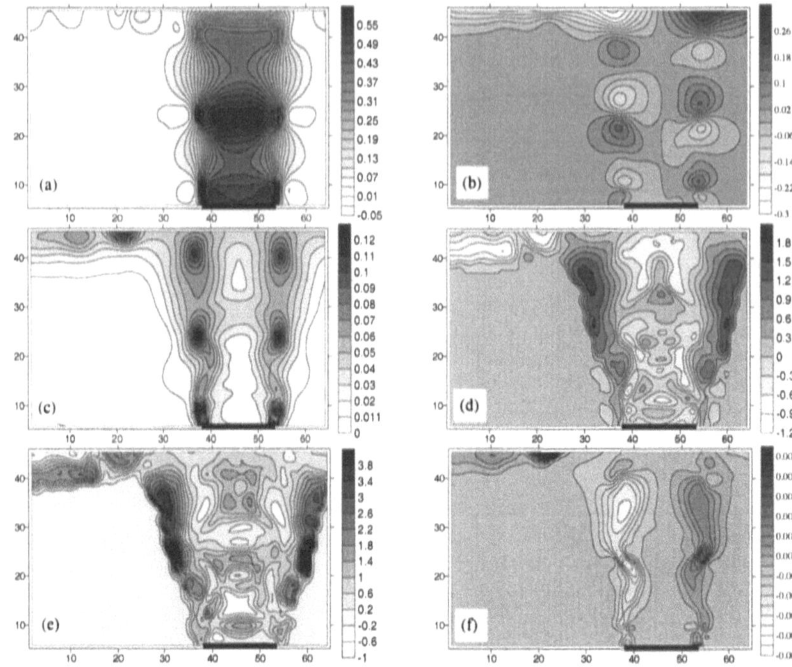

Fig. 3. Spatial distributions of conditionally averaged characteristics for forced impinging jet flow field. Re = 7600, H/d = 3, Sh = 0.5. (a), (b) – mean longitudinal (v) and radial (u) velocity components correspondingly, (c) – RMS value (v) velocity component, (d) skewness factor, (e) – kurtosis, (f) – Reynolds stresses uv. Averaging over 9000 instant frames of velocity field recorded at the same phase.

near field of the jet and well pronounced large scale vortex structures develop in the shear layer (Alekseenko et al., 1997). This fact allowed us to perform the triple decomposition of velocity and wall shear stress pulsations in order to separate the coherent and broad-band components of physical quantities. According to the common approach (Hussain, Reynolds, 1972) the triple decomposition for instant value of generalised variable a is:

$$a(\vec{x},t)=\bar{a}(\vec{x})+\tilde{a}(\vec{x},t)+a'(\vec{x},t),$$

where first, second and third terms in the left part of equation are correspondingly time average, coherent (periodic) and stochastic components of pulsations. In the present work the velocity and wall shear stress components were considered as partial cases of a. Phase averaging gives the following:

$$\langle a(\vec{x},t_i)\rangle=\bar{a}(\vec{x})+\tilde{a}(\vec{x},t_i)=\lim_{N\to\infty}\frac{1}{N}\sum_{k=1}^{N}a_k(t_i)$$

It is equivalent to the average value a for each phase of the flow. In EDM experiments the probe signal was separated into cycles according excitation frequency. For N pulsation cycles, the ensemble average at time instant t_i from the beginning of a cy-

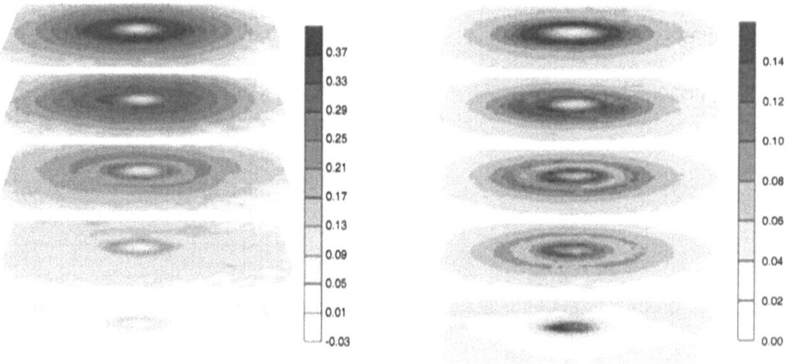

(a) (b)

Fig. 4. Distributions of radial velocity component in near-wall horizontal cross-sections. Re = 7600, $H/d = 3$, $d = 15$ mm, Sh = 0.5. Upper cross-section: $y = 1$ mm, lower - 5 mm. (a) - mean radial velocity V_r , (m/s), (b) - RMS pulsations of radial velocity $\sqrt{\overline{V_r'^2}}$, (m/s).

cle was calculated for velocity and friction components and their squares. $a_k(t_i)$ is the physical value reading in the kth cycle at time t_i. For PIV measurements the conditional sampling was provided by synchronization of excitant signal with laser pulses and setting corresponding phase delay. Both centered and non-centered (relative to time average values) characteristics were calculated.

The set of statistical characteristics including 2nd, 3rd and 4th order moments has been obtained for a number of flow phases: 100 phases for EDM wall shear stress and velocity measurements and 20 phases for whole velocity field (PIV experiments) were tested. The number of samples, processed for each phase was varied from 1000 to 9000 in different experiments. Figure 2 shows the phase-averaged distributions of wall shear stress along impingement surface (solid lines). Dashed lines correspond to the whole averaged value. Six equidistant phases from 100 are plotted. It is clearly seen from the Figure that local values of skin friction can be extremely large owing to coherent structures influence. In the same Figure the decomposition of pulsations level on coherent and stochastic parts is presented. In the vicinity of stagnation point the coherent component of skin friction pulsations dominates whereas in the far field of the radial wall jet the broad-band component yields a major contribution into whole turbulence level.

Figure 3 presents the phase averaged distributions of velocity components, dispersion, skewness factor, kurtosis of longitudinal velocity component pulsations and also Reynolds stress. Presented distributions show the "frozen structure" of the flow, averaged only over the stochastic part of turbulent pulsations. One can clearly observe the periodic spatial flow structure caused by large vortices. The areas with maximum values of second moment of velocity pulsations (Fig. 3, c) correspond to the regions with large velocity gradients (Fig. 3, a, b). The distribution of Reynolds stress (Fig. 3, f) is also strongly nonuniform with locations of maximums in the centre of mixing layer. The pictures with higher moments distributions (Fig. 3, d, e) can manifest the areas

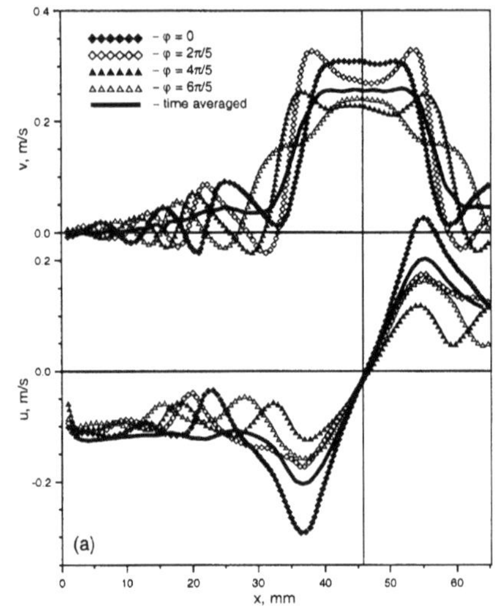

(a)

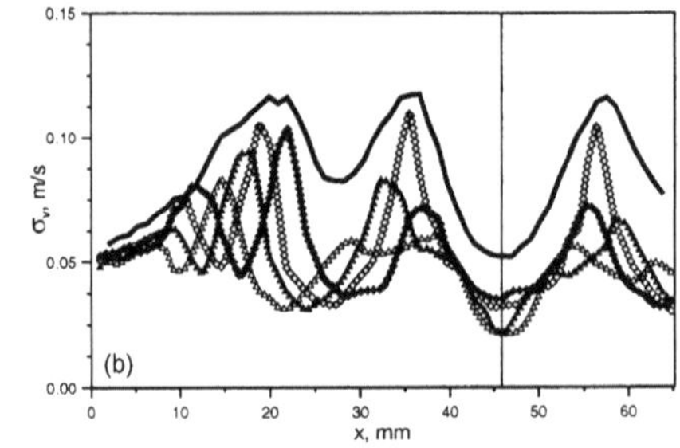

(b)

Fig. 5. Phase averaged distributions of velocity components (a) and dispersion σ_v (b) in horisontal cross-section of the flow: $z = 3.1$ mm. PIV measurements. Flow conditions as in Fig. 3.

with largest asymmetry of pulsations and largest degree of "spottiness" for given location of "frozen" large-scale vortex. Zero values of 3rd and 4th moments correspond to Gauss PDF. The greatest deviation from the normal distribution takes place at the edge of the mixing layer in the areas where two neighbouring large vortices are located.

The analysis of flow structure was performed also on the basis of measurements of local flow velocities in the horizontal near-wall cross sections. These results are presented in Fig. 4, a for time averaged radial velocity component and in Fig. 4, b for its RMS deviation. In the horizontal cross-sections the polar coordinates are used for data processing. The typical nonmonotonical shape of velocity profiles can be observed in different cross-sections of flow. Along radial coordinate the velocity initially grows from zero at critical point up to maximum at some radial distance and then decreases with further increase of r. However at some intermediate range of axial coordinate y (point of origin - impingement plane) one can observe the appearance of second local maximum of velocity and, correspondingly, maximum of pulsations level (Figure 4, b, third cross section, $y = 3$ mm) which is more pronounced. Such nonmonotonical behaviour of these distributions can be explained on the basis of analysis of nonlinear contribution of large-scale structures, propagating along the wall, into flow characteristics. The tangential component of velocity is negligible both for instant and whole averaged distributions and so these data are not presented in the paper. The example of conditionally averaged values of velocity components and fluctuation level for different flow phases is shown in Fig. 5, a, b for near-wall horizontal cross section. This Figure presents the distributions of "frozen" velocity components. The periodical spatial structure of phase-locked flow is demonstrated. Similar to Fig. 4, b the double-maximum of whole averaged distributions for pulsations level can be also observed.

The next part of the work is devoted to the study of the fine structure of the im-

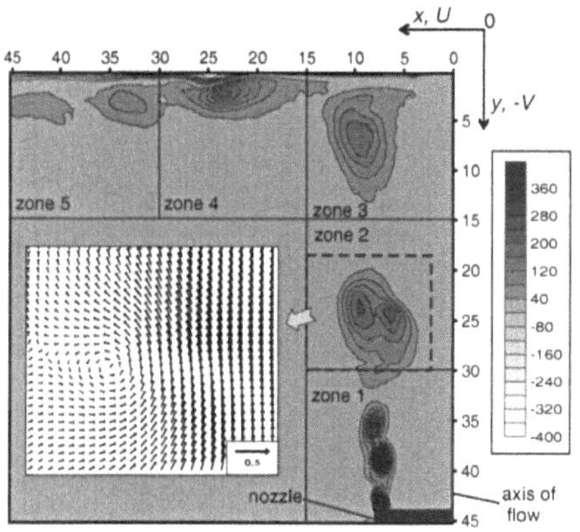

Fig. 6. Vorticity distribution and flow field for impinging jet. Re = 7600, $H/d = 3$, $d = 15$ mm, Sh = 0.5. Spatial resolution is 0.26 mm. Phase locked measurements. Averaging over 1000 instant frames.

pinging jet with the aid of high resolution PIV approach. The flow field was separated into 5 zones with dimensions of 15x15 mm^2 where measurements have been performed with applying the conditional sampling technique. For the results presented below the spatial resolution was equal to 0.26 mm. In Figure 6 the vorticity field is shown for some a priori chosen phase of flow. The distributions in five tested zones are combined into one spatial distribution. The development of vortex structures in the mixing layer of the jet can be clearly observed beginning from the nozzle exit and up to area where these structures interact with the impingement wall.

The use of conditional sampling technique together with high resolution PIV approach has allowed obtaining the principally novel information about fine structure of turbulence in the core of the jet impinging onto the wall. It has been found that the distributions of statistical moments calculated using conditionally sampled ensemble of instant velocity fields had pronounced periodic character. Figure 7 shows the spatial distributions of second and third statistical moments correspondingly. Note that these distributions represent the values averaged only over stochastic part of turbulent pulsations. Thus, for each phase of coherent structure propagation cycle one can observe periodical distribution of statistical moments. At the same time the statistical characteristics stored and processed without conditional sampling do not show any peculiarities. Described above effect has been observed at two essentially different experimental conditions. Parameters of these experiments are the following: Re = 7600, $H/d = 3$, Sh = 0.5 – for the first case and Re = 9000, $H/d = 2$, Sh = 0.6 – for the second case. The spatial resolution of PIV was 0.26 mm and 0.17 mm for these two cases correspondingly. As a result, for this phenomenon we might say that a certain new kind of instability is detected in a sense of "wave structure" of intensity of turbulent pulsations for the conditions when the general flow field is influenced by periodically propagating large-scale vortices. The physical nature of such periodical pattern of pulsation field is not completely clear at the moment, however the analysis of obtained phase-locked pulsations fields points that such a behavior can't be described in the frames of the closure models of gradient type.

4. Conclusions

The statistical analysis was performed for the characteristics of axisymmetric submerged impinging jet under the action of external forcing with the frequencies from the range of most jet's sensitivity. On the basis of PIV and electrodiffusion measure-

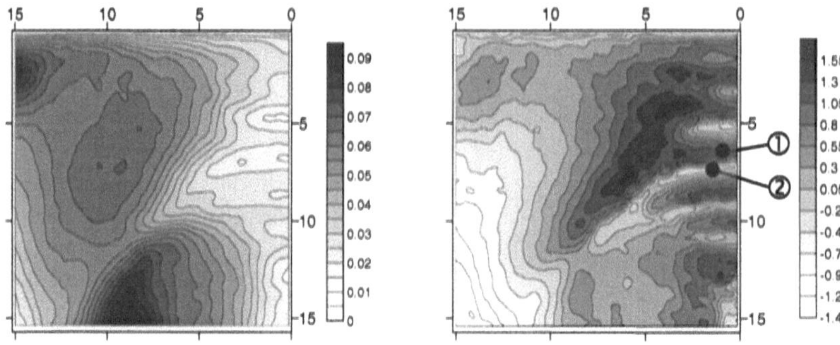

(a) (b)

Fig. 7. Distributions of second (a) and third (b) statistical moments for longitudinal velocity component measured at phase-locked conditions. Zone 3 (see Fig. 6.)

ments and conditional sampling technique the phase averaged flow characteristics were determined. It was shown that coherent component provides the major contribution into main flow characteristics. The strong anisotropy of phase-locked pulsations was observed whereas whole statistical quantities can have substantially more isotropic character. The novel kind of instability is detected which manifests in a wave (periodical) structure of distributions of higher statistical moments for turbulent pulsations.

Acknowledgments

This work was supported by RFBR Foundation, grants N 01-02-17662, 00-15-96810, 02-02-08081-inno and INTAS foundation, grant N YSF 2002-240.

References

Alekseenko, S.V. and Markovich, D.M., 1994. "Electrodiffusion diagnostic of wall shear stresses in impinging jets", J. Appl. Electrochemistry, vol. 24, pp. 626-631.

Alekseenko, S.V., Markovich, D.M. and Semenov, V.I., 1997, "Effect of external disturbances on the impinging jet structure", Proc. of the 4th World Conference on Experimental Heat Transfer, Fluid Mech. and Thermodynamics, Brussels, June 2-6, vol. 3, pp. 1815-1822.

Cooper, D., Jackson, D. C., Launder, B. E. and Liao, G. X., 1993, "Impinging jet studies for turbulence model assessment - I. Flow-field experiments", Int. J. Heat Mass Transfer, vol. 36, pp. 2675-2684.

Didden, N. and Ho, C.-M., 1985, "Unsteady separation in a boundary layer produced by an impinging jet", J. Fluid Mech., vol. 160, pp. 235-256.

Hussain, A. K. F. M., and Reynolds, W. C., 1972, "The mechanics of an organized wave in turbulent shear flow. Part 2, experimental results", J. of Fluid Mech., vol. 54, pp. 241-261.

Landreth, C.C. and Adrian, R.J., 1990, "Impinge-ment of a low Reynolds number turbulent circular jet onto a flat plate at normal incidence", Experiments in Fluids, vol. 9, pp. 74-84.

Nishino, K., Samada, M., Kasuya, K. and Torii, K., 1996. "Turbulence statistics in the stagnation region of an axisymmetric impinging jet flow", Int. J. Heat and Fluid Flow, vol. 17, pp. 193-201.

RESUSPENSION BY DROPLETS

STUART B. DALZIEL & MARTIN D. SEATON
Department of Applied Mathematics and Theoretical Physics
The University of Cambridge
Silver Street, Cambridge CB3 9EW, ENGLAND
s.dalziel@damtp.cam.ac.uk

1. Introduction

Most of the work on hydrodynamic resuspension has concentrated on the lift (and drag) forces generated by motion in the turbulent boundary layer beneath some mean flow. In the fluvial context, this hydrodynamic mechanism represents the dominant component of resuspension for most naturally occurring particle-laden flows. However, in the aeolian context, the hydrodynamic mechanism has great difficultly resuspending fine dust particles, although may often resuspend larger 'sand' particles. It is then the saltating of these sand particles (which perform a series of approximately parabolic hops along the ground) that drives the ballistic mechanism of direct mechanical interaction that is capable of resuspending the finer dust. Further details may be found in the many review articles on both these processes (*e.g.* Willets 1998). An additional hydrodynamic mechanism, whereby the wake of a colliding particle can play a role, has explored recently by Eames & Dalziel (2000). The resuspension mechanism driven by droplets, however, turns out to be quite different.

This paper presents some first preliminary results from a set of laboratory experiments studying the resuspension produced by a liquid droplet colliding with a bed of fine particles. The resuspension mechanism observed has some similarities with the classical ballistic mechanism, but the hydrodynamics of the droplet also plays a substantial role, leading to a more efficient resuspension than is found for the collision of a solid particle. There is some similarity with the collision of a droplet with a solid surface. Collisions of a droplet with a solid boundary has long been studied (*e.g.* Bussmann *et al.* 2000), but not collisions with a bed of particles.

2. Experimental setup

Droplets were produced by a standard laboratory pipette (hand-held and operated) above a bed of 120μm hollow glass microspheres. The region of impact was illuminated by a 1kW halogen photographic lamp, and visualised by a 262 frame per second, 512×512 pixel digital video camera. The four-channel LVD output from the camera (around 68MByte/s) was fed into a digital framegrabber then stored in real time onto a three-disk SCSI array in the host computer. In this way sequences of arbitrary length could be recorded, thus removing any need to synchronise the recording process with the droplet formation.

The glass microspheres were selected due to their relatively low density (940kgm^{-3}) and the almost complete absence of cohesive forces. This combination of characteristics made them relatively easy to resuspend, and they moved relatively freely in response to ballistic impacts with the bed.

A. Gyr and W. Kinzelbach (eds.), Sedimentation and Sediment Transport, 101–104.

3. Results

Figure 1 shows a sequence of a typical droplet colliding with a bed of particles. In this case the particle layer was sufficiently thick that the presence of a rigid boundary below did not play a role. The release height of the droplet was around 100mm above the bed, and the diameter of the droplet was around 4.3mm. The sequence shows images at a spacing of 1/262s from just before the point of collision.

As expected for a droplet of this size, the droplet is approximately spherical (Figure 1a) before the collision. There is some evidence of capillary waves causing a slight shape oscillation, the origin of which is the process of droplet formation and detachment at the tip of the pipette. The presence of these oscillations is not important, however, for the subsequent collision.

Prior to impact, the air between the droplet and the particle bed must be removed. This will occur partly through the porous particle bed, and partly as a radial flow away from the droplet. While it is possible that the radial flow component may resuspend some material, the current visualisations suggest this is unlikely.

The initial impact of the droplet (Figure 1b) is very similar to that of a solid particle. There is a direct transfer of vertical momentum from the droplet to the particle bed. However, unlike the collision of a rigid particle, deformation of the droplet means that the droplet does not decelerate uniformly, and so the peak vertical impulse to the bed is somewhat lower but extending over a longer period. Nevertheless, the initial response of the particle bed is to a point loading, providing the initial lateral redistribution and burst of resuspension.

Although the total vertical momentum of the droplet is the same as that of an equivalent sized solid particle, the lower vertical force on the particle bed means that

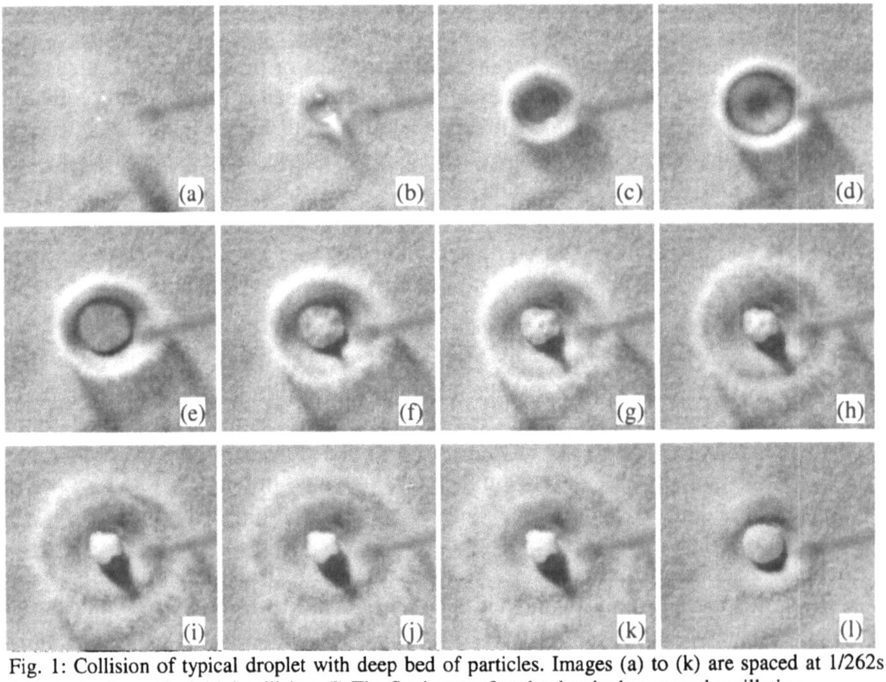

Fig. 1: Collision of typical droplet with deep bed of particles. Images (a) to (k) are spaced at 1/262s from just prior to the initial collision. (l) The final state after the droplet has ceased oscillating.

if the droplet penetrates to the same depth, less of its impact kinetic energy is converted to work within the particle bed. Instead a significant fraction of this impact energy is converted to radial motion – the motion that would cause the splattering of a droplet impacting a solid surface. This radial motion, which peaks about the time the droplet achieves its maximum penetration (at which time its vertical deceleration will be greatest) pushes laterally against the particles (Figure 1c), throwing more particles into the air. Due to the stresses within the particle bed making it easier to move the particles nearer the surface, the deforming droplet is accelerated upwards at the same time as it moves out, leading to the characteristic bowl shape that may be seen in Figure 1c. The net effect is that not only are the particles thrown outwards by the distorting droplet, but also upwards.

The rate of deformation of the droplet decelerates rapidly due not only to the loss of energy to the resuspended particles and those remaining in the bed, but also to increase in surface energy that results from deforming the droplet away from its initial spherical shape. Soon all the impact kinetic energy has been converted, and surface tension begins to draw the droplet back towards a lower energy state (Figure 1d). If the surface of the droplet had remained 'clean', the droplet would approximately recover its original spherical shape (although would overshoot this). However, during its expansion, the surface of the droplet picked up a coating of particles while the surface area of the droplet was increasing. The distribution of this coating – purely on the lower side of the droplet at maximum extent – suggests that the surface divergence occurred primarily on the parts of the surface not in contact with the bed. During the collapsing, the particle-free surface of the droplet is able to contract, but the particle-coated surface is not. The presence of the particles provides the compressive stresses that balance the tension in the surface. The crinkling of this semi-rigid particle-covered surface may be seen clearly in Figures 1e and f.

The collapsed particle-covered droplet continues to deform due to its inward inertia, even after all the particle-free surface has been removed, as can be seen in Figures 1g to j. During this time there is also a gradual rearrangement of the particles on the surface, removing the crinkles, that continues as the droplet dissipates its remaining energy and relaxes (Figure 1k) to eventually reach a smooth equilibrium state (Figure 1l). This final state is not spherical as the surface area covered by particles is greater than that of the initial spherical droplet.

The initial deformation of the droplet is governed by inertia, surface tension, and the pressure forces exerted by the particles. Even though it is likely that the particles in contact with the droplet provide effectively a no-slip boundary, viscous diffusion is too slow for this to play a significant role during this initial very rapid expansion. In contrast, viscous diffusion, enhanced by the droplet's particle coating, plays an important role during the subsequent collapse of the droplet, rapidly dissipating the surface energy that was recovered during the collapse. Seepage into the particle bed did not occur in these experiments; it is not yet clear if this would play a role if the particles were more strongly hydrophilic.

The late-time behaviour of the droplet is a little different if the dimensionless impact energy is significantly lower, not allowing the droplet to expand to the same extent and thus preventing its entire spherical surface to be covered by particles. Figure 2a shows a 6.2mm droplet just prior to impact, while Figure 2b shows the droplet as it is beginning to collapse back. The lower energy level has reduced the maximum extent of the droplet and reduced the time to reach that extent, although significant re-

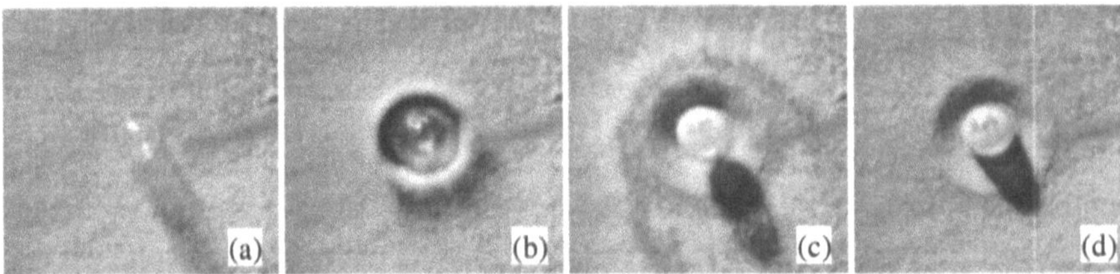

Fig. 2: Low energy collision of a droplet showing droplet bounce after subsequent collapse. Time of images relative to impact: (a) 0s, (b) $t = 0.008$s, (c) $t = 0.036$s, (d) $t = 0.080$s.

suspension still occurs. The droplet collapse is only weakly hindered by a dispersed partial covering of particles, allowing an approximately spherical shape to be recovered and permitting more of the recovered surface energy to be converted to kinetic energy. The collapse is sufficiently fast that the associated capillary waves converging on the lower side of the droplet cause the droplet to bounce back off the particle bed, as seen by the detached shadow in Figure 2c. The final state of the droplet (Figure 2d) is approximately spherical with a dispersed particle covering over part of the droplet.

If the impact is with a thin layer of particles (or the layer of particles is relatively immobile), then the deforming droplet is able to escape from the particle bed before its impact kinetic energy has all been transferred to the particles or converted to surface energy. In such a case the droplet can shatter before the main droplet and satellite droplets collapse back into their equilibrium forms.

4. Conclusions

A droplet colliding with a fine particle bed can be an efficient mechanism for resuspending particles due to the rapid distortion of the droplet providing a direct transfer of the impact kinetic energy to the particles. Surface tension and inertia play dominant roles in this process, while viscous effects are not seen until the subsequent collapse of the distorted droplet.

The mechanisms described here will apply to other types of particles, but in many cases higher collision velocities/energies would be required, and the shorter associated timescales would be too fast for the current camera system. Moreover, splattering of the droplet would become more important with less mobile particle beds.

The environmental implications of this resuspension mechanism may, however, be limited. While raindrops can clearly resuspend dry material, they also act to wash dust from the atmosphere. Moreover, once the soil has been wet fine particles are much more difficult to resuspend, and this mechanism is essentially turned off.

References

Bussmann, M. Chandra, S. & Mostaghimi, J. 2000 Modelling the splash of a droplet impacting a solid surface. *Phys. Fluids* **12**, 3121-3132.

Eames, I. & Dalziel, S.B. 2000 Resuspension of dust by the flow around a sphere impacting a wall *J. Fluid Mech.* **403**, 305-328.

Willets, B., 1998 Aeolian and fluvial grain transport. *Phil. Trans. Roy. Soc. A* **356**, 2497-2513.

Particle resuspension by an impacting vortex ring

RICK J. MUNRO & STUART B. DALZIEL
Department of Applied Mathematics and Theoretical Physics
The University of Cambridge
Silver Street, Cambridge CB3 9EW, ENGLAND
r.j.munro@damtp.cam.ac.uk & s.dalziel@damtp.cam.ac.uk

1. Introduction

Coherent vortex structures play a significant and important role in the dynamics of many commonly occurring natural flows, for example turbulent boundary layers and channel flows. One particularly important feature of these coherent vortices is how they interact with flow boundaries, and the subsequent generation of new coherent vortex structures.

The results presented here are from a set of visualization experiments performed as the initial stage of an on-going project. The overall aim of this project is to attempt to identify the fundamental mechanisms by which a vortex ring impacting on a thick sediment layer is able to resuspend and redeposit particles.

The experiments identify two principle mechanisms for the resuspension and redeposition of sediment. First, particles are entrained into the primary, or generated, ring and the ejected secondary vortex ring produced by the unsteady separation of the boundary layer induced by the primary. Second, the high concentration particle *splash* produced at the point of separation results in a debris flow, which deposits the sediment in a thin radial layer outside of the initial crater.

It is hoped that the understanding obtained from simple experiments such as these will provide valuable insight into the role of coherent vortices in the suspension and deposition of sediment in more complicated flows.

2. Experimental setup

Figure 1(a) shows the basic experimental setup used for the visualisations. A small Perspex tank of volume 36 litres was filled with tap water to a level 6 cm below the top of the tank. Small glass ballotini, with diameter range 0.01-0.02 cm and relative density $\rho_r = 1.4$ g cm^{-3}, were added to form a smooth, even layer of sediment 1 cm thick. The depth of the sediment layer was chosen so that the presence of the solid tank bottom would not affect the sediment suspension. The vortex ring generator consisted of a circular tube of diameter 5 cm with one end open and the other connected to a bicycle pump. The tube was clamped into a vertical position at the centre of the tank, with the open end submerged 8 cm below the surface of the water. When the pump is compressed, the slug of water occupying the tube is forced out producing an axisymmetric vortex ring (see Maxworthy (1972) for details of vortex ring formation). The vortex ring was fully developed a distance of about 5 cm (*i.e.* one ring diameter) away from the exit plane, and travelled a vertical distance of about 20 cm before making impact with the sediment layer. The impact area was illuminated from the side, and the images were captured at a rate of 45 frames per second using a digital video camera with field of view aligned normal to the light

A. Gyr and W. Kinzelbach (eds.), Sedimentation and Sediment Transport, 105–108.

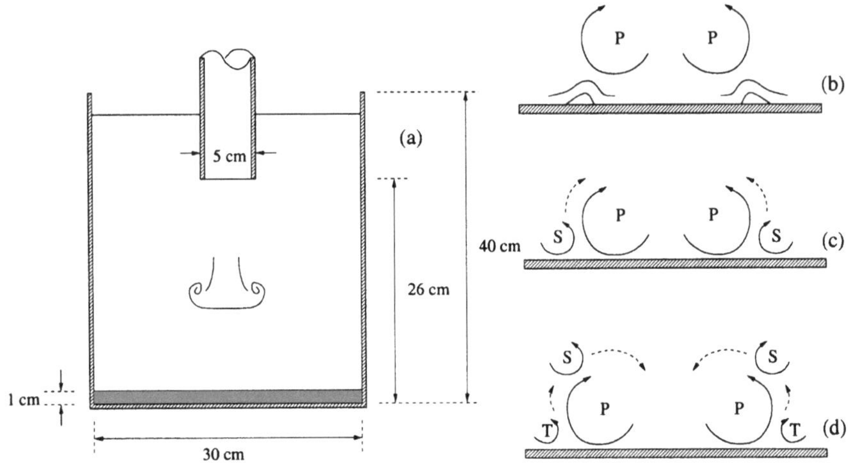

Figure 1: (a) Experimental setup, (b)-(d) schematic showing the important stages of a vortex ring colliding with a solid boundary.

source. Images were captured from two different camera perspectives, the first looking down at an angle of about 30°, and the second looking horizontally across the sediment bed.

The Reynolds number $Re = D_0U_0/\nu$ of the vortex rings used in the experiments was approximately 10^4. Here $D_0 = 5$ cm is the diameter of the tube, U_0 is the averaged velocity of the ring over the distance travelled, and ν is the kinematic viscosity.

3. Results

To understand what happens when a vortex ring impacts a sediment layer, it is worth describing the process of a vortex ring colliding with a solid surface. The fundamental characteristics of this process are shown schematically in figure 1(b)-(d). For a more detailed description, see Walker *et al.* (1987) and Doligalski *et al.* (1994).

As a vortex ring approaches the surface, a boundary layer is induced. Directly below the vortex core a velocity maximum occurs near the wall, corresponding to an instantaneous pressure minimum. As a result, an annular region, with vorticity of opposite sign to that of the vortex ring, develops adjacent to the wall. This causes rapid thickening of the boundary layer in a region local to this eddy (figure 1(b)). Provided the vortex ring is sufficiently strong, boundary-layer separation occurs in which a secondary vortex ring of opposite rotation is ejected from the boundary layer. The mutual influence of the primary and secondary vortices causes the secondary to orbit around the outside of the primary towards the interior (figure 1(c)). Provided the primary vortex ring is sufficiently strong a tertiary vortex ring, again of opposite rotation, can be ejected in much the same manner (figure 1(d)).

The sequences of images in figure 2, taken from two different perspectives, shows the initial stages of the vortex ring impacting with a sediment layer. The images have been extracted from separate experiments; however the conditions are essentially identical. The rows are aligned to approximately correspond to the same stage of the impact, and each sequence corresponds to a period of about 0.2 s.

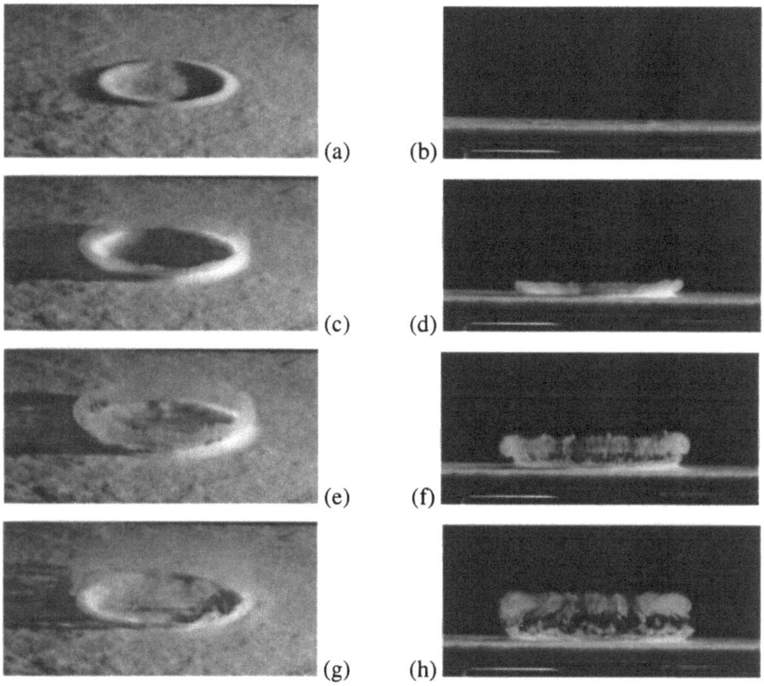

Figure 2: Two sequences of images showing different stages of the initial impact of the vortex ring. Both sequences correspond to a period of about 0.2 s.

When the vortex ring is far from the boundary, the tangential velocity induced near the wall is small. However, as the ring gets close to the wall the unsteady boundary layer has developed, and the tangential velocities are sufficient to displace the sediment (figures 2(a) and 2(b)). The initial displacement of sediment seems to occur when the vortex ring is approximately one ring radius above the layer. At the point of separation of the boundary layer, a high concentration sediment *splash* is produced (figures 2(c) and 2(d)), a small proportion of which is entrained into the primary and ejected secondary vortices. The entrained sediment in the secondary vortex structure can be seen clearly at the extreme right of the splash in figure 2(d). As with solid surface impact, the secondary vortex orbits the primary vortex and so the entrained sediment is suspended in the fluid above the point of impact. Figures 2(e) to 2(h), particularly 2(f) and 2(h), show the orbiting of the secondary vortex as it carries the entrained sediment around the primary. The entrainment of sediment in the primary and secondary vortices is the predominant mechanism for resuspending the sediment. When the secondary vortex orbits into the interior of the primary vortex, instabilities are produced due to vortex compression. These instabilities develop smaller and smaller scales, and eventually a turbulent state is reached.

Another important mechanism for the redeposition of sediment was identified from these images. At the point of separation, the majority of sediment displaced in the high concentration splash is not entrained into the primary and secondary vortices. As a result, the remaining sediment develops into a debris flow redepositing the sediment in a thin radial layer outside of the initial crater produced by the splash. The sequence of images in figure 3 shows different stages of the development of this flow.

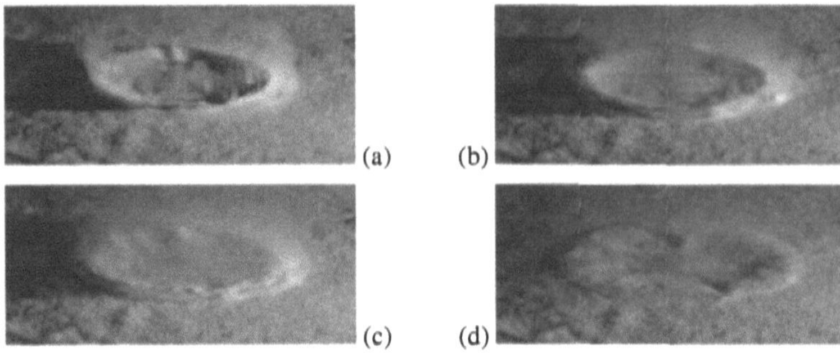

Figure 3: A sequence of images, corresponding to a period of about 2 s, showing different stages of the debris flow.

The debris flow also seems to interact with primary vortex ring as its diameter increases due to the presence of the boundary. Note that the sequence of images shown in figure 3 corresponds to a period of about 2 s. However, the images are not taken at equal intervals, and the first three correspond to a period of about 0.8 s.

4. Discussion

The initial experiments have identified two fundamental mechanisms by which particles are resuspended and redeposited by the impact of a vortex ring on a thick sediment layer. The initial displacement of sediment is caused by the induced boundary layer due to the primary vortex ring, which eventually separates causing a high concentration splash. The predominant mechanism for suspension is the entrainment of particles from the splash in the primary and ejected secondary vortices. However, the majority of the redeposition of sediment in this high concentration splash is produced by the resulting debris flow.

The results obtained from these initial visualisation experiments can now be used to design more accurate and controlled experiments to analyse these mechanisms in greater detail.

References

T. L. Doligalski, C. R. Smith and J. D. A. Walker, 1994. Vortex interactions with walls. *Annu. Rev. Fluid Mech.*, **26**, 573-616.

T. Maxworthy, 1972. The structure and stability of vortex rings. *J. Fluid Mech.*, **51**,15-32.

J. D. A. Walker, C. R. Smith, A. W. Cerra and T. L. Doligalski, 1987. The Impact of a vortex ring on a wall. *J. Fluid Mech.*, **181**, 99-140.

IV

Two-Phase Flows

ON THE EFFECTS OF STOKES, RICHARDSON, AND STABILITY NUMBERS IN PERSISTENT AND ACCELERATING VORTICES

ALAIN BURGISSER,[1] GEORGE W. BERGANTZ,[2] and ROBERT E. BREIDENTHAL[3]
[1]*University of Alaska Fairbanks - Geophysical Institute, Alaska Volcano Observatory, Fairbanks, AK 99775-7320 USA, alain@gi.alaska.edu*
[2]*University of Washington, Dept. of Geological Sciences, Box 351310 Seattle, WA 98195-1310 USA, bergantz@u.washington.edu*
[3]*University of Washington, Department of Aeronautics and Astronautics, Box 352400 Seattle, WA 98195-2400 USA, breident@aa.washington.edu*

1. Introduction

1.1 PERSISTENT VORTICES

Based on observations in stratified flow, Cotel [1,2] proposed a new model for stratified entrainment, in which a new parameter was asserted to be necessary to explain the observations. When a vortex is near an interface, the fluxes across the interface are determined by conventional parameters such as Richardson and Reynolds numbers as well as a new "persistence parameter." A measure of the stationarity of the vortex, the persistence parameter is essentially the ratio of the rotational to the translational speed of the vortex with respect to the interface. It plays as large a roll in stratified entrainment as any other parameter.

According to the theory, the fluxes across the interface would be reduced to their laminar values, independent of the fine-scale structure of the turbulence, if the interface is sufficiently stratified to remain essentially flat, $Ri > \mathrm{Re}^{1/4}$. If a vortex cannot engulf across the interface, all fluxes there must be purely diffusive. This is also true if the interface is a solid wall. In recent water tunnel experiments [3], vortex generators added strong, streamwise vortices to the flow at the upstream end of a corrugated plate. When each streamwise groove contained a single vortex in the optimum location, the vortices were stabilized. The wall heat flux measured in the bottom of a groove was reduced to a laminar value when the vortices were stable.

It is counter-intuitive that adding strong vortices to the flow would reduce the wall flux. It is also surprising that a turbulent flow would exhibit a laminar wall flux. There are two different kinds of turbulence in wall flows, the ordinary nonpersistent flow with relatively high, "turbulent" wall fluxes and the persistent case with low, "laminar" wall fluxes. With such unusual flux behavior, a natural question is the effect of persistence on two-phase flow.

1.2 ACCELERATING VORTICES

Self-similarity is one of the few powerful tools in turbulence research. Early on, Richardson and Obukhoff used the concept to deduce the behavior of the energy cascade in the inertial range of eddies in turbulence [4,5]. Because there is no distinguishable eddy size in this range, each eddy must behave like any other, except for a stretching in length and time scales. While self-similarity in space has received considerable attention, self-similarity in time has been relatively neglected.

A. Gyr and W. Kinzelbach (eds.), Sedimentation and Sediment Transport, 111–120.

Consider ordinary unforced turbulence at high Reynolds number. Viscosity sets the size of the smallest eddy, given by the Kolmogorov microscale λ_0 [6,7]. The ratio of the smallest to the largest eddy is

$$\lambda_0/\delta = Re_\delta^{-3/4}, \tag{1}$$

where δ is the largest eddy size and

$$Re_\delta = \Delta U\delta/\nu \tag{2}$$

is the Reynolds number of the largest eddies, whose characteristic velocity is ΔU. At large Re_δ, the microscale is much smaller than the large scale vortices. According to Richardson and Obukhoff, energy cascades from the largest eddies down towards the smallest in an essentially inviscid process without dissipation, until the Kolmogorov microscale is reached. The eddy Reynolds number at this point is

$$Re_{\lambda_0} \equiv \frac{v_{\lambda_0}\lambda_0}{\nu} \cong 1, \tag{3}$$

where v_{λ_0} is the characteristic velocity of scale λ_0. The kinetic energy is rapidly converted to thermal energy by viscous stresses.

At large Re_δ, there is a wide range of eddies larger than λ_0 but smaller than δ. Kinetic energy is transferred from one eddy to the next by inertial forces without appreciable energy loss in the inertial subrange. The kinetic energy flux, e, is constant so that

$$e = v_\lambda^3/\lambda = \Delta U^3/\delta. \tag{4}$$

Within the inertial subrange, an eddy of size λ and speed v_λ presumably hands over a large fraction of its energy in one turnover time given by

$$\tau_\lambda = \lambda/v_\lambda = \lambda^{2/3}/e^{1/3} = (\lambda/\delta)^{2/3}\,\tau_\delta, \tag{5}$$

where $\tau_\delta = \delta/\Delta U$ is the large scale vortex rotation period. A different question is how long it takes to reach scale λ from the start of the cascade. It has been suggested that self-similarity requires

$$dt = const.\lambda^{2/3}d\lambda/(e^{1/3}\lambda), \tag{6}$$

where dt is the differential time interval during which energy is transferred from scale λ to $\lambda - d\lambda$ [5]. Integration of (6) and the use of (4) provides the elapsed time $t(\lambda)$ to reach scale λ starting from scale δ or

$$t(\lambda) = [1 - (\lambda/\delta)^{2/3}]\,\tau_\delta. \tag{7}$$

By combining equations (5) and (7),

$$\tau_\lambda/\tau_8 = 1 - t(\lambda)/\tau_8. \tag{8}$$

The rotation period at scale λ is a linearly decreasing function of the elapsed time since the start of the cascade. This seems reasonable since there is no other time scale available. The radius of curvature of the function is therefore infinite, and therefore the relationship must be linear. The initial and final conditions of the energy cascade require the function to decline linearly in t.

2. Super-exponential forced turbulence

A new class of forced turbulence has recently been proposed [9]. The forcing is of form $\exp(t/\tau_v)$ where the e-folding time imposed on the vortices $\tau_v = \tau_v(t)$ is itself a function of time. Self-similar flow implies that there is no distinguished scale, either spatial or temporal. Hence the vortex rotation period $\tau_v(t)$ at any time t is always a constant fraction of the next rotation period, $\tau_v(t + \tau_v(t))$.

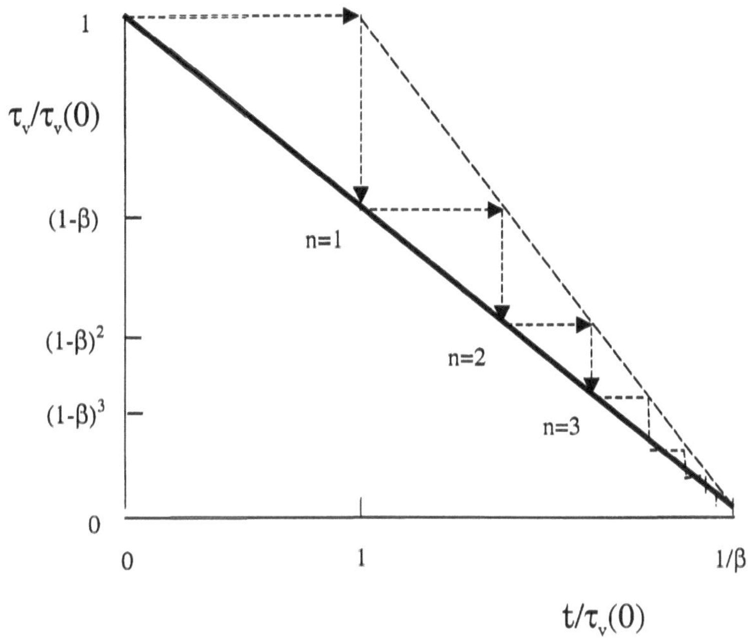

Fig. 1 Temporal evolution of the vortex rotation period τ_v for 0<β<1 in self-similar flow.

Every rotation looks like every other one, except for a constant proportional change in scale. Expressed another way, the decrease in rotation period per rotation is a constant, independent of time, namely

$$[\tau_v(t) - \tau_v(t + \tau_v(t))]/\tau_v(t) = \beta, \qquad (9)$$

where β is a constant. Suppose $\tau_v(t)$ is a general polynomial of order n given by

$$\tau_v(t) = \tau_v(0) + a_1 t + a_2 t^2 + \ldots\ a_n t^n. \qquad (10)$$

After substituting equation (10) into equation (9) and some algebra, it is found that

$$[a_1(\tau_0+a_1t+a_2t^2+\ldots a_nt^n)+O(t^{n^2})]/(\tau_0+a_1t+a_2t^2+\ldots at^n) = \beta. \qquad (11)$$

The left hand side has terms of order t raised to the $(n^2\text{-}n)$ power while the right-hand-side is independent of t. This can only be satisfied if $n^2\text{-}n=0$. The solutions are n = 0 or n = 1. Thusa$_i$ = 0 for all i >1, and $a_1 = -\beta$. For this to be true, the only possible self-similar form is

$$\tau_v(t) = \tau_v(0) - \beta t, \qquad (12)$$

where $\tau_v(0)$ is the rotation period at t = 0. In dimensionless form,

$$\tau_v(t)/\tau_v(0) = 1 - \beta t/\tau_v(0). \qquad (13)$$

If all self-similar vortices follow equation (13), the self-similar vortices in the inertial subrange must also. From a comparison between equations (7) and (13), evidently β = 1 following the energy cascade in unforced turbulence. This suggests that dissipation always vanishes at β = 1.

Figure 1 illustrates in graphical form the self-similar evolution for the general case of β > 0. At the end of the first rotation (n=1), the rotation period is reduced by a constant factor (1 - β). At the end of the second rotation (n=2), τ_v is again reduced by the same factor, and so on. At finite time $t/\tau_v(0) = 1/\beta$, τ_v vanishes.

To an observer following the cascading energy in the inertial subrange, β = 1. However, to an observer following the large scale vortices in the same unforced flow, β = -1 [6,7]. The rotation period is approximately equal to the vortex chronological age, since there is no other time scale available. The value of β thus depends on the observer. There seem to be two different canonical coordinate frames, in analogy with Eulerian and Lagrangian viewpoints of fluid motion, that yield different values of β for the same flow. An observer may either remain at a fixed eddy scale or follow the energy. In ordinary, unforced turbulence, the largest eddies extract energy from the mean flow via instabilities and hand it off to the top of the cascade in the inertial subrange. Energy flows through the largest eddies. An observer remaining at the largest vortices could not follow the energy in unforced turbulence.

The reference frame of the observer can be defined in terms of the dimensionless flux of energy past him. Define γ to be the rate of energy flux e' past the observer normalized by the energy cascade flux e through the largest eddies so that

$$\gamma = e'/e = e'\delta/\Delta U^3. \qquad (14)$$

For an observer fixed at scale $\lambda = \delta$, $\gamma = 1$ and $\beta = -1$ for unforced turbulence. For an observer moving down the cascade with the energy flux, $\gamma = 0$ and $\beta = 1$. For intermediate cases, a reasonable conjecture is a linear relation, where $\beta = 1 - 2\gamma$ or $\gamma = (1-\beta)/2$.

In the limit of $\beta = 1$, the implication is that the entrainment rate vanishes [9]. This implies that the cascade process is interrupted, shutting off the flow of energy from the largest eddies into the cascade. Such a situation is plausible in light of the fact that the hand-off process of energy requires something like a full rotation of the vortex. For $\beta = 1$, however, the time for a vortex to rotate or for a vortex sheet to roll up is just equal to the interval remaining before the singular time.

3. Vorticity

The self-similar evolution represented by equation (13) is also the solution to the scalar equation

$$D\omega/Dt = \beta\omega^2, \tag{14}$$

if $\omega = 1/\tau_v$. The vector vorticity equation for constant density flow in the absence of body forces is

$$D\omega/Dt = (\nabla\mathbf{u})\,\omega + \nu\nabla^2\omega. \tag{15}$$

This suggests that, for inviscid self-similar flow, the magnitude of the nonlinear vortex stretching term $(\nabla\mathbf{u})\,\omega$ corresponds to $\beta\omega^2$. Thus, there is a kind of symmetry between the rate of deformation tensor and the vorticity, such that their inner product is always proportional to the square of the magnitude of the vorticity. The rate of deformation becomes proportional to the vorticity magnitude, and the proportionality coefficient is simply β.

4. Related experiments

4.1 EXPONENTIAL JET

A flow with a constant imposed time scale is the exponential jet [11,12], which exhibits a constant vortex rotation period for the largest vortices. Figure 2 illustrates the evolution of the large-scale rotation period as a function of time. All known turbulent self-similar are represented as a straight line on this figure, differing only in their slope, $-\beta$. The exponential jet entrains and mixes at a somewhat lower rate than the unforced jet.

4.2 TRANSVERSE EXPONENTIAL JET

In an exponential transverse jet, vortices are formed in the near field with an initial rotation period controlled by the nozzle time, the ratio of local nozzle diameter to speed [13]. However, as the vortices advect downstream, they are subjected to subsequent nozzles that increase in size and speed in an exponential way.

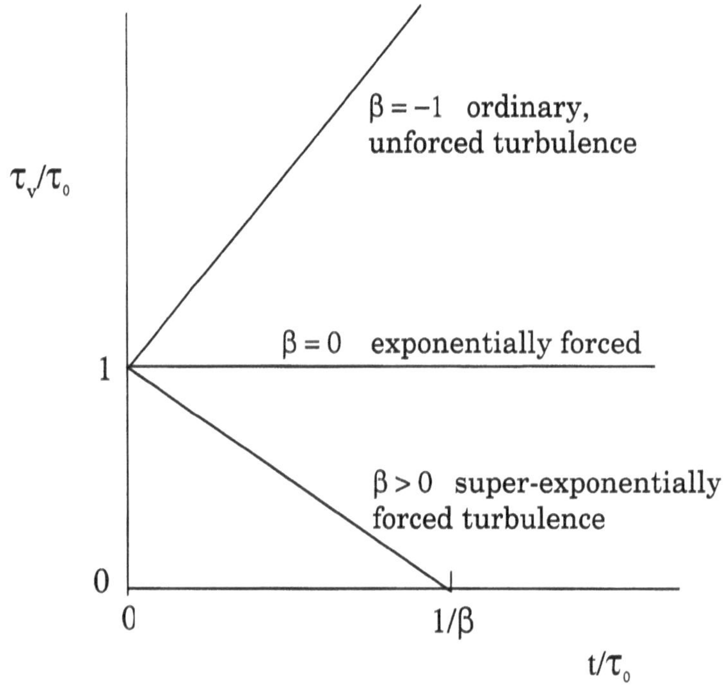

Fig. 2 Vortex rotation period $\tau_v(t)$ for all self-similar flows

The corresponding e-folding time introduces a second time scale on top of the first one. The ratio of these two times is an acceleration parameter α, a measure of the forcing. As α increases above the value of one, vortex rollup and entrainment are reduced. There is less time available for the vortices to complete a rotation.

4.3 GENERALIZED RAYLEIGH-TAYLOR FLOW

In order to test the conjecture that entrainment should vanish in the limit of $\beta \rightarrow 1$ in super-exponential forcing, a generalized Rayleigh-Taylor flow was simulated in two-dimensional inviscid calculations using the CLAWPACK code [14,15]. The density ratio across the interface was two, and the reduced acceleration g' was given a super-exponential profile in time with an initial dimensionless value of 0.1. The initial dimensionless e-folding time was one. When β was negative or zero, the interface rapidly formed into a mushroom shape as baroclinic vortex sheets roll up into discrete vortices. However, when β is one, the vortex sheet does not roll up at all. This is anticipated for all initial disturbances larger than about $g_0'\tau_0^2$. A plot of the vortex size δ_v, normalized by the total layer thickness δ, as a function of β is shown in figure

3. Here, the ratio δ_v/δ was evaluated at fixed δ. As $\beta \rightarrow 1$, the ratio δ_v/δ decreases to zero. Vortex sheet rollup is then completely inhibited.

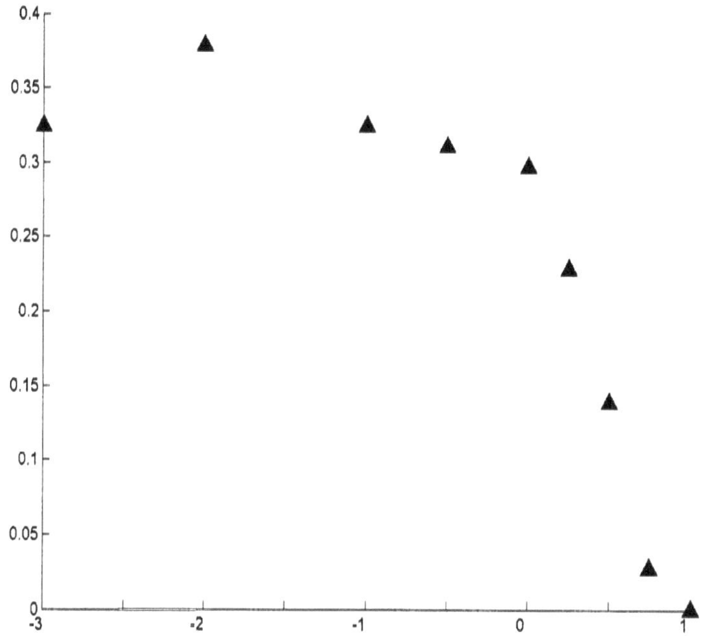

Fig. 3 Vortex size as a function of β.

The chaotic mixing regions associated with entrainment in accelerating turbulence have recently been explored by Govindarajan [16]. The connection between the chaotic mixing regions and β has not yet been determined, but the prime suspect is the motion of the saddle point on the separatrix.

When $\beta = 1$, the vortex sheet does not roll up, even though the remaining time is equal to the instantaneous rotation period, see figure 4. Furthermore, there is some inhibiting effect on vortex growth even when $\beta < 1$. J. Jimenez (personal communication) has pointed out that this implies the flow must be stabilized by the acceleration, in contrast to merely not having sufficient available time for rollup. Future work will analyze the stability.

5. Two-phase flow

5.1 PERSISTENT VORTICES

Define the persistence parameter T to be the number of rotations a vortex makes before it moves a distance with respect to the nearby interface equal to its diameter. In the persistent

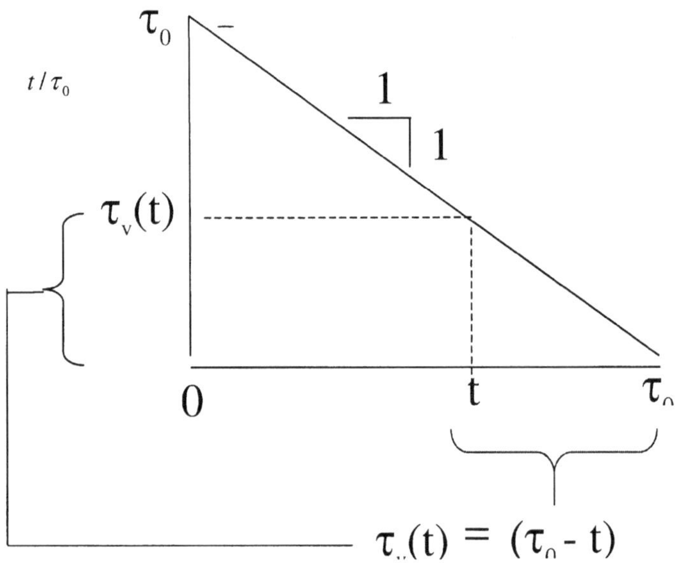

Fig. 4 At every instant, the current rotation period is just equal to the remaining available time for $\beta = 1$. The vortex sheet never rolls up.

limit of $T >> 1$, we expect particle separation for Stokes numbers $\frac{1}{T} \le St \le 1$, since the vortex makes many rotations while stationary, like a centrifuge. This contrasts with ordinary, nonpersistent turbulence, where particle separation only occurs for $St \cong 1$ [17].

5.2 ACCELERATING VORTICES

For a self-similar, accelerating vortex, $\tau_v = 1 - \beta t$, so the Stokes number $St \equiv \frac{\tau_p}{\tau_v}$ increases without limit as $t \to \frac{\tau_v}{\beta}$, where τ_p is the particle relaxation time. If the Stokes number is initially greater than one, it remains so for all time. The particles are always uncoupled from the fluid. However, if the Stokes number is initially less than one, it

increases to unity at time $t = \frac{\tau_0}{\beta}\left(1 - \beta\frac{\tau_p}{\tau_0}\right)$. The particles are then slung out of the vortex as it spins up.
As an example of the stability parameter, consider a forced Rayleigh-Taylor flow, in which the acceleration goes as $g = \frac{g_0}{\left(1-\beta\frac{t}{\tau_0}\right)^2}$, and the density contrast is solely due to dense particles. The stability parameter is then $StRi = \frac{g\tau_p\tau_v}{\delta} = \frac{g_0\tau_p\tau_0}{\delta}\left(1-\beta\frac{t}{\tau_0}\right)$, where δ is the scale of the vortex. The stability parameter decreases with time. So if $g_0 < \frac{\delta}{\tau_p\tau_0}$, then $StRi < 1$ for all time, and the particles immediately fall out. On the other hand, if $g_0 > \frac{\delta}{\tau_p\tau_0}$, then the particles do not fall out initially. Only at time $t = \frac{\tau_0}{\beta}\left(1 - \frac{\delta}{g_0\tau_p\tau_0}\right)$, the stability parameter becomes unity and the particles fall out.
However, depending on the value of β, the particles may first be slung out radially. If $\beta > \frac{\delta}{g_0\tau_p^2}$, the particles will reach unity Stokes number before unity stability parameter, and they will be centrifuged out instead. The particle motion may thus influence the flow stability.

6. Conclusions

Two new classes of turbulent flows have been recently recognized. In the first, the large-scale vortices are stationary near an interface. If the interface is strongly stratified or a solid wall, the fluxes are laminar. A vortex in the persistent limit will centrifuge out particles over a wide range of the Stokes number, since the vortex is stable and completes many rotations in place.
In the second, self-similar acceleration requires a vortex rotation period to decrease linearly with time. The entrainment and dissipation rates depend solely on the logarithmic rate of decrease in rotation period. If the latter is large, vortex rollup is strongly inhibited. For vortices in an accelerating Rayleigh-Taylor flow, particles will eventually fall out unless the rotation period decreases at a sufficient rate, in which case the particles will be slung out radially.

ACKNOWLEDGMENTS
The application to Rayleigh-Taylor flows was originally stimulated by discussions with Guy Dimonte. Hamid Johari suggested the use of the CLAWPACK software, which

was written by Randall LeVeque and his group. Cristian Anitei performed the numerical simulations of the Rayleigh-Taylor flow. The authors are also grateful to Derek Bale and James Rossmanith for their help in running the code and to Greg Balle for discussions about the vorticity equation. William Criminale and Hamid Johari suggested many improvements in the manuscript.

7. References

[1] A.J. Cotel and R.E. Breidenthal, Persistence effects in stratified entrainment, *Applied Scientific Research* **57** (1997) 349-366.

[2] R.E. Breidenthal, Turbulent stratified entrainment and a new parameter for surface fluxes, *Recent Research Developments in Geophysical Research*, ed. S.G. Pandalai, Research Signpost, Trivandrum, India (1999).

[3] G.J. Balle and R.E. Breidenthal, Stationary vortices and persistent turbulence in Karman grooves, *J. Turbulence* **3** (2002) 033.

[4] L.F. Richardson, *Proc. Roy. Soc. A* **110** (1922) 709.

[5] A.M. Obukhoff, *C.R. Acad. Sci. URSS* **32**, 19 and *Izv. Akad. Nauk. SSSR, Ser. Geogr. i. Geofiz.* **5**, (1941) 453.

[6] A.N. Kolmogorov, *Dok. Akad. Nauk SSSR* **30** (1941) 301.

[7] L.D. Landau and E.M. Lifshitz, *Fluid Mechanics*, Pergamon (1959) 121.

[8] J.E. Broadwell and R.E. Breidenthal, A simple model of mixing and chemical reaction in a turbulent shear layer *J. Fluid Mech.* **125** (1982) 397-410.

[9] G.W. Bergantz and R.E. Breidenthal, Non-stationary entrainment and tunneling eruptions: A dynamic template for eruption processes and magma mixing *Geophys. Res. Let.* **28** (2001) 3075-3078.

[10] J.E. Broadwell and R.E. Breidenthal, A simple model of mixing and chemical reaction in a turbulent shear layer, Structure and mixing of a transverse jet in incompressible flow, *J. Fluid Mech.* **148** (1984) 405-412.

[11] Q. Zhang and H. Johari, Effects of acceleration in turbulent jets, *Phys. Fluids* **8** (1996) 2185-2195.

[12] H. Johari and R. Paduano, Dilution and mixing in an unsteady turbulent jet *Exp. Fluids* **23** (1997) 272-280.

[13] A. Eroglu and R.E. Breidenthal, Exponentially accelerating jet in cross-flow *AIAA J.* **36** (1998) 1002-1009.

[14] R.J. LeVeque, *Finite Volume Methods for Hyperbolic Problems*, Cambridge in press (2002).

[15] C. Anitei and R.E. Breidenthal, Accelerating Rayleigh-Taylor flow, Project Report, Department of Aeronautics and Astronautics, University of Washington, 2002.

[16] R. Govindarajan, *Phys. Rev. Let.* in press (2002).

[17] C.T. Crowe, J.N. Chung, and T.R. Troutt, Particle dispersion by organized turbulent structures, *Particulate Two-Phase Flow*, ed. M.C. Roco, Butterworths, New York, (1993).

CENTRIFUGAL SEDIMENTATION PROCESSES IN SUSPENSIONS - FUNDAMENTALS AND CHALLENGES OF EFFICIENT SIMULATION

MARIUS UNGARISH
Computer Science Department, Technion, Haifa 32000, Israel

1. Introduction

The sedimentation (separation) process which is induced by centrifugal-Coriolis forces in a mixture of buoyant particles (droplets, bubbles) in a suspending fluid, remains very much on the frontier of fluid flow research. The motivations are:

1. *Practical* These flows are essential ingredients in separation/mixing technologies and environmental control, e.g., (a) separation of cream-milk, fuel - water droplets, and blood plasma - cells; (b) mixing of micro-carriers of animal cells and nourishing fluid ; (c) propagation of turbidity currents and volcanic ashes (here a combined centrifugal and gravitational driving may appear). The practical objective is to improve the design and control of these processes.
2. *Academic* These flows display interesting and novel phenomena in the realm of multi-phase flow. The objective is to enhance the theoretical and experimental knowledge concerning accurate formulation, efficient solution and reliable interpretation.

In recent years, many fundamental aspects of centrifugal separation of suspensions have been formulated, analyzed, numerically simulated and experimentally verified. The concomitant advance of computer power opens the possibility for implementing these achievements in large scale parametric studies and engineering-oriented simulations. However, there are gaps of knowledge concerning strong rotation, concentrated suspensions and motion of the sediment, which require additional theoretical and experimental investigations.

In a typical system the number of suspended particles is very large (say 10^{10}) and hence a practical description must rely on some kind of averaging. The physico-mathematical framework used in most of the recent studies on suspensions is the continuum-mechanics approach.

The numerous dispersed macroscopic particles lose their identities and are envisaged as a continuum— "the dispersed phase", which co-exists with another continuum—"the continuous phase", representing the embedding fluid. The "concentration" of the phases is usually represented by the volume fraction of the dispersed particles, α; evidently, the continuous phase occupies the volume fraction $1-\alpha$. We emphasize that even for quite dilute suspensions the inter-particle distance, e, is of the order of magnitude of the particle size. In particular, for a suspension of spheres of radius a, simple volume considerations show that

$$e \approx 2.0 a \alpha^{-\frac{1}{3}}. \tag{1}$$

The combination of the dispersed and continuous phases is regarded as the "mix-

A. Gyr and W. Kinzelbach (eds.), Sedimentation and Sediment Transport, 121–126.

ture fluid".

It is endeavored that the flow fields of these "phases" and "mixture" models reproduce the averaged motion of the physical system. To this end, the "averaged" equations are derived from basic principles. The resulting "two-fluid" model consists of two coupled sets of conservation equations. A more common variant, the "mixture" or "diffusion" model, uses one set of conservation equations for the whole mixture, supplemented by a "diffusion" equation to account for the internal changes of volume fraction (concentration) of the dispersed phase. Both models still require some constitutive closure assumptions: for non-concentrated suspensions, it is usually postulated that the generalized stresses are Newtonian with an "effective" viscosity, the interfacial "drift" force is given by a modified Stokesian drag formula, the pressures of the "phases" are related by the capillary law, etc. The final set of conservation equations resembles a Navier-Stokes systems, see [4]. Obviously, the hopelessly complicated original particle-fluid system has been cast in a tractable form by the continuum approach.

However, it is not evident, a-priori, how and if this formulation can be applied to rotating flows, because various Coriolis couplings, Ekman layers and Taylor columns, which may violate the postulated closure assumptions. The practical strategy is to attempt the solution and evaluate critically the results. Solutions pertinent to typical configurations, in particular centrifugal separation from solid body rotation, spin-up from rest, and Ekman/von-Kármán boundary layers, are presented and discussed in [4]. We conclude that the continuum approach produces useful results for these problems, and is able to point out essential differences between the gravitational and centrifugal processes.

An obvious parameter in separation processes is the relative density difference,

$$\varepsilon = \frac{\rho_D - \rho_C}{\rho_C}, \tag{2}$$

where the subscripts C and D denote the continuous and dispersed "phases". Typically, ε is small. A key parameter in centrifugal separation is the (modified) particle Taylor number,

$$\beta = \frac{2}{9}\frac{\Omega a^2}{\nu_0}, \tag{3}$$

which expresses the ratio of Coriolis to viscous forces on a particle (ν_0 is the kinematic viscosity of the fluid, a the radius of the particle, and Ω the angular velocity of the centrifuge). When $\beta \ll 1$ the velocity of the dispersed particles relative to the fluid can be calculated from the balance between the centrifugal buoyancy and the Stokes drag, which yields

$$\mathbf{v}_R = \mathbf{v}_D - \mathbf{v}_C = (\varepsilon\beta\Omega)\left[\frac{1-\alpha}{\mu(\alpha)}\right] r\hat{r}, \tag{4}$$

where r is the radial coordinate in a cylindrical system $\{r, \theta, z\}$, and z is the axis of rotation. The classical Stokes result is modified by the α-dependent hindrance function, where $\mu(\alpha)$ represents the increase of the effective viscosity due to the particles. The particle Reynolds number is assumed small. When β is not very small, significant changes of the hydrodynamic "drag" forces on the particle and of the mixture rheology expressed by $\mu(\alpha)$ are expected, see [4] §7.5 and [3].

2. Sediment transport considerations

An important assumption in the analysis of the abovementioned problems is that the sediment plays a passive role, i.e., the rate of sedimentation is determined by the flow-field inside the container (centrifuge), but this flow-field is not influenced by the presence and motion of the sediment on the boundaries of the container. The validity of this assumption, and the needed modifications of the analysis when this assumption is violated, are expected to be of interest to the participants of this symposium.

One rotating suspension problem where the sediment may become significant is the von-Kármán layer, see [4] §5.4. The theoretical results indicate that some of the particles which enter the boundary layer hit the rotating disk - but it is not clear what happens afterwards. Do these particles accumulate on the boundary, or are they expelled to the periphery ($r \to \infty$) by the main boundary layer circulation? If particles accumulate in a sediment layer, does a steady-state develop? A direct experiment for this boundary-layer flow is an important challenge for the validation and extension of the theory.

Another important case where the behavior of the sediment may play an essential role concerns the understanding of the similarities and differences between the long axisymmetric cylindrical centrifuge and the tube centrifuge, see Figure 1. The latter centrifuge is widely used in biological and medical laboratories. In both centrifuges the inner and outer radius are r_i and r_o, respectively. The detailed analyses are presented in [4] §4.2 and in [5], and some essential consequences are discussed below.

We assume that the suspension consists of small mondispersed particles, heavier than the fluid, of initial volume fraction $\alpha(0)$, the Reynolds and Taylor numbers of the particle are small, the flow is stable, and the typical spin-up time is much shorter than the separation time. Thus, the relative velocity is given by (4).

In the (long) cylindrical centrifuge a region of pure fluid, $r_i \le r \le r_P(t)$ and of sediment, $r_S(t) \le r \le r_o$ appear. The suspension in the shrinking domain $r_P(t) \le r \le r_S(t)$ has a volume fraction with decaying $\alpha(t)$ according to

$$\frac{d\alpha}{dt} \approx -2\,(\epsilon\beta\Omega)\,\alpha(1-\alpha)^{2+n}; \quad \alpha(0) \ \text{given}; \tag{5}$$

where $n \approx 3.1$ is the semi-empirical power of the effective viscosity increase correlation, $\mu(\alpha) \approx (1-\alpha)^{-n}$. The integration of (5) is straightforward. The jump conditions between the various regions yield the compact results

$$r_P(t) = r_i \left[\frac{\alpha(0)}{\alpha(t)}\right]^{1/2}; \tag{6}$$

$$r_S(t) = r_o \left[\frac{\alpha_M - \alpha(0)}{\alpha_M - \alpha(t)}\right]^{1/2}. \tag{7}$$

Here we assume that the sediment has a constant volume fraction, α_M (typically 0.65) and sticks to the outer wall. In this case there is, in our opinion, no reason to question the validity of this assumption, at least as a good approximation. Some resuspension may occur because there is some relative angular motion between the suspension and the cylinder, but because of the axial symmetry this effect is restricted to a thin annulus and cannot penetrate into the core. (For large $\alpha(0)$ the sediment shock may be unstable but this case is beyond the scope of the present discussion).

What is the difference between the process in the tube centrifuge and in the cylindrical centrifuge? The centrifugal buoyancy, and hence the driving separative velocity of the particles, is the same in both devices. But in the tube centrifuge some of the particles which move on radial outward trajectories encounter the side walls, and settle there; only a portion of the initially suspended particles can settle directly on the outer wall, see Figure 2.

Because of the similar (practically, identical) driving force on the particles in the cylindrical and tube centrifuges, (5) is valid in both cases. This is not surprising, because the suspension in a small control volume in the interior of the centrifuge "does not know" what is the shape of the boundaries, and the behavior of the local concentration of the particles, $\alpha(t)$, is governed by the local motion of the particles, according to the local centrifugal-drag balance which yielded (4). Moreover, the analysis indicates that, when the sediment layer on the side walls of the tube centrifuge is thin (relative to the typical width D of the tube) and immobile, the propagation of the pure fluid and outer wall sediment interfaces is also given by (6)-(7), as in the axisymmetric cylinder case, independent of the shape of the cross section (but this shape is assumed constant). This interesting result may be, however, affected by the details of the sediment accumulation and motion on the side-walls. Consider three possibilities:

1. The sediment slips on the side-walls towards the outer wall. In this case the theoretical solution for $\alpha(t)$ and $r_P(t)$ remain valid, but the interface $r_S(t)$ propagates faster inwardly than predicted by (7).

2. The sediment produced on the side-walls sticks to the boundary. As long as this sediment layer is thin, the theoretical results (5)-(7) remain valid. Let the cross-section area of the sediment and tube be $\delta A(x,t)$ and A. The deviation from the theoretical solution is expected to be of the order of magnitude $\delta A/A$. Thus, for non-concentrated suspensions, or for an initial period of time, the theoretical results (5)-(7) remain valid. A more detailed analysis, which takes into account the change of the geometry as a result of the sediment is feasible, but, to our best knowledge, has not been done yet.

3. The sediment on the side walls re-suspends and re-mixes with the suspensions. The results (5)-(7) are evidently not valid in this case, and a detailed analysis has not been attempted. However, the qualitative behavior in this case bears some similarity to the assumption introduced by [1] in their analysis of the tube centrifuge, namely, that there is no particle settling on the side walls, and the centrifugal driving is parallel to the side-walls. This "one dimensional" (in a Cartesian coordinate system) approach predicts, essentially, the following modification: (a) the leading coefficient 2 in the

RHS of (5) is replaced by 1, i.e., the decay of $\alpha(t)$ is slower (consistent with re-mixing), and (b) the power 1/2 in (6)-(7) is also replaced by 1.

Which of these possibilities occurs in practical cases may depend on the properties of the suspension and centrifuge side-wall. Our expectation is that the suspension is stable and hence the first two possibilities prevail. Experimental verifications and further research are an interesting challenge, and further progress is bound to be beneficial to various applications.

We note in passing that some experimental investigations of a tube centrifuge were reported by [2], but the interpretation of the results is problematic. Unfortunately, no comparisons with the predictions (5)-(6) of [4, 5] were made, and only $r_P(t)$ (but not $\alpha(t)$) was measured. The paper [2] shows that the "one dimensional" theory of [1] can predict accurately $r_P(t)$ if a special, ad-hoc evaluated, value of n (typically, 3.5) is used. But there is no physical justification for this value of n; moreover, the possibility that with this special value of n the "one dimensional" theory recovers well the measured $r_P(t)$, but miss-predicts $\alpha(t)$ has not been excluded. There are indications that the predictions (5)-(6) of [5] provide a more acceptable interpretation (both mathematically and physically) to the experimental data, but this comparison is left for future work.

3. Concluding Remarks

(1) The foundations of the theory of rotating and separating suspensions seem reliable and productive. The body of available solutions provides confidence in the equations, boundary conditions and methods of approach, as well as understanding of basic results and indications on the capability of treating complex flow fields.

(2) Application of the theory still requires a great deal of "insight" for the proper formulation of the problem and interpretation of the results. The available results may serve as good tests and calibration runs for computer codes, in particular for commercial packages (some of which may be provided as a "black-box" module without access to the source program).

(3) There are some fundamental differences between gravitational and centrifugal separation. "Extrapolations" from the former case to the latter should be carefully checked. The rotating gravity current, that combines both effects, requires further investigation.

(4) The "state of the art" in gravity settling is more advanced than in centrifugal separation. Studies on resuspension, high concentration and stability are unavailable for rotating suspensions. Progress in the analysis of rotating suspensions — in particular for rapid separation with non-small Taylor and Rossby numbers — requires a much better understanding of some "classic" problems in single-phase rotating fluids.

(5) It is both useful and interesting to promote this field by applications, simulations, experiments and further research.

Acknowledgment The work was supported by the Fund for promotion of research, Technion, Israel.

References

[1] G. Anestis and W. Schneider. Application of the theory of kinematic waves to the centrifugation of suspensions. *Ingenieur-Archiv*, 53:399–407, 1983.

[2] D. Frömer and D. Lerche. An experimental approach to the study of sedimentation of dispersed particles in a centrifugal field. *Arch. Appl. Mech.*, 72:85–95, 2002.

[3] E. Minkov, M. Ungarish, and M. Israeli. The motion generated by a rising particle in a rotating fluid - numerical solutions. Part 2: The long container case. *J. Fluid Mech.*, 454:345–364, 2002.

[4] M. Ungarish. *Hydrodynamics of Suspensions: Fundamentals of Centrifugal and Gravity Separation.* Springer, 1993.

[5] M. Ungarish. On the separation of a suspension in a tube centrifuge. *Int. J. Multiphase Flow*, 27:1285–1291, 2001.

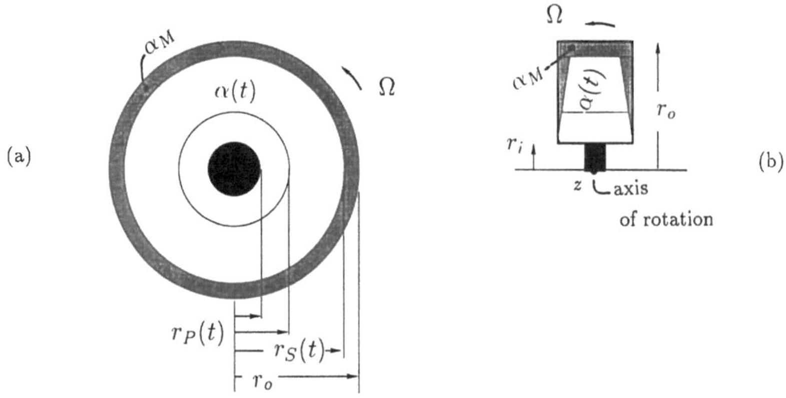

Figure 1: Centrifugal separation of a suspension of heavy particles (a) in a straight axisymmetric container rotating about its axis of symmetry, z; (b) in a tube rotating about the axis z which is perpendicular to the tube centerline x. The regions of pure fluid (white), suspension (grey) and sediment (dark grey) are shown.

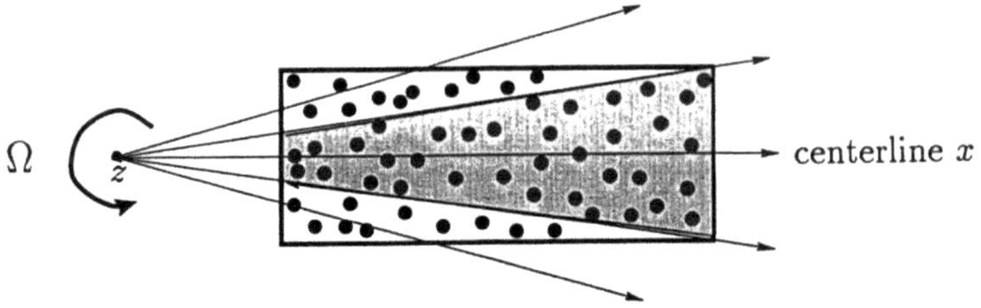

Figure 2: Schematic settling behavior in a tube centrifuge with particles in initial position. The rays indicate the direction of the centrifugal force, and hence the particles in the white domain are expected to settle on the side wall, while these in the grey domain are expected to settle on the outer wall.

ON THE ONE-DIMENSIONAL FLOW APPROXIMATION IN SEDIMENTATION PROCESSES

WILHELM SCHNEIDER
Institut für Strömungslehre und Wärmeübertragung, TU Wien
Resselgasse 3, A-1040 Vienna, Austria

1. Introduction

Following the pioneering work by G. J. Kynch [7], the theory of one-dimensional kinematic waves has been applied to a variety of settling processes due to gravity and centrifugal forces, respectively, cf. [6], [10], [12] and [14] for surveys. While most of the work on gravity settling is concerned with constant cross sections, cross sectional areas changing with the distance from the axis of rotation play an important role in centrifugation processes. In particular, in cylindrical centrifuges without compartments the cross-sectional area is proportional to the radial coordinate, giving rise to rather peculiar kinematic-wave phenomena [2]. It is obvious that the formulation of the problem in terms of only one spatial coordinate ("one-dimensional flow") is strictly valid in that case.

In most other cases, however, the assumption of one-dimensional flow is, at best, an approximation. Among others, a conventional one-dimensional flow analysis of sedimentation processes does not account for the following phenomena.

(a) *Wall friction.* As has been shown in [11], the no-slip condition at the walls of a settling vessel may substantially affect the (two-dimensional) flow of the suspension. Wall friction also gives rise to a flow of the clear liquid on top, which is not predicted by the one-dimensional theory [7]. The effects of wall friction are particularly pronounced in very slender vessels [5].

(b) *Particle-free layers at side walls.* At downward-facing walls of gravity settlers there are thin layers of clear liquid that give rise to enhanced settling (Boycott effect). To analyse this effect, the theory of kinematic waves has been generalized, taking into account the two-dimensional or three-dimensional bulk flow of the suspension [9]. An analogous effect has been observed in centrifuges with radial, or nearly radial, walls [8], cf. also the more general analysis given in [10].

(c) *Particle settling at side walls.* It has been pointed out recently by M. Ungarish [13] that the one-dimensional flow approximation, as applied in [2], may lead to severe errors if particles settle not only at the bottom of the vessel but also at side walls that are inclined with respect to the force acting on the particles. The particular problem considered in [13] is the tube centrifuge, i.e. a rotating tube of constant cross section, with the axis of rotation being normal to the tube's centreline. In case of gravity settling, an analogous problem arises for vessels with upward-facing side walls.

Although recently published experimental data [4] appear to be in good agreement with predictions based on the one-dimensional flow analysis [2], the present author accepts Ungarish's criticism [13] concerning the *conventional* one-dimensional flow approximation as basically correct. However, the analysis given in [13] appears to be incomplete. In particular, mass conservation requires a flow of the suspension to

A. Gyr and W. Kinzelbach (eds.), Sedimentation and Sediment Transport, 127–130.

compensate for the loss of particles that are gliding along the side walls, but this induced bulk flow is not taken into account in [13]. It is the aim of the present paper to predict the induced flow field and discuss how the settling time is affected. For simplicity, gravity settling is considered in what follows, while the more elaborate analysis of centrifugal settling is reserved for future work.

2. Two-dimensional kinematic wave analysis

As indicated in Figure 1, batch sedimentation of particles of uniform shape and size in a liquid-filled vessel with upward-facing plane or conical side walls is considered. The inclination angle of the side walls with respect to the horizontal plane is Θ. The flow is assumed to be either plane or axisymmetric, with x,z denoting Cartesian and cylindrical coordinates, respectively. The z axis is pointing downwards, i.e. in the direction of the gravity acceleration g. The lengths are referred to the vessel hight, H, the velocities to the terminal settling velocity of a single particle, U, and the time is scaled with H/U.

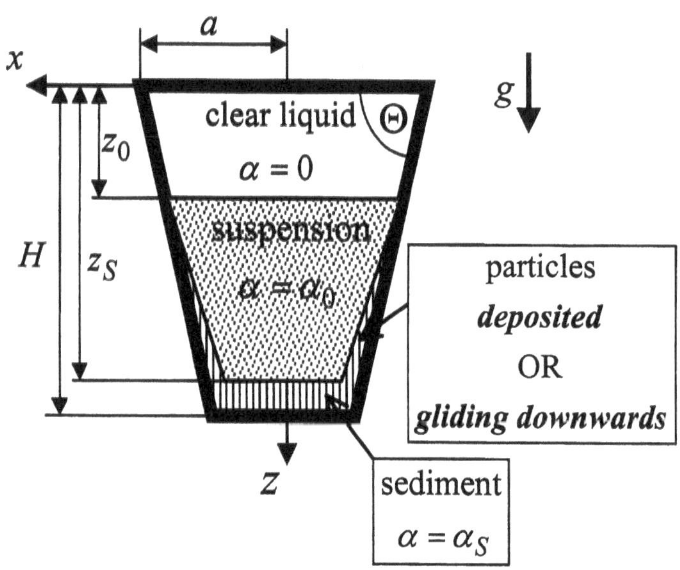

Fig. 1. Gravity settling in vessels with inclined side walls.

The present analysis is based on the kinematic-wave theory as given in [9], taking two-dimensional (plane or axisymmetric) flow into account. Following common practice in the theory of sedimentation, the viscosity of the suspension as well as inertial effects are neglected. This essentially implies sufficiently small particles settling in sufficiently large vessels. More precise conditions of validity, formulated in terms of appropriate non-dimensional parameters, can be found in [9]. As a consequence of dropping both viscous and inertial terms, the momentum equation for the mixture reduces to a balance of pressure and gravity. Thus the pressure is the hydrostatic one and the particle volume fraction α does not depend on the horizontal coordinate x, i.e. $\alpha = \alpha(t, z)$. Furthermore, the momentum equation associated with the relative motion of particles and liquid reduces to the drift-flux relation $\vec{j}_p = \alpha\vec{j} + f(\alpha)\vec{e}_z$, where $\vec{j} = \vec{j}_p + \vec{j}_l$ is the total volume flux, with $\vec{j}_p$ and $\vec{j}_l$ denoting the volume fluxes of the particle and liquid phases, respectively, while $\vec{e}_z$ is the unit vector in z direction. The so-called drift-flux function $f(\alpha)$ is to be determined from suspension mechanics or by experiments. Often an empirical correlation due to Richardson and Zaki is used for $f(\alpha)$, cf. [14], p.178.

With the use of the drift-flux relation the equation of continuity for the particle phase becomes

$$\frac{\partial \alpha}{\partial t} + \left[j_z + f'(\alpha)\right]\frac{\partial \alpha}{\partial z} = 0, \tag{1}$$

where j_z is the z-component of $\vec{j}$. Differentiating Eq.(1) with respect to x shows that j_z (like α) is independent of the horizontal coordinate, i.e. $j_z = j_z(t, z)$. This also holds for the special case $\alpha = \text{const}$, as was shown in [9]. Hence Eq.(1) describes *one*-dimensional kinematic waves embedded in a *two*-dimensional field of total volume flux. The latter is governed by the continuity equation for the (incompressible) mixture subject to appropriate boundary conditions at the bottom of the vessel and at the side walls, respectively.

While it is natural to assume that the sediment (with particle volume fraction α_S) is at rest at the bottom, the boundary conditions at the side walls are less obvious, cf. [1]. Two limiting cases may be distinguished: all particles being deposited at the wall; or all particles gliding downwards in a thin boundary layer, in which the particle volume fraction has a constant value, say α_B, and the drift flux is negligible. In the former case there is no total volume flux, and the time required for complete settling is independent of the inclination of the side walls. This resembles Ungarish's result [13] for the tube centrifuge. In case of particles gliding downwards at the walls, however, a non-vanishing total volume flux is required to satisfy the boundary condition $\alpha_B/(1-\alpha_B) = j_p^{(n)}/j_l^{(n)}$, which follows from the mass balance for the boundary layer. The superscript $^{(n)}$ refers to components normal to the wall.

Following the procedure outlined in [9], a solution of the set of basic equations and boundary conditions can easily be obtained. Provided the initial value of the particle volume fraction, α_0, is not too large, the particle volume fraction remains constant in the suspension during the settling process, and horizontal kinematic shocks separate the suspension from the clear liquid on top and from the sediment at the bottom, respectively. The position of the upper shock, as a function of time, is found to be

$$z_0 = A_0 t, \tag{2}$$

with $A_0 = f(\alpha_0)/\alpha_0$, whereas the motion of the lower shock is governed by the differential equation

$$\frac{\mathrm{d}z_S}{\mathrm{d}t} = -\left[A_{S1} j_z(z_S) + A_{S2}\right], \tag{3}$$

with

$$j_z(z) = -A_B\left[\left(\frac{\bar{a} - A_0 t}{\bar{a} - z}\right)^{1+\sigma} - 1\right] \qquad \text{(for } z \geq A_0 t\text{)}. \tag{4}$$

To simplify the notation, the following constants have been introduced: $\bar{a} = a\tan\Theta$, $A_{S1} = \alpha_0 /(\alpha_S - \alpha_0)$, $A_{S2} = f(\alpha_0)/(\alpha_S - \alpha_0)$, $A_B = f(\alpha_0)/(\alpha_B - \alpha_0)$, $\sigma = 0$ for plane flow, and $\sigma = 1$ for axisymmetric flow. Apart from the first term within the brackets on the right-hand side of Eq.(3), Eqs. (2) and (3) resemble the classical result for a vessel with vertical walls. Since the discriminating term is negative, the lower shock moves more slowly, and the time required for complete settling increases, as a consequence of the particles gliding downwards the inclined side walls.

3. Conclusions

Walls that are not in alignment with the driving force may affect the over-all settling process in various ways, depending on whether the particles are deposited at, or glide along, those walls. A theory of one-dimensional kinematic waves that are embedded in a two-dimensional (or three-dimensional) bulk flow of the mixture is capable of describing these effects.

4. References

[1] *Amberg, G., Dahlkild, A.*, and *Bark, F.*: Sediment transport in some unsteady settling processes. Poster 783, XVIth IUTAM Congress, Lyngby, Denmark, 1984.

[2] *Anestis, G.*, and *Schneider, W.*: Application of the theory of kinematic waves to the centrifugation of suspensions. Ing.-Archiv **53**(1983), 399-407.

[3] *Baron, G.*, and *Wajc, S.*: Behinderte Sedimentation in Zentrifugen. Chem.-Ing.-Techn. **51**(1979), 333.

[4] *Frömer, D.*, and *Lerche, D.*: An experimental approach for studying the sedimentation of dispersed particles in a centrifugal field. Arch. Appl. Mech. **72**(2002), 85-95.

[5] *Herbolzheimer, E.*, and *Acrivos, A.*: Enhanced sedimentation in narrow tilted channels. J. Fluid Mech. **108**(1981), 485-499.

[6] *Kluwick, A.*: Kinematische Wellen. Acta Mechanica **26**(1977), 15-46.

[7] *Kynch, G. J.*: A theory of sedimentation. Trans. Faraday Soc. **48**(1952), 166-176.

[8] *Schaflinger, U., Köppl, A.*, and *Filipczak, G.*: Sedimentation in cylindrical centrifuges with compartments. Ing.-Archiv **56**(1986), 321-331.

[9] *Schneider, W.*: Kinematic-wave theory of sedimentation beneath inclined walls. J. Fluid Mech. **120**(1982), 323-346.

[10] *Schneider, W.*: Kinematic wave description of sedimentation and centrifugation processes. In: Flow of Real Fluids (*G. E. A. Meier* and *I. Obermeier*, Eds.), Lecture Notes in Physics Vol. **235**, Springer-Berlin, 1985, pp.326-337.

[11] *Smek, E.*: Two-dimensional viscous flow of a suspension due to particle settling in a vertical vessel. Acta Mechanica **55**(1985), 21-31.

[12] *Ungarish, M.*: Hydrodynamics of Suspensions. Springer, Berlin 1993.

[13] *Ungarish, M.*: On the separation of a suspension in a tube centrifuge. Int. J. Multiphase Flow **27** (2001), 1285-1291.

[14] *Wallis, G. B.*: One-dimensional Two-phase Flow. McGraw-Hill, New York 1969.

PARTICLE SWARMS SETTLING IN CALM WATER

J. BUEHLER[1], D.A. PAPANTONIOU[2]
[1] *Institut für Hydromechanik und Wasserwirtschaft, ETH Zurich, 8093 Zurich, Switzerland, buhler@ihw.baug.ethz.ch*
[2] *Performance Technologies S.A. 48 Agisilaou St., 17341 Agios Dimitrios, Greece, dap@performance.gr*

1. Introduction

Rubble from tunnel or road construction is increasingly used to create wildlife habitats and artificial islands in designated areas of lakes and coastal waters. Provided that the grain size is small enough, the collective velocity which the particles have as members of the cloud is initially much greater than the one they would have by settling alone. In this case their predominant role is to provide the excess mass of the cloud relative to its surroundings. The cloud then acquires a circulation similar to that in a vortex ring, and the particles closely follow the corresponding toroidal motion of the fluid (Slack, 1963). The resulting flow is dynamically similar to that of thermals of warm air rising in the atmosphere, and they are called suspension thermals. The loss of momentum and buoyancy to a wake of the cloud is usually neglected in analogy to classical thermal theory. A review of work on thermals as well as on puffs, their nonbuoyant counterparts, is due to Scorer (1997).

Slack (1963) further noted that a quite different type of flow arises when a load of larger particles is released in air. The particle clouds then assumed a bowl-like appearance as shown in his Fig. 1. Boothroyd (1971) further suggested that the circulation within flows of this type is too weak to return particles from the base of the cloud back to the top, and that they settle more like if they were alone, i.e. with their individual settling velocity. Bühler and Papantoniou (1991, 2001) referred to the bowl-shaped clouds as swarms to distinguish them from suspension thermals, and from clusters of a few particles only. They concluded that suspension thermals eventually turn into particle swarms, because they slow down as they sink, and swarms settle with approximately constant velocity. The corresponding transition distance was computed from thermal theory as the distance from the source at which the two velocities are about the same. By assuming nearly self-preserving flow in swarms, they also derived power laws for their widening rate. The relevant flow constants were obtained from experiments with 1.5 – 2 and 2 - 3 mm sand. The authors also reviewed previous work on thermals and particle clouds in their latter paper, which will be referred to as BP. Recent studies on particle clouds are due to Noh (2000), McNamee(2000), and to Bush and Thurber (2002).

The growth rate of thermals, suspension thermals, and of puffs due to an instantaneous source of momentum, is usually specified by making use of the entrainment principle proposed by Morton et al. (1956), which implies that the temporal rate of increase of the volume of a cloud is proportional to its surface area and its velocity. A disadvantage of the entrainment principle is, however, that it is inconvenient for nonspherical thermals, unless their surface area can be readily specified. Moreover, it cannot be applied to

A. Gyr and W. Kinzelbach (eds.), Sedimentation and Sediment Transport, 131–136.

thermals with a large density difference without making additional assumptions. To avoid these limitation, BP proposed two alternative formulations inspired by classical closure assumptions for steady shear flows due to Prandtl (1926), and to Ricou and Spalding (1961). The second one of these formulations is also suitable for dense flows, and it will be outlined in the following section. The resulting momentum equation is then extended to swarms in section 3, and in the last section the results are compared with experimental data on swarms of bronze particles settling in water.

2. Dimensional analysis of dense, thermal – like flows.

Ricou and Spalding (1961) measured the entrainment rate of dense and light axisymmetric gas jets in air. Measurements were made at moderate distances from the source, where buoyancy effects were still negligible, such that the momentum flux $\dot{M}$ in the jets was preserved. They proposed that the mass flux $\dot{m}$ through a cross-section of the jet depends only on the momentum flux, the uniform ambient density ρ_a, and the downstream distance, which leaves the relation

$$\dot{m} = c_j(\rho_a \dot{M})^{1/2} x \tag{1}$$

where c_j is a constant which was found to have a mean value of 0.282 for all experiments.

In analogy to (1), the mass m of nonbuoyant puffs which conserve their momentum M in an unstratified environment can be related to the elapsed time t as

$$m^{4/3} = c_p \rho_a^{1/3} M t \tag{2}$$

Where c_p is a growth coefficient for puffs. A dependence on t instead of x is necessary here for dimensional reasons. The growth rate of puffs can be specified as the derivative

$$\frac{d(m^{4/3})}{dt} = c_p \rho_a^{1/3} M, \tag{3}$$

and it will be shown that the spatial widening rate of puffs and thermals are identical if (3) is retained for the buoyant case as well.

Escudier and Maxworthy (1973) applied the concept of added mass to thermals. Specifically, they accounted for the forward momentum associated with the induced motion in the ambient fluid by assuming that a shell of this fluid – the added mass - surrounds the thermal and moves along with it, essentially without being entrained. Since the momentum of a fluid mass is normally given as the product of a mass and its velocity, one may argue that the added mass of ambient fluid should be a part of the total mass of the cloud. This view is also in line with the approach of Prandtl (1926) if the "added mass" of thermals is considered as containing turbulence, as both the cloud and its shell can then be considered as growing due to turbulent diffusion of mass and momentum. We shall, therefore, include the added mass in m, such the average velocity u of the cloud in the thermal stage is

$$u = M / m = dx / dt \tag{4}$$

The conservation equations for mass, momentum and buoyancy B for a spheroidal puff or thermal of mean density ρ, height h, and width b can then be stated as

$$\frac{\pi}{6}\frac{d}{dt}\left((1+k_v\frac{\rho_a}{\rho})\rho b^2 h\right)=\frac{3}{4}c_p\rho_a^{1/3}u\left(\frac{\pi}{6}(1+k_v\frac{\rho_a}{\rho})\rho b^2 h\right)^{2/3} \tag{5}$$

$$\frac{\pi}{6}\frac{d}{dt}\left((1+k_v\frac{\rho_a}{\rho})u\rho b^2 h\right)=\frac{\pi}{6}(\rho-\rho_a)gb^2h=B \tag{6}$$

$$\frac{\pi}{6}\frac{d}{dt}\left((\rho-\rho_a)gb^2h\right)=0 \tag{7}$$

Scorer (1957) expressed the volume V of thermals as $V= c_V\,(b/2)^3$. For a spheroidal thermal $h = 3c_V\, b/(4\pi)$; and this relation completes the set (4) to (7) unless changes of the thermal shape in time are modelled as well. With d/dt = u d/dx relation (5) can be written in terms of the spatial derivative, and after integrating the resulting expression from a virtual origin we obtain

$$\left(b^2h\right)^{1/3}=c_p\left(\frac{3}{32\pi}\frac{\rho_a}{\rho}\right)^{1/3}(1+k_v\frac{\rho_a}{\rho})^{-1/3}x \tag{8}$$

This result is valid for puffs and thermals, and in case their spreading rates are indeed the same in the Boussinesq limit, as suggested by Scorer (1997), the same value of the growth coefficient c_p applies for both flows. Scorer (1957) carried out experiments on salt water thermals in fresh water, and measured the position x_f of their front, and their half-width, in time. He recommended a value of 4 for the ratio $n = x_f/(b/2)$, and of 3 for the shape factor c_v. Escudier and Maxworthy (1973) assigned the value of 0.5 to the added mass coefficient k_v which is applicable for a sphere in potential flow. k_v depends slightly on the ratio h/b for spheroids, but the value for spheres will be used here throughout. The position x of the centre of a spheroidal thermal is $x = x_f - h/2$, and the above recommendations for n, c_v and k lead to $c_p = 2.01$.

Scorer, and later experimenters, determined the extent and velocity of thermals and puffs from visualisations, i.e. as mass-based flow scales derived from the distribution of a dye or a stratifying agent. This choice is consistent with the concept of added mass, and his scales are the ones adopted here. Within this framework, k_v can also be considered as a (poorly known) shape constant which ties properties of the velocity distribution in the entire domain of motion to the fairly well determinable mass-based flow scales. A similar approach was proposed by Bühler et al. (1991) for jetlike flows. The analysis of swarms in the following section requires an additional velocity scale, the mean velocity u_v of the interstitial fluid within a particle cloud. The index v refers to the fact that this scale is derived from the velocity distribution, and not from the particle distribution like the remaining flow scales. As outlined above, the two velocity scales u and u_v are essentially equal for suspension thermals. Eqs. (5) and (6) will, therefore, be generalized by replacing u by u_v to facilitate their extension to swarms.

3. Swarms

The average vertical velocity u of particles in a suspension cloud is the sum of the average velocity u_v of the interstitial fluid, and of the slip velocity, or, approximately, $u = u_v + u_t$, where u_t denotes the individual terminal settling velocity of the particles in calm fluid. In contrast to suspension thermals, where $u_t \ll u_v$, the particles in swarms sink at a rate close to their individual settling velocity u_t, and $u_v \ll u_t$.

In terms of the concept of added mass one may consider the lower part of the moving water mass of a dilute swarm as consisting of a hemisphere of diameter $b_v = b\,(1+k_v)^{1/3}$. An estimate for the outflow rate of water through the upper rim of the swarm is then $\rho_a u_t (1+k_v)^{2/3} b^2 \pi/4$. This fluid leaving the cloud has a small downward velocity u_v. A good part of the momentum generated by the buoyancy force is thus lost through the upper rim of the swarm, forming a wake, and this loss is approximately $\rho_a u_v u_t (1+k_v)^{2/3} b^2 \pi/4$.

If it is assumed that the generalized form of equation (5) remains applicable to swarms as well, this momentum loss is proportional to u_t dm/dt. Since u_t is constant, the momentum equation (6) can then be extended to the swarm stage by restating it as

$$\frac{\pi}{6}\frac{d}{dt}\left((u_v + c_M u_t)(1+k_v\frac{\rho_a}{\rho})\rho b^2 h\right) = B, \quad \text{with } c_M \approx \frac{1}{c_p}\frac{4\pi}{3c_v^{2/3}} \approx 1 \tag{9}$$

Variations of the local particle velocity over the height of a swarm are of order u_v and can be neglected in comparison to u_t. The entire cloud can thus be taken to descend at a velocity close to the individual settling velocity u_t, i.e.

$$u_f = c_S u_t, \tag{10}$$

where c_S should be close to one for particles of uniform size and shape. The cloud velocity over the entire range from thermals to swarms is then

$$\frac{dx}{dt} = c_S u_t + u_v \tag{11}$$

To focus on the motion of dilute swarms, we shall simply neglect u_v in eq. (9) in comparison with u_t, and set $\rho_a/\rho \cong 1$. An integration from the virtual time origin then leads to

$$b^3 = \frac{8Bt}{c_M c_V (1+k_v)\rho_a u_t} \tag{12}$$

where h was replaced by $3c_V\, b/(4\pi)$. As $c_S\, u_t \cong x_f/t$, we also find

$$b^3 = \frac{8Bx_f}{c_M c_S c_V (1+k_v)\rho_a u_t^2} = c_b \frac{Bx_f}{\rho_a u_t^2} \tag{13}$$

The swarm stage may come to an end when the particles have drifted so far apart that their wakes no longer interact within the cloud, or when the flow within the swarm becomes laminar.

4. Experiments on swarms of bronze particles

A number of experiments were carried out in a 1.5x3 m tank of 1.1 m depth (Figure 1). The tank was partly glass-walled for visualizations, and previously wetted bronze particles were released by dropping them from a height of about 10 cm through a funnel with a nozzle of 2.5 cm diameter, which protruded an equal distance below the water surface. The funnel was positioned in a corner of the tank at 50 cm from the side walls. The experimental arrangement and procedures were the same as those described in BP for experiments with coarse sand. The bronze particles had a mean diameter of 0.475 mm, a density ρ_p of 8.61 g/cm^3, and a settling velocity u_t of 21 cm/s.

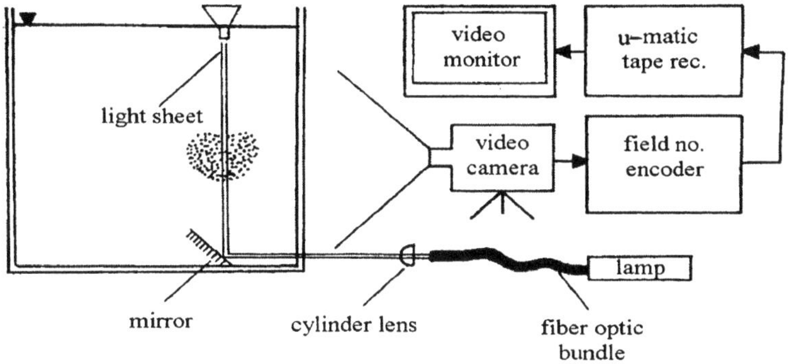

Fig.1: Experimental arrangement for particle clouds settling in water.

Tests for four values of the thermal buoyancy B (the submerged weight of the load) were made, corresponding to 0.28, 0.14 and 0.07 kg m/s^2. Results for the bronze particles are presented in Figure 2, which explains the symbols, and Figure 3. Length scales are divided by $x_0 = (B/\rho_a)^{1/2}/u_t$. Figure 2 is a plot of the ratio of the settling velocity u_t and the front velocity u_f of thermals against the nondimensional distance xf' of the front from the end of the funnel. In the swarm stage the front velocity remains constant far downstream, and somewhat larger than the mean settling velocity u_t. This is to be expected, as the larger particles would tend to sink faster and congregate near the front of the swarms. The constant c_S in (10) can be determined as 1.4, which agrees with the results reported by BP for 1.5-2 mm and dry 2-3 mm sand. In Figure 3 the third power of the width b of the clouds is plotted against their frontal position, as suggested by eq. (13). The virtual origin of the swarm stage is clearly located below the source. Two tests were conducted for B = 0.007 kgm/s^2. The corresponding values of $b/(x_f x_0^2)^{1/3} = c_b^{1/3}$ in the swarm stage are about 1.65, and 1.35. The values reported by BP for sand swarms are within this range as well, and c_M is obtained as 0.4 from eq. (13), i.e. as less than 1. One of the possible reasons is that the assumption of a hemispheroidal shape might be more adequate for swarms.

5. REFERENCES

Boothroyd, R.G. Flowing gas-solids and suspensions. Chapman and Hall, 1971.

Bühler, J., Wright, S.J., and Kim, Y. Gravity currents advancing into a coflowing fluid. Jour. Hydraulic Res. **29**, 2, 1991, 243-257.

Bühler, J., and Papantoniou, D.A. Swarms of coarse particles falling through a fluid. Environmental Hydraulics, Lee and Cheung, eds., **1**, Balkema, 1991, 196-201.

Bühler, J., and Papantoniou, D.A. On the motion of suspension thermals and particle swarms. Jour. Hydraulic Res., **39**, 6, 2001, 643-653 (see on www.iahr.org).

Bush, J.W.M and Thurber, B.A. Particle clouds in homogeneous and stratified environments (submitted to JFM).

Escudier, M.P. and Maxworthy, T. On the motion of turbulent thermals. J. Fluid Mech. **61**, 541, 1973.

McNamee, Michael R. An experimental study of axisymmetric turbulent puffs: their motion, dilution, and sedimentation. MS Thesis, Arizona State University, May 2000.

Morton, B.R., Taylor, G.I. and Turner, J.S. Turbulent gravitational convection from maintained and instantaneous sources. Proc. Royal Society, A234, 1-23, 1956.

Noh, Y. Sedimentation of a particle cloud across a density interface. Fluid Dynamics Research **27**, 2000,129-142.

Prandtl, L. Ueber die ausgebildete Turbulenz. Proc. Int. Congr. of Appl. Mech., Sept. 12-17 Zurich,1926, 62-74 (see also Schlichting, H. Boundary layer theory, McGraw-Hill).

Ricou, F.P. and Spalding, D.B. Measurement of entrainment by asymmetrical turbulent jets. J. of Fluid Mech., **11**, Part. 1, 21-32, 1961.

Scorer, R.S. Experiments on convection of isolated masses of buoyant fluid. J. Fluid Mech. **2,** 583-594, (1957).

Scorer, R.S. Dynamics of metrology and climate. Wiley - Praxis Series in Atmospheric Physics, 1997.

Slack, G.W. Sedimentation of a large number of particles as a cluster in air. Nature, **200**, 1306, 1963.

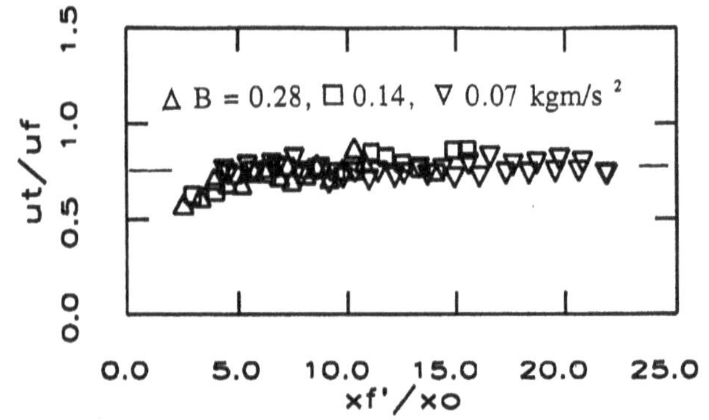

Fig. 2: Decay of the velocity u_f of the front with its distance x_f' from the source

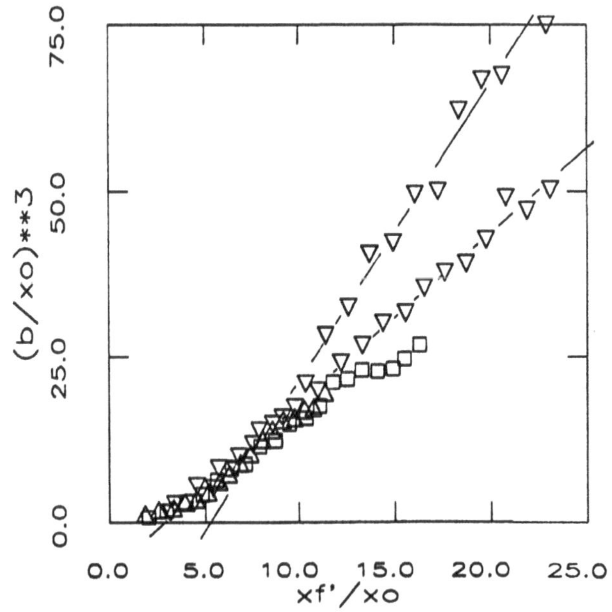

Fig. 3: Cube root growth of the swarm width b with the distance x_f' to their fronts.

FLOW OF MEDIA WITH HIGH NANOPARTICLES' CONCENTRATION

SERGEY P. BARDAKHANOV
Novosibirsk State Technical University and Institute of Theoretical and Applied Mechanics Novosibirsk 630090 RUSSIA E-mail, bard@itam.nsc.ru

1. Introduction

In the last decade the "Nanotechnology" as an area of modern science is developing. The key point is the consideration of properties of substances, consisting of particles with the sizes less then 100 nanometers, as such objects become comparable to the sizes of some fundamental values (size of the magnetic domain, dislocation, etc.) it leads to quantum mechanical effects. From the point of view of the mechanics, probably, the applicability of the continuum representation for the description of such a media, in particular of nanopowders can be formulated.

The powders are widely used in manufacturing processes as either final product or intermediate constituents. The theories of chemistry and physics had been successfully applied in the different are of powder technology, but a clear analysis of the powder behavior is still far from being understood. From the scientific point of view the powders, they are a granular media, frequently considered as a fourth state of matter, the mechanical properties of which are understood more intuitively. In particular, the formulation of constitutive laws, reflecting the rheological properties of the powder flow, is the problem, which is needed for solving the transport and other mechanical problems.

There is no clear agreement on how this matter should be treated and many visions are opposed, according to which the different theories have been proposed. Authors of [1] suggested an analogy with the kinetic theory of gases, where the viscosity increases with temperature. In paper [2] a grain media is considered as interstitial liquid of high density and viscosity. A fluid mechanical approach to slow powder flows is proposed in [3]. Accordingly, the number of experimental devices were designed to obtain data for powders of comparatively large particles. But the main question in the research of granular media of the exact measurement of macroscopic quantities, especially with the intention to compare them with fluid mechanics.

The author assumes, that with the development of the nanotechnology the mechanical behavior of "nanopowders", consisting of primary nanoparticles, can become important. In connection with the above mentioned, we propose the idea, that a nanopowder is a new kind of continuum, which is similar to liquid or gas. At the same time a nanopowder can be considered also as a granular material, but with small particle comparable to molecular size.

The purpose of this work was to obtain experimental data of the flow properties of nanopowders, focusing to the analogy of the fluid mechanics and the dynamic properties of nano- granular flow.

A. Gyr and W. Kinzelbach (eds.), Sedimentation and Sediment Transport, 137–142.

2. Experimental procedure

2.1. THE POWDERS

The following oxide powders were used as the working media:
nanopowders – pyrogenic SiO_2 (silica) and Al_2O_3 (alumina) of Aerosil series A-380 (as of Kalush Plant, Ukraine, as of Degussa-Huls, Germany), A-90, Aluminium oxide C (Degussa-Huls) with average primary particle size d = 7, 13 and 20 nanometers and corresponding specific surface S = 380, 100 and 90 square meters per gram (m^2/g); plasma-chemical alumina with d = 40 nm and S = 40 m^2/g (Siberian plant); physical vapor deposited silica "Tarkosil" with d = 60 nm and S = 30 m^2/g [4].
precipitated silica micropowders of Sipernat series with d = 3000-50000 nm and S = 50-750 m^2/g (Degussa-Huls), of "white black" series BS-50, BS-100, BS-120 (different Russian plants, number corresponds to S in m^2/g) and of Zeolex-15, Tixosil-73, Zeosil-85, Zeosil –175 (Rhone-Poulenc, France).
The powders had also a large difference in bulk density, the lowest one was about 40 gram per liter. The most of them (except Al_2O_3) were amorphous with different contents of crystalline phase.
In "normal" granular theories grains are taken to be of the same size. In divergence with this and unlike molecules, the nano- and micro- powder primary particles sizes are not identical, and always have a distribution. In addition, the specific powder characteristics, depending on the method of producing of primary particles, as cohesion and agglomeration of different kind play a significant role. In latter case it is not easy to distinguish the characteristic size of the particle, especially because the agglomerates can change, for example, in process of motion due to their interaction. So, the particle size is not only a parameter concerning the applicability of the continuum hypothesis to fine powders, but at this stage we choose them as the main value for their characterization. In this continuum we assume the number of primary particles at rest, for example, in one cubic centimeter, only two or three orders lower then the number of air molecules. (Certainly, the nanopowder in motion should be considered as compressible fluid, because the concentration of particles in some volume changes, but still remains high.) The air molecules are distributed between particles, and their energy exchange, at least partly, is prevented by particles, because their mean free path is of the same size as the one of the air molecules. The total density is many times larger then the density of the air. Such an approach allows us to apply fluid mechanics ideas, including hot-wire anemometry.

2.2. THE FACILITIES

The facilities of different complexity modeled the problems of classical fluid mechanics: heavy sphere falling under the gravity force in still nanopowder [5, 11, 12]; plane and round Couette flow [5-7, 11, 12]; powder flow through the pipes of different diameter under the action of pressure [6, 7], gravity force [5-8, 11, 12] or driven by ventilator [6, 7]; wake behind cylinder in nanopowder flow [8]; pulsating jet in nanopowder [6, 7]; sound excitation and wave propagation in nanopowder [9, 10].

The general scheme of an experiment with nanopowder is the flow in a vertical pipe under gravity as presented in Figure 1a). A glass tube is filled with powder of nanoparticles. After the plug opens the powder starts to move and accelerates under the gravity force, the hot-wire probe detects the motion. The inner diameter of the pipes varied from 30 to 95 mm, height from 0.4 to 1.3 m, loaded at bulk density (for alumina) from 30 to 400 g. In modern facilities the operation is controled, which allows to regulate the speed of the flow in a wide range.

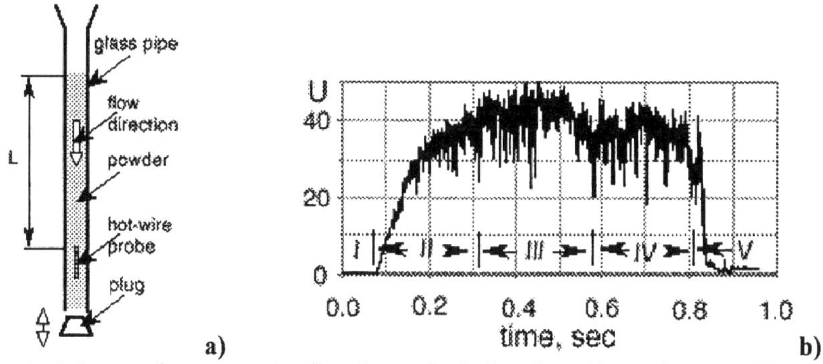

Figure 1. Scheme of nanopowder flow in vertical pipe a) and hot-wire trace b).

2.3. THE EXPERIMENTAL TECHNIQUES

The dynamic behavior of granular media, specifically of nanopowders has been studied. For this purpose a hot-wire anemometry was used for the first time [9] and a video recording were employed. The electromagnetic, impact and optical probes were also used in different cases. Due to instationary nature of the processes computer data acquisition systems were widely used.

Hot-wire method is used in liquid and gas flow measurements (and even in solid bodies, usiong transient hot-wire), as well for the velocity, as for heat transfer measurements. However, the application of conventional hot-wire probes for the direct measurements of the flow velocity in nanopowder (and of powders of larger scales) was not found in the literature, including the textbooks on hot-wire anemometry. In different experiments we employed the probes of different design, but always the diameter of golden tungsten wire was 4-6 micron. Surprisingly the wire of such small diameters did not break, even at nanopowder speed of 1 m/s.

3. Results and Discussion

In the present paper we will discuss only the results obtained by the hot-wire method for the nanopowder flow in vertical pipes.

First, the heat transfer of the hot-wire was compared for most of the above listed powders at rest by using the voltage E_0, which is analog to the free convection voltage of the hot-wire in air). It was different, but certain correlation had not been found with primary particles size, specific surface, substance, method of production. The difference helped to calibrate the probe taking the value in air and comparing them with the velocity in the nanopowder flow of known speed. The corresponding hot-wire signal is presented at Figure 1b). U shows the speed as if the probe would be in an air flow at 40 m/s, while the actual speed of the powder in the column was 0.5 m/s. It means that the heat transfer of hot-wire in nanopowder flow is much more intensive then in air. It is necessary to point out, that the usual problem of contaminating the probe does not exist, because all substances are refractory oxides with a melting point higher then 1000 °C, and most nanopowders contained more then 99.8% of the main substance.

It is possible to classify the signal by a few characteristic regimes. Prior to the beginning of a powder motion the speed is equal 0 (part I on the graph). At incipient motion the powder is quickly accelerates (II). It is important to note, that in the subsequent range two regimes III and IV were observed. The majority of the signals showed the given level of a pipe filling and in part III the average flow speed was a little bit higher than in subsequent part IV. Probably, it happened due to formation of an average velocity distribution through the section of a pipe. The end of part IV shows the moment at which the whole powder has already passed through the probe. In part V mostly air with particles went through, and then – air alone.

Comparing the signal in part II with visual observations it is possible to say that at first the powder in this regime moves similarly to a solid body or by analogy to the fluid mechanics in a "laminar" regime. However one can recognize the occurrence of essential velocity fluctuations. Strong fluctuations in parts III and IV were observed visually as intensive chaotic motions of particles.

Hot-wire allows measurements of the fastest occurring flow changes, what is very difficult in the common experimental studies of granular media. The velocity of the whole powder column was measured by shooting video pictures of the motion of the nanopowder free surface. An example of velocity measurements by video and its corresponding hot-wire voltage record is presented at Figure 2.

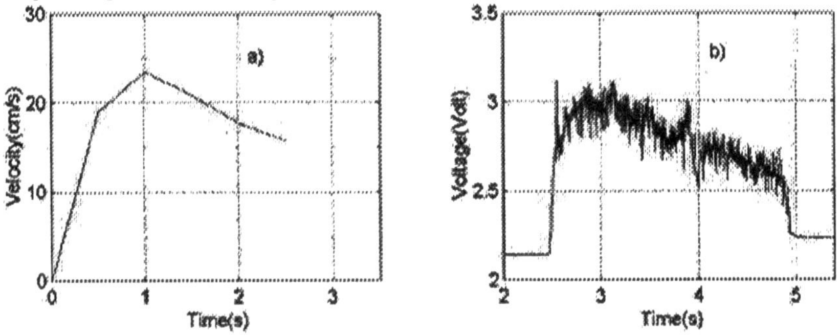

Figure 2. Change of velocity from video a) and voltage from hot-wire b) data.

An example of the calibration results from the same hot-wire measurements in air (according the standard procedure used for Pitot-Prandtl tube) and in powder (Aluminium Oxide C), using as the reference the video data (which plays the role of Pitot-Prandtl tube), are presented in Figure 3. It can be clearly seen the two order of magnitude difference in the velocity of air and of nanopowder at the same hot-wire voltage. It again supports the idea that the heat transfer from hot-wire is significantly different in air and in nanopowder flow.

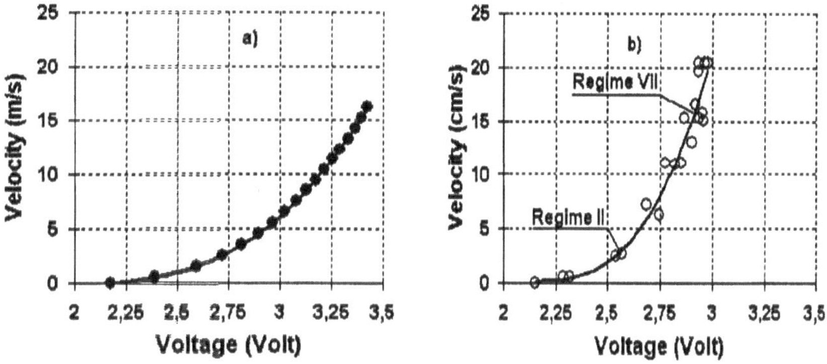

Figure 3. Calibration curves of a hot-wire probe in air a) and in powder b).

The calibrated hot-wire had been used for the measurements of the mean velocity and the velocity fluctuations distributions across the pipe. As an examples for the velocity regimes II and VII (see Fig. 3) measured over the half pipe section are shown in Figure 4. It can be seen that the mean velocity and fluctuations are approximately uniform in the core of the flow, and the level of fluctuations in both cases deviates by about 20% only. The last one can be treated as turbulence level in "nanomedia" flow or as a "granular temperature" in nano- granular flow.

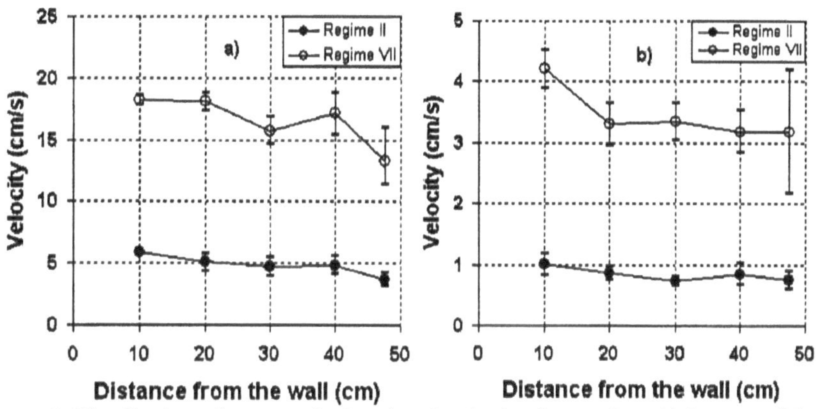

Figure 4. Distribution of mean velocity a) and velocity fluctuations b) in core of flow.

The corresponding measurements in boundary layer flows near the wall of the pipe, that means the mean velocity and velocity fluctuations distribution and their peculiarities will be discussed in a separate paper.

4. Conclusions

Experimental studies concerning the rheology of nanopowder flow in conditions where the concentration of the powder is kept "close" to bulk density had been conducted. The developed techniques, especially the applied hot-wire method, can be used for the estimation of dynamical properties of nanopowder flow. The results shows that nano- granular media can be considered as a new kind of continuum. For the first time the estimation of the flow behavior in the boundary layer of nanogranular media had been obtained. Probably, the concept of laminar and turbulent flow, certainly with necessary restricts, can be used in the description of such flows. As far as the author knows, a similar concept was expressed in [13] for granular media with large particles. Apparently, for such a flow another problems appeare , for example, concerning its stability, development of disturbances, etc.

The author expresses his thanks for the help to his collaborators M.E. Djuraeva, S.W. Joo, Y.U. Han, O.Y. Korkina, V.V. Larichkin, D.Y. Lee, Y.H. Lim and M.Y. Won. This work was partially supported by grant No.00-15-96164 from the Russian Foundation for Basic Research and under the Brain Korea 21 Program of Republic of Korea Government.

5. References

1. J.T. Jenkins, S.B. Savage, A theory for the rapid flow of identical, smooth, nearly elastic, spherical particles,Journal of Fluid Mechanics, **130**, 1983, 187-202.
2. Haff, P.K., Grain flow as a fluid-mechanical phenomenon, Journal of Fluid Mechanics, **134**, 1983, 401-430.
3. Tardos, G.I., A fluid mechanic approach to slow, frictional flow of powders, Powder Technology, **92**, 1997, 61-74.
4. S.P. Bardakhanov, A.I. Korchagin, N.K. Kuksanov, A.V.Lavrukhin, S.N. Fadeev, R.A. Salimov. Fine particle production by electron beam in air. Proc. of International Congress for Particle Technology (PARTEC 2001), Nuremberg, Germany, 27-29 March 2001.
5. Bardakhanov, S.P., Djuraeva, M.E., Korkina, O.Y., Lim, H.Y., Won, M.Y., Rheology of Nanopowders, in: Nanotechnology in Mechanical Engineering, S.P.Bardakhanov (Ed.), School of Mechanical Engineering, Yeungnam University, 2002, 2-29.
6. Bardakhanov, S.P., Larichkin, V.V., Rudyak, V.Ya., Experimental studies of nanopowder flows, Abs. of 8th International Conference "Stability of Homogeneous and Heterogeneous Fluids", April 2001, Novosibirsk, ITAM, 2001, 18-19.
7. Bardakhanov, S.P., Larichkin, V.V., Rudyak, V.Ya., Investigation of Nanopowder Flow, Proc. of First Russian-Korean International Symposium on Applied Mechanics ("RUSKO-AM-2001"), October 2-4, 2001, Novosibirsk State Technical University, 2001, 6 pages.
8. Bardakhanov, S.P., New way of nanomaterials production and new method of nanopowder dynamic flow measurements, Proc. of Brain Korea 21 Workshop "Dynamic Flow Measurements", January 31 – February 2, 2002, School of Mechanical Engineering, Yeungnam University, 1-35.
9. Bardakhanov, S.P., Kozlov, S.A., Hot-wire measurements in nanopowder flow, Abs. of 8th International Conference "Stability of Homogeneous and Heterogeneous Fluids", April 2001, Novosibirsk, ITAM, 2001, 16-17.
10. Bardakhanov, S.P., Ivanov, E.G., Sound propagation in nano- granular media, in: Nanotechnology in Mechanical Engineering, S.P.Bardakhanov, Ed., School of Mechanical Engineering, Yeungnam University, 2002, 30-45.
11. Bardakhanov, S.P., Djuraeva, M.E., Joo, S.W., Han, Y.U., Korkina, O.Y., Larichkin, V.V., Lee, D.Y., Lim, H.Y., Rheology of ultrafine powders, Proc. of JSSUME-2002, Kyongu, August 2002 (in press).
12. Bardakhanov, S.P., Djuraeva, M.E., Joo, S.W., Han, Y.U., Korkina, O.Y., Larichkin, V.V., Lee, D.Y., Lim, H.Y., Rheology of nano- and micro- powders, Proc. of ISMANAM 2002, Seoul, September 2002 (in press).
13. Savage, S. B., Disorder, diffusion and structure formation in granular flows and granular media, in: Disorder and Granular Media, D. Bideau, A. Hansen, Eds., North-Holland, Amsterdam, 1993, 255-285.

V

Self Organization in Sedimentation Processes

THE EFFECT OF HOMOGENEOUS ISOTROPIC TURBULENCE ON THE SETTLING OF HEAVY PARTICLES

A. ALISEDA, A.CARTELLIER[1] & J.C. LASHERAS
Department of Mechanical and Aerospace Engineering,, University of California San Diego, La Jolla, CA 92093, USA

1 Introduction

The experiments were performed in a horizontal wind tunnel, and a grid was used both to induce the turbulent and to inject the water droplets. The water droplets were injected through an array of atomizers embedded in the grid that produced a polydispersed spray uniformly distributed along the central portion of the tunnel. The turbulent Weber number for the droplets was always very small, so the droplets behave as heavy spherical particles throughout the experiments.
The carrier flow turbulence was studied using a hot-wire anemometer. Both the mean and the fluctuating velocity were found to be uniform in planes perpendicular to the mean flow, outside of the boundary layers. The decay of the turbulent intensity in the test section is well inside the near region, [1] [7], and following the former, a linear fit can be applied to our data, $(U/u')^2 = 38.41(x/M - 10.17)$. The turbulence characteristics obtained from the hot-wire measurements are shown in table 1. It is important to notice that the Kolmogorov length scale in the carrier flow is an order of magnitude larger than the droplet's diameter (both the arithmetic and Sauter mean diameters). Two different experimental techniques were used to characterize the velocity and concentration field of the particles in the flow. Phase Doppler Particle Analyzer was used to measure the horizontal and vertical velocities, as well as the diameter, of the particles. Flow visualizations were used to characterize the spatial distribution of the particles.

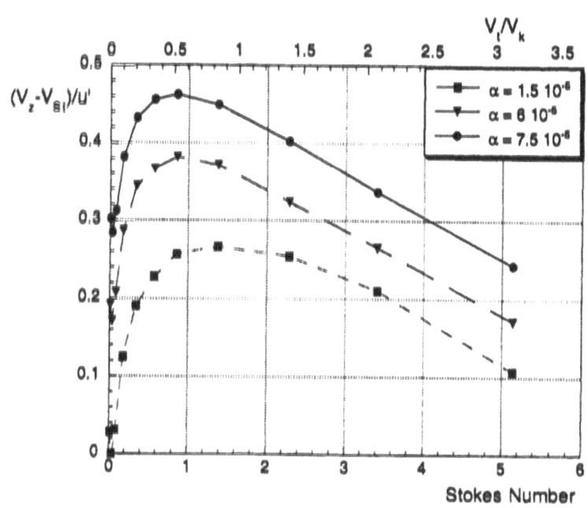

Figure 1: Increase in the settling velocity of particles as a function of the Stokes number, for dfferent volume fractions.

[1] On leave from: LEGI/CNRS/UJF/INPG, Grenoble, France

A. Gyr and W. Kinzelbach (eds.), Sedimentation and Sediment Transport, 145–148.

2 Experimental Results

The settling velocity of the particles was found to be significantly higher than that they would have in still fluid. The difference between them was found to depend on the particles inertia, characterized by the Stokes number $S_t = \tau_p/\tau_k$, with a maximum value for particles with Stokes number around one. This result had been predicted from both analytical studies and numerical simulations, [4] and [6]. The experiments confirmed those predictions but also showed an important dependency on the particle volume fraction, figure 1.
The concentration field of particles showed strong inhomogeneities which had been found in simulations, [5] [6], as well as in experiments, [3]. The images taken of slices of the flow were analyzed statistically to find a set of parameters that could characterize the concentration field. By dividing the images into windows of a certain size and counting the number of particles in each window, a pdf of the number of particles in a window can be build and compared with the one produced by a random process (Poisson distribution). The comparison can be quantified by using different parameters. When the ones introduced by [3] and [6] were tried, they both yielded the same result. Deviation from randomness is maximum for a certain window size, in both cases equal to approximately 10 times the Kolmogorov length scale of the flow, figure 2. This is, then, a characteristic length scale for these regions of high particle concentration that we refer to as clusters.
With the knowledge of the cluster structure gained from the analysis of the flow visualizations, the velocity data obtained from the PDPA was reprocessed. Since the arrival time of each particle was recorded, together with the diameter and the velocity, and the mean horizontal velocity is independent of the particle size, one can relate inter- velocity of a particle can be computed as the sum of two terms: $V_z = V_{is} + V_{cluster}$.arrival time to inter-particle distance, and thus obtain a value of the local concentration around each particle. The velocity can then be averaged conditioned to size and local concentration, those averages are plotted in figure 3. What they show is that the average settling velocity of a particle in turbulence depends on the local concentration around it. This result supports the finding from the unconditional velocity averages that showed a dependency on the volume fraction: if the volume fraction is higher a larger percentage of particles will be in regions where the local concentration is high. Moreover, the dependency is very similar for particles of all sizes, and is almost linear with the local concentration.

3 The effect of preferential accumulation on the settling velocity of the particles.

A simple phenomenological model was proposed to explain this collective effect on the settling velocity, and tested against the experimental results. The idea behind the model is that once clusters are formed by the preferential accumulation mechanism, they evolve as a somewhat individual structure, until the interaction with the large energetic scales in the flow destroys it, dispersing the particles that formed it. The vertical velocity of a particle can be computed as the sum of two terms: $V_z = V_{is} + V_{cluster}$.

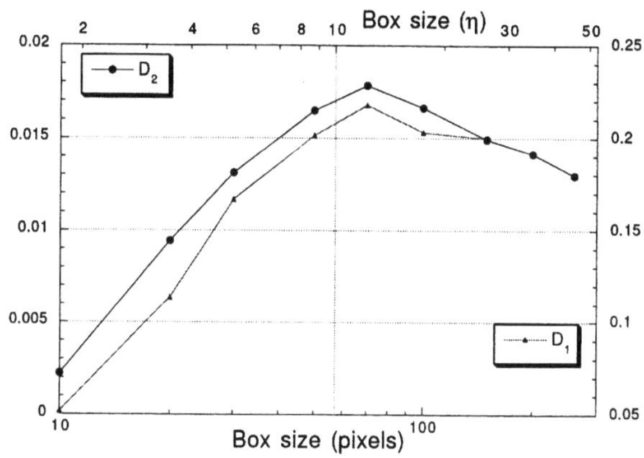

Figure 2: Two estimates of the distance between the PDF of particles in the flow and that of a random process. Evolution with the length scale.

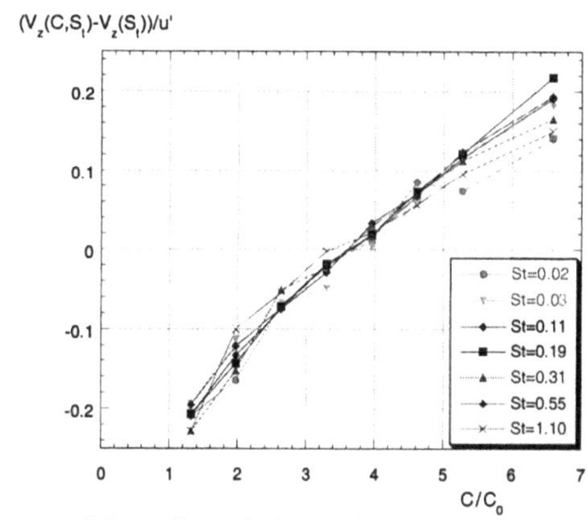

Figure 3: Enhancement of the settling velocity as a function of the local concentration

The first one is due to buoyancy and interaction with the turbulence (in isolation). The second is the one due to the cluster, if the particle is inside a cluster at that time. Averaging over time, the average settling velocity of particles of a certain size can be expressed as: $\overline{V_z(S_t,\alpha)} = V_{is}(S_t) + \int \gamma(S_t,\alpha,C) \bullet V_{cl}(C)dC$. The effect of the clusters can be computed as the percentage of time that particles spend inside clusters (γ intermittency factor) times the velocity of those clusters. That velocity can be evaluated by a very simple balance between buoyancy and viscous drag, resulting in: $V_{cl} = \frac{K_T}{18}\frac{\rho_p}{\rho_{air}}\frac{g}{\nu_{air}} C l_{cl}^2$ We can see that this expression has the linear dependency on the local concentration that was found from the experiments. The comparison of the slopes obtained from the model to the ones from the ex periments is shown in table 2. What has not been possible to model, however, is the intermittency factor, which would allow to recover the unconditional velocity from the conditional data.

x(cm)	u' (cm/s)	$e(m^2/s^3)$	L (mm)	λ (mm)	η(mm)	τ_k(ms)	v_k(cm/s)
83	26.2	1.75	37.7	5.06	0.210	2.92	7.16
101	21.1	1.00	43.0	5.37	0.241	3.87	6.23
138	19.6	0.88	48.4	5.26	0.249	4.13	6.03
168	17.6	0.76	50.5	4.80	0.258	4.43	5.82
187	16.3	0.68	56.5	4.64	0.265	4.69	5.66
207	15.6	0.61	56.2	4.66	0.273	4.96	5.50

Table 1: Downstream evolution of the turbulence characteristics for the single phase air flow.

		From the model	From figure 3	
x(cm)	a	$\dfrac{V_{cl}/u'}{c/c_0}$	$\dfrac{dV_z/u'}{dC/C_0}$	$C/C_0(V_z=\overline{V_z})$
100	$6\ 10^{-5}$	0.051	0.052	2.81
	$7.5\ 10^{-5}$	0.063	0.065	3.05
200	$5\ 10^{-5}$	0.053	0.068	2.95
	$7.5\ 10^{-5}$	0.074	0.102	3.84

References

[1] Comte-Bellot, G. & Corrsin, S. The use of a contraction to improve the isotropy of grid-generated turbulence. J. Fluid Mech. **25**, 657–682, 1966.

[2] Eaton, J.K. & Fessler, J.R. Preferential concentration of particles by turbulence. Intl J. Multiphase Flow **20**, Suppl., 169–209, 1994.

[3] Fessler, J.R., Kulick, J.D. & Eaton, J.K. Preferential concentration of heavy particles in turbulent channel flow. Phys. Fluids **6**, 3742–3749, 1994.

[4] Maxey, M.R. The motion of small spherical particles in a cellular flow field. Phys. Fluids **30**, 1915–1928, 1987a.

[5] Squires, K.D. & Eaton, J.K. Preferential concentration of particles by turbulence. Phys. Fluids **A 3**, 1169–1178, 1991b.

[6] Wang, L.P. & Maxey, M.R. Settling velocity and concentration distribution of heavy particles in homogeneous isotropic turbulence. J. Fluid Mech. **256**, 27–68, 1993a.

[7] Wells, M.R. & Stock, D.E. The effect of crossing trajectories on the dispersion of particles in a turbulent flow. J. Fluid Mech. **136**, 31–62, 1983.

NUMERICAL INVESTIGATION OF TWO-WAY COUPLING MECHANISMS IN DILUTE, PARTICLE LADEN FLOWS

E. MEIBURG
Department of Mechanical and Environmental Engineering, UCSB
Santa Barbara, CA 93106, USA

1. INTRODUCTION

Over the last decade, two-way coupled simulations of particle or droplet laden flows have contributed significantly to our insight into the interaction mechanisms between the continuous and the dispersed phase, respectively. The effects of this two-way coupling on the statistical properties of both isotropic and homogenous turbulence have been addressed in the numerical simulations of several authors (Squires and Eaton 1990, Elghobashi and Truesdell 1993, Truesdell and Elghobashi 1994, Maxey, Patel, Chang and Wang 1997, Sundaram and Collins 1999), while other investigators have focused on the modification of wall turbulence (Pan and Banerjee 1996) and fully developed channel flows (Kulick, Fessler and Eaton 1994) by heavy particles. Druzhinin (1995) conducts an analysis of two-way coupling effects in several simplified flowfields such as free stagnation points, as well as constant vorticity and Stuart vortices, in order to obtain improved physical insight. By using a small Stokes number expansion, he makes the interesting observations that two-way coupling effects lead to a reduction in the vorticity near vortex centers, while the strain is enhanced at hyperbolic free stagnation points. Two-way coupling effects in transitional flows have been investigated by Saffman (1961), as well as Chen and Chung (1995). Recently, Dimas and Kiger (1998) extended the above linear stability analyses. Their focus is on the impact of the particle Stokes number and the mass loading, and they observe that an increase in either one of these parameters decreases the growth rate of the instability. Under certain conditions, they find the appearance of a second, low-frequency instability mode.

We will briefly review the case of a unidirectional base flow of a mixing layer, in the presence of gravitational sedimentation. Subsequently, the focus will be on two-dimensional two-way coupled mixing layers. The importance of the vorticity variable will be emphasized. More detailed discussions can be found in Meiburg et al. (2000), as well as Wallner and Meiburg (2002).

2. UNIDIRECTIONAL BASE FLOW

Consider the case in which the fluid velocity initially is given by a unidirectional, horizontal base flow. Let us furthermore assume that the initial conditions for the particle velocity and number density are independent of the streamwise location as well. Then, in the absence of two-dimensional perturbations, the particle velocity field can remain single-valued under certain conditions, cf. Meiburg et al. (2000), so that one can define a particle vorticity field As shown by Meiburg et al. (2000), the coupled and rescaled fluid

A. Gyr and W. Kinzelbach (eds.), Sedimentation and Sediment Transport, 149–154.

and particle vorticity fields are then governed by a set of coupled partial differential equations with interesting properties. Depending on the values of the governing dimensionless parameters D (particle loading ratio), S (particle Stokes number), Re (Reynolds number), and Fr (Froude number), numerical simulations of this system of equations show the formation of a downward traveling wave solution (Meiburg et al. 2000). Interestingly, the downward propagation of the fluid vorticity field is not accomplished through convection, but rather by the production and loss of vorticity on opposite sides of the mixing layer. For moderate settling velocities, the simulation results reveal an optimal coupling mechanism between the fluid and particle vorticities at intermediate values of the mass loading parameter. For large settling velocities and intermediate mass loadings, more than one local maximum is seen to evolve in the vorticity field.

3. NUMERICAL APPROACH FOR THE TWO-DIMENSIONAL PROBLEM

In the following, we will focus on the temporal evolution of the above base flow in the presence of two-dimensional, spatially periodic perturbations. In order to avoid the high computational cost of a phase space based approach, we pursue a less expensive Lagrangian numerical method for the particulate phase. The field equations for the fluid phase are solved in an Eulerian fashion, by applying a spectral Fourier series expansion in the streamwise, periodic direction (Gottlieb and Orszag 1977), and sixth order compact finite differences in the transverse direction (Lele 1992). Time advancement is accomplished by means of a third order Runge-Kutta method (Wray 1991). An initial perturbation is applied to the fluid velocity field that has the form of the eigenfunction of the most unstable mode according to inviscid theory. Slip conditions are applied at the upper and lower boundaries of the control volume. At the beginning of the simulation, the particle velocities are set equal to the local fluid velocity.

Our numerical approach for modeling the particulate phase and its effects on the fluid motion is similar in spirit to the techniques employed by Squires and Eaton (1990), Elghobashi and Truesdell (1993) and Maxey et al. (1997) in that it tracks individual particles with inertia in a Lagrangian fashion. However, while those authors focused on the two-way coupling mechanisms in a velocity-pressure formulation of the equations governing the fluid motion, our focus will be on gaining a vorticity-based understanding of the two-way coupling effects by which the particulate phase affects the fluid motion. Consequently, the simulations to be discussed in the following will be based on the vorticity formulation of the fluid equations of motion. Since the vorticity equation for the fluid involves derivatives of the particle number density field (Meiburg et al. 2000), care has to be taken in reconstructing this field at every time level from the Lagrangian information of the individual particle locations. The numerical approach was first introduced by Wallner and Meiburg (1998), and its key features are reviewed here. The dispersed particles are tracked in a Lagrangian way, with each of the computational particles representing a cluster of physical particles located in the same neighborhood. The number density field is then obtained by summing over the computational particles. In this process, a radially symmetric Gaussian shape function is employed. All computational clusters initially are of identical size and strength. Their initial separation is typically equal to the grid size of the Eulerian mesh, while their core size is twice the mesh spacing. In order to maintain a well-resolved particle concentration field for long times, a remeshing procedure is employed. In order to evaluate the instantaneous acceleration of each Lagrangian cluster, the flow velocity at its location is determined by means of a fourth-order, two-dimensional Lagrangian interpolation scheme (Martin and Meiburg 1994).

4. RESULTS OF TWO-DIMENSIONAL SIMULATIONS

In the following, some of the qualitative differences in the vorticity and concentration fields between the one-way coupled and the two-way coupled cases will be described, at first in the absence of gravitational settling effects. To this end, we will focus on the case of St=1, for which the coupling between the two phases is quite pronounced, as will be seen below. For an initially uniform seeding with particles, and for a mass loading of D=0.5, significant differences in the evolution of the mixing layer at Re=200 are visible, cf. figure 1. Both in the braids, as well as in the regions to the immediate left and right of the core, an increase in the magnitude of the vorticity due to the two-way coupling can be observed. In contrast, the vorticity within the core, and in the regions immediately above and below it, is reduced in magnitude. Furthermore, we observe that the braids remain somewhat thicker under two-way coupling, and that the vortex generally is less circular than in the passive case.

The particle concentration field is affected by the two-way coupling as well. As the coupling reduces the core vorticity, the centrifugal forces acting on the particles become weaker. As a result, for two-way coupling we observe a slight increase in particle concentration in the core, with a corresponding decrease in the braids. Due to the less pronounced ejection of particles from the core, the concentration field appears somewhat compressed in the cross-stream direction, when compared to the passive case. This behavior is again consistent with the predictions by the linear stability analysis of Dimas and Kiger (1998).

A more detailed understanding of the underlying vorticity production and destruction mechanisms due to particle loading can be obtained by analyzing the source term in the fluid vorticity equation. Upon expanding

$$\nabla\times\left(n(\vec{u}_p-\vec{u})\right)=\nabla n\times\left(\vec{u}_p-\vec{u}\right)+n\left(\nabla\times\vec{u}_p-\nabla\times\vec{u}\right), \tag{1}$$

two potential mechanisms for vorticity generation/loss are identified. The first term indicates that the vorticity will be altered if the concentration gradient is misaligned with the direction of the slip velocity. The second term describes the influence of a difference between the particle and fluid vorticities. The simulations usually show that in the vicinity of the core, the second term is largest, due to the particles' inertia. Keeping in mind that the fluid vorticity is negative everywhere, the source term will thus be positive in the core, thereby reducing the fluid vorticity. In the braids, on the other hand, the first term on the right hand side of equation (1) becomes more significant due to the large variations of the particle concentration field, and the relatively weak vorticity. Overall, this leads to a quadrupole structure of the vorticity source term near the stagnation point, cf. Meiburg et al. (2000). The vorticity generation and destruction caused by the two-way coupling is seen to slow the transport of vorticity from the braids into the core region, thereby weakening the Kelvin-Helmholtz instability mechanism of the fluid phase.

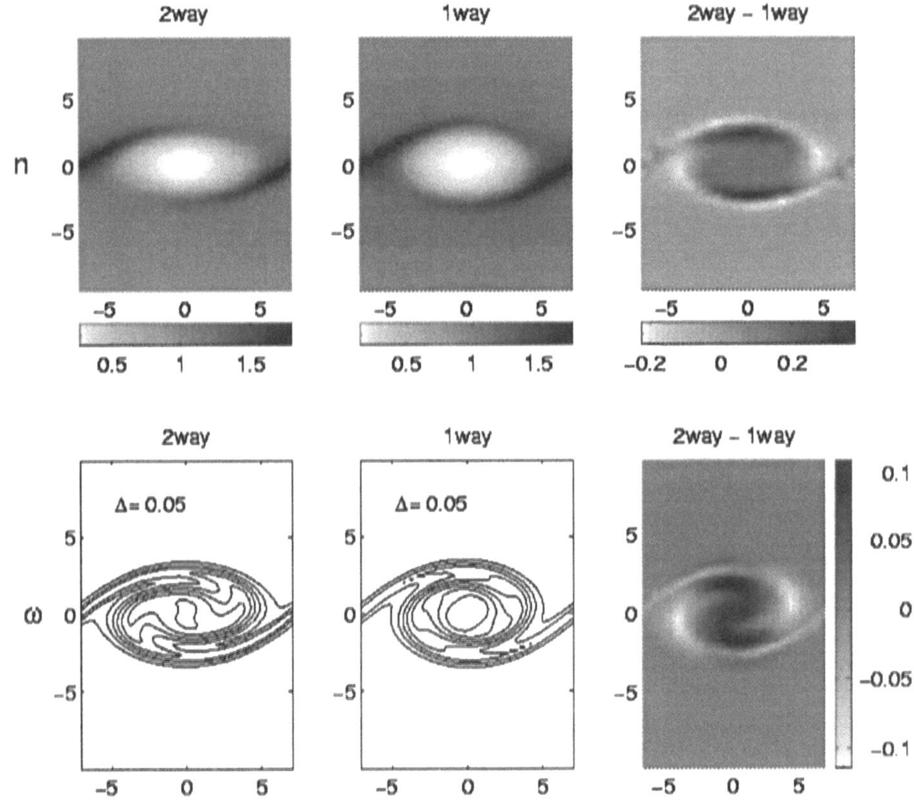

Fig. 1: Concentration of the two-way coupled and the one-way coupled cases, respectively. The right column shows the differences between the two flows. In the vorticity contour plots, Δ gives increments between contour levels. The vorticity difference is indicated by shading, with dark (light) shading visualizing an increase (decrease) in the vorticity magnitude due to coupling. The particle number density is seen to increase in the core, while the vorticity decreases.

4.1 GRAVITATIONAL SETTLING AND VORTICITY DYNAMICS

Inspection of the governing equations shows that gravity does not directly enter into the conservation equation for the vorticity. Rather, there are two ways in which gravitational forces affect the vorticity production term indirectly. Firstly, by causing the particles to settle through the fluid, gravity tends to convectively modify the gradient field of the particle number density, thereby altering the locations of vorticity production and cancellation. Secondly, the gravitational acceleration of the particles affects their slip velocity, which again affects the vorticity source term. Here these two mechanisms will be briefly discussed for uniformly seeded particle laden mixing layers.

The case of $St=Fr=3$, $D=0.1$, and $Re=200$ is well suited for demonstrating the qualitative changes caused by the two-way coupling. Figure 2 shows that two-way coupling results in a global shift of the vorticity field in both the $+x$- and the $-y$-directions. This is effected by the vorticity source term, which tends to cancel the vorticity on the upper side of the mixing layer, while amplifying it on the lower side. A series of corresponding simulations, cf. Meiburg et al. (2000), suggest that after an initial transient the vortical structure assumes nearly constant velocities in the horizontal and vertical directions, which can be estimated from least squares fits of the displacement data.

The present investigation has addressed only two-dimensional two-way coupling mechanisms between the two phases, i.e., the effect of the suspended phase on the convection and diffusion of the carrier phase vorticity. For the evolution of fully three-dimensional two-phase flows, an important aspect will be the influence of the particulate phase on the vortex stretching mechanism. This is currently under investigation.

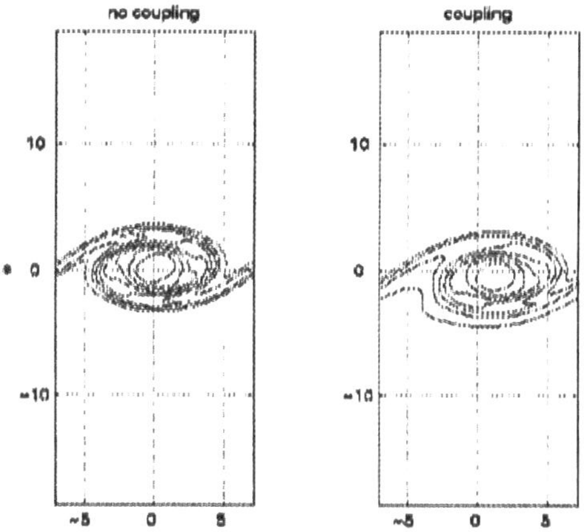

Fig. 2: Uniformly seeded flow. Shown are the respective vorticity fields for one-way and two-way coupling. The two-way coupling results in a net horizontal displacement of the large scale vortical structure in the direction of the upper stream. At the same time, it is shifted downward in the direction of the gravitational acceleration.

REFERENCES

CHEN, Y.-C. and CHUNG, J.N. 1995 The linear stability of an oscillatory two-phase channel flow in the limit of small Stokes number. Phys. Fluids 7, 1510.

DIMAS, A.A. and KIGER, K.T. 1998 Linear instability of a particle-laden mixing layer with a dynamic dispersed phase. Phys. Fluids 10, 2539--2557.

DRUZHININ, O.A. 1995b On the two-way interaction in two-dimensional particle-laden flows: The accumulation of particles and flow modification. J. Fluid Mech. 297, 49--76.

ELGHOBASHI, S. and TRUESDELL, G.C. 1993 On the two-way interaction between homogeneous turbulence and dispersed solid particles. I: Turbulence modification. Phys. Fluids A 5, 1790--1801.

GOTTLIEB, D. and ORSZAG, S.A. 1977 Numerical Analysis of Spactral Methods: Theory and Applications. SIAM, Philadelphia.

KULICK, J.D., FESSLER, J.R., and EATON, J.K. 1994 Particle response and turbulence modification in fully developed channel flow. J. Fluid Mech. 277, 109--134.

LELE, S.K. 1992 Compact finite difference schemes with spectral-like resolution. J. Comput. Phys. 103, 16--42.

MARTIN, J.E. and MEIBURG, E. 1994 The accumulation and dispersion of heavy particles in forced two-dimensional mixing layers. I. The fundamental and subharmonic case. Phys. Fluids 6, 1116--1132.

MAXEY, M.R., PATEL, B.K., CHANG, E.J., and WANG, L.-P. 1997 Simulations of dispersed turbulent multiphase flow. Fluid Dyn. Res. 20, 143-156.

MEIBURG, E., WALLNER, E., PAGELLA, A., RIAZ, A., HAERTEL, C. and NECKER, F. 2000 Vorticity dynamics of dilute two-way-coupled particle-laden mixing layers. J. Fluid Mech. 421, 185--227.

PAN, Y. and BANERJEE, S. 1996 Numerical simulation of particle interactions with wall turbulence. Phys. Fluids 8, 2733--2755.

SAFFMAN, P.G. 1961 On the stability of laminar flow of a dusty gas. J. Fluid Mech. 22, 120--128.

SQUIRES, K.D. and EATON, J.K. 1990 Particle response and turbulence modification in isotropic turbulence. Phys. Fluids A 2, 1191--1203.

SUNDARAM, S. and COLLINS, L. 1999 A numerical study of the modulation of isotropic turbulence by suspended particles. J. Fluid Mech. 379, 105--143.

TRUESDELL, G.C. and ELGHOBASHI, S. 1994 On the two-way interaction between homogeneous turbulence and dispersed solid particles. II: Particle dispersion. Phys. Fluids 6, 1405--1407.

WALLNER, E. and MEIBURG, E. 1998 Numerical simulation of two-way coupled particle laden mixing layers. Proceedings of the Third International Workshop on Vortex Flows and Related Numerical Methods, held at Toulouse, France, Aug. 24-27, 1998.

WALLNER, E. and MEIBURG, E. 2002 Vortex pairing in two-way coupled, particle-laden mixing layers. Int. J. Multiphase Flow 28, 325--346.

WRAY, A.A. 1991 Minimal storage time-advancement schemes for spectral methods. Preprint.

STATISTICS OF TURBULENCE-INDUCED FLUCTUATIONS OF PARTICLE CONCENTRATION

G. Falkovich, A. Fouxon and M. Stepanov
Physics of Complex Systems, Weizmann Institute of Science, Rehovot 76100, Israel

The motion of heavy/light particles is retarded/advanced with respect to the ambient fluid. Fluid vortices act as little centrifuges spinning heavy particles out and sucking light particles in, creating inhomogeneities in their concentration. We describe the concentration fluctuations of inertial particles in a random flow. We show that the concentration is inhomogeneous at the scales less than the correlation scale of the velocity gradients η (few times the viscous scale in turbulence). In the interval of scales between η and the particle size a, the particles are distributed very nonuniformly forming a fractal set. We describe the pair correlation function of concentration as a function of particle and flow parameters. We consider polydisperse set of particles in both low-Reynolds random flows and high-Reynolds turbulence and account for gravity. We show how small inertial effects are enhanced by turbulence intermittency and by the presence of a wide interval of scales available for clustering.

If there is no flow produced by external factor, then a single small spherical particle moves vertically with the velocity $u_g = g\tau(\rho_0 - \rho)/\rho_0$ where the Stokes time is as follows $\tau = (2/9)(\rho_0/\rho)(a^2/\nu)$. Here ρ_0, ρ are particle and fluid densities respectively and the Reynolds number of the flow around the particle, $Re_a \equiv (2/9)(\rho_0/\rho)(ga^3/\nu^2)$, is assumed small. The velocity may be increased by hydrodynamic interaction between different particles which can be estimated by summing up the perturbations of the flow that decay as inverse distances [1]:

$$\delta u_g(a, \mathbf{r}) \simeq \sum_{\mathbf{a}'} (\mathbf{3}/\mathbf{4})\mathbf{a}'\mathbf{u_g}(\mathbf{a}') \int \mathbf{d^3r}'\, \mathbf{n}(\mathbf{a}', \mathbf{r}')(1 + \cos^2 \vartheta)|\mathbf{r} - \mathbf{r}'|^{-1} . \quad (1)$$

Here, ϑ is the angle between $\mathbf{r} - \mathbf{r}'$ and the vertical. To estimate the integral, one cuts it at the Oseen scale ν/u_g and concludes that n-dependent corrections to u_g are negligible when $na^3 Re_a^{-2} \ll 1$, which is assumed here.

Consider now particles in a flow. The particle velocity $\mathbf{v}$ is related to the fluid velocity $\mathbf{u}$ by the equation [1,2]

$$\frac{d\mathbf{v}}{dt} - \beta\frac{d\mathbf{u}}{dt} = \frac{\mathbf{u} - \mathbf{v}}{\tau} + \mathbf{g} . \quad (2)$$

here $\beta = 3\rho/(\rho + 2\rho_0)$. Both $\mathbf{u}, \mathbf{v}$ are evaluated on the particle trajectory. To illustrate the effect of preferential concentration, let us consider the limit when fluid velocity encountered by particle change slightly during τ. In this case, one can approximate

$$\mathbf{v} = \mathbf{u} + (\beta - 1)\tau[\partial_t \mathbf{u} + (\mathbf{u} \cdot \nabla)\mathbf{u}] + \mathbf{g}\tau . \quad (3)$$

If we now consider the set of particles suspended in the flow then (2,3) define the field $\mathbf{v}(\mathbf{r})$. Because of inertial retardation, the particles' velocity (3) is compressible even if the air flow is not [3]: $\nabla \cdot \mathbf{v} = (\beta - 1)\tau\nabla[(\mathbf{u} \cdot \nabla)\mathbf{u}]$. Since $\nabla[(\mathbf{u} \cdot \nabla)\mathbf{u}] = S^2 - \Omega^2$, where S, Ω are respectively symmetric and antisymmetric parts of the velocity derivative tensor, then it is indeed clear that heavy particles ($\beta < 1$) leave vorticity-dominated elliptic regions of the flow ($\nabla \cdot \mathbf{v} > 0$) and concentrate in the strain-dominated hyperbolic regions

A. Gyr and W. Kinzelbach (eds.), Sedimentation and Sediment Transport, 155–158.

while light particles behave in an opposite way. Compressibility means that the particle distribution is inhomogeneous; this effect is long-known and used, for instance, for flow visualization, it has been directly observed in laboratory experiments and in numerical simulations (see,[4,5,6,7] and the references therein). Of course, turbulence also provides mixing. To characterize the relative significance of the three factors (flow, particle inertia and gravity) one introduces two dimensionless parameters. To characterize the relative strength of segregation versus mixing, consider the concentration $\tilde{n}(r)$ coarse-grained over the scale r which satisfies the equation $\partial\tilde{n}/\partial t + \tilde{n} div\, v(r) + (v\nabla)\tilde{n} = D(r)\Delta\tilde{n}$ where $div\, v(r) \simeq (\beta-1)\tau[\delta u(r)/r]^2$ and $D(r) = r\delta u(r)$. The relative strength of inertia (that leads to segregation) versus eddy diffusivity (that tends to make the concentration uniform) is measured by the ratio of the second term to the fourth one which is the scale-dependent Stokes number $St(r) = t(r) div\, v(r)$ where the eddy turnover time is $t(r) = r/\delta u(r)$ (here we neglect gravitational settling). In the viscous interval, $St(r) = St = |\beta - 1|\lambda\tau$ is scale-independent while in the inertial interval it decreases with the scale, $St(r) = St(\eta/r)^{2/3}$. We denote the rms fluid velocity gradient by $\lambda = \sqrt{\langle|\nabla u|^2\rangle}$ and the correlation scale of velocity gradients by η assuming everywhere $\eta \gg a$. The effects of compressibility are thus negligible and the concentration is almost uniform on the scales from the inertial interval of turbulence. Quantitatively, the pair correlation function of the concentration was shown to decay as a stretch-exponential in the inertial interval [9]. Turbulence effectively mixes particles at the scales exceeding η while inertial effects lead to segregation and concentration inhomogeneities at the scales below η. In other words, the clusters of particles are created with sizes less than η as it follows from theory [9] and seen in numerics [10] and observations [11,12]. Note in passing that it follows from [9] that it is enough to coarse-grain the concentration field over the scale larger than $\eta St^{-3/2}$ (at $St < 1$) to make the clusters unobservable. The effect of the flow relative to gravity is characterized by $\epsilon = \lambda\eta/u_g$ which can be called small-scale Froude number. We also introduce $\tilde{\epsilon} = \epsilon(1+\epsilon^2)^{-1/2}$. The below theory is valid for $St\tilde{\epsilon} < 1$ which physically means that either particle inertia is small ($St < 1$) or turbulent acceleration $\lambda^2\eta$ is much less than g [4].

The initial stage of clustering can be readily described since one can show that any compressible flow with a Lagrangian chaos creates a fractal concentration field (so-called Sinai-Ruelle-Bowen measure) [13]. Indeed, as far as one can neglect diffusion due to Brownian motion of the particles, the concentration n satisfies the equation $d\ln n/dt = -\mathrm{div}\,\mathbf{v}$ in a Lagrangian frame of reference. If $\mathrm{div}\,\mathbf{v}$ is some random function of time independent of n, one gets an exponential growth of the moments of n with an asymptotically lognormal statistics. That corresponds to an uninterrupted clustering of inertial particles in a smaller and smaller regions accompanied by a growth of voids. The growth of concentration fluctuations and the validity time of diffusionless description were described for a white in time velocity [14]. An attempt to account for diffusion was made in [15] by writing the equation for the pair correlation function of concentration within the framework of the same white in time model. Unfortunately, due to calculational error, the authors of [15] concluded that diffusion does not stop fluctuation growth at sufficiently high Reynolds numbers (a correct solution for the pair correlation function can be found in [9]). Physically natural statement that finite-size effects do stop the clustering of inertial particles at the scale $r_d = \max(a, (\kappa/\lambda)^{1/2})$ (κ is the particle diffusivity) have been shown in [9]. Back reaction of particles on the fluid velocity may also stop compression. If one replaces u_g by u_{in} in (1) and cuts the integral at $|\mathbf{r} - \mathbf{r}'| \simeq \eta$ (since particles move coherently within the viscous scale) then one sees that the role of interaction in creating concentration inhomogeneities can be neglected as long as $\delta u \ll u$ which requires $(9\pi/4)St\bar{n}a\nu\lambda^{-1} \simeq (\rho_0/\rho)\bar{n}a^3 \ll 1$. Therefore, as long as the mean mass content of the particles in the fluid is small, the concentration n can be considered passive, i.e. $\mathbf{v}$ is independent of n. If one assumes that the interaction stops compression when n is locally so large that $\delta u \simeq u$, then the minimal size due to interaction is estimated as $r_i \simeq 9\pi\eta\bar{n}a^3\rho_0/8\rho$. Thus for equal-size particles the minimal scale of clusters is $r_d = \max\{a, (\kappa/\lambda)^{1/2}, r_i\}$. It was shown in [9] that for Gaussian (non-intermittent) velocity field the pair correlation function of concentration at $\eta \gg r \gg r_d$ is as follows: $\langle n(a,0)n(a,\mathbf{r})\rangle \simeq \langle n(a)\rangle^2(\eta/r)^{St^2/2}$. The correlation function grows as a power law when r decreases, the growth is saturated at $r \simeq r_d$. The meaning of the result is that spatial average is determined by small fractions of volume occupied by clusters: particles contained in the volume η^3 are compressed by velocity gradients into the clusters for which the concentraion is increased by a power of volume ratios.

A more general setting was considered in [4]: a polydisperse set of particles in a general velocity field and under the action of gravity. Let us define a general pair correlation function between the concentrations of particles with different sizes: $k_{12}(r) \equiv \langle n(a_1,0)n(a_2,\mathbf{r})\rangle/\langle n(a_1)\rangle\langle n(a_2)\rangle$. Here and below subscripts denote sizes of the particles. At the scales $\eta \gg r \gg r_d(a_1) + r_d(a_2)$, the correlation function is as follows [4]:

$$k_{12}(r) = \left(\frac{\eta}{r + r_{12}}\right)^{\alpha}, \qquad r_{12} = (\eta + g\lambda^{-2})(St_1 - St_2)\tilde{\epsilon}^{-1} . \tag{4}$$

At smaller scales, $r < r_d(a_1) + r_d(a_2)$, the correlation function saturates: $k_{12}(0) \sim [\eta/(r_{d1} + r_{d2} + r_{12})]^{\alpha}$. Particles of different sizes have additional relative velocity $|\tau_1 - \tau_2|(g + \lambda^2\eta)$ that stops clustering at r_{12}. We see that inertia and gravity make k_{12} a decaying function of $|a_1 - a_2|$ that is concentrations of particles at close points are uncorrelated if sizes are substantially different ($\eta \lesssim r_{12}$). To put it simply, particles of different sizes have their smallest clusters in different places.

The power α is a complicated function of St, ϵ and the Reynolds number of the flow. It can be estimated from below by the expression that uses the single-time probability distribution $\mathcal{P}(\sigma)$ of the velocity gradients $\sigma = \nabla v$ [4]:

$$\alpha \simeq (\tau^2/\lambda\tilde{\epsilon}) \int_{\tau|\sigma|<1} d\sigma\sigma^4\tau_c(\sigma)P(|\sigma|) . \tag{5}$$

To relate $P(\sigma)$ to $\mathcal{P}(s)$ of the fluid velocity gradients, $s = \nabla u$, one uses the extrapolation formulas: $P(|\sigma|) = (1+\sigma^2/s_*^2)\mathcal{P}(|s| = |\sigma| + \sigma^2/s_*)$ and $\tau_c(|\sigma|) = \tau + (|\sigma| + \lambda^{1/2}|\sigma|^{1/2}\epsilon^{-1})^{-1}$. For sufficiently small particles ($St < 1$, $\epsilon > 1$ and not very high Re), one has $\alpha \simeq St^2F_3$, where $F_3 = \lambda^{-3}\langle|s^3|\rangle$ which is a growing function of Re. That describes how turbulence intermittency amplifies the effect of small particle inertia [4], see also [16,17]. For not very high Reynolds numbers, both α and k_{12} have their maximum for particle size that correspond to $St \simeq 1$. Particular prediction of (5) is that the maximum shifts to lower St as the Reynolds number increases (that shift seems to be observed experimentally [12]). At very high Re (typical for developed turbulence), α has a maximum at $St \simeq \epsilon^2$ which corresponds to the balance between inertia and gravity when the Stokes time τ is the universal value $(\nu/g^2)^{1/3}$. Particles of such size take time τ to fall through the vortex with the turnover time τ. The effect of centrifugal force is less both for smaller particles (which are less inertial) and for larger ones (which spend less time inside the vortex). We see that concentration fluctuations may be significant ($k_{12} \gg 1$) even when inertia is small ($St \ll 1$). On the other hand, the interplay between gravity and turbulence intermittency makes concentration fluctuations significant only in the restricted interval of particle sizes.

Turbulence-induced fluctuations of concentration influence both the mean and the fluctuations of the settling velocity which will be described elsewhere.

References

[1] Pruppacher, H. and Klett, J. (1998) *Microphysics of Clouds and Precipitation*, Kluwer, Dordrecht.

[2] Maxey, M.R., (1987) The gravitational settling of aerosol particles in homogeneous turbulence and random flow field, *J. Fluid Mech.*, **174**, 441-465.

[3] Maxey, M.R. and Riley, J.J. (1983) Equation of motion for a small rigid sphere in a nonuniform flow, *Phys. Fluids*, **26,** 883–889.

[4] G. Falkovich, A. Fouxon and M. Stepanov (2002) Acceleration of rain initiation by cloud turbulence, *Nature*, **419**, 151–154.

[5] Squires, K. and Eaton, J. (1991) Measurements of particle dispersion from direct numerical simulations of isotropic turbulence, *J. Fluid Mech.*, **226**, 1–35.

[6] Wang, L. P. and Maxey (1993) M. Settling velocity and concentration distribution of heavy particles in homogeneous isotropic turbulence, *J. Fluid Mech.*, **256**, 27–68.

[7] Pinsky, M. and Khain, A. (1997) Turbulence effects on droplet growth and size distribution in clouds - a review, *J. Aerosol. Sci.*, **28**, 1177–1214.

[8] Zhou, Y., Wexler, A. and Wang, L.-P. (1998) On the collision rate of small particles in isotropic turbulence, *Phys. Fluids,* **10**, 1206–1216.

[9] E. Balkovsky, G., Falkovich, and Fouxon (2001) A. Intermittent distribution of inertial particles in turbulent flows, *Phys. Rev. Lett.*, **86**, 2790–2793.

[10] Grits, B., Pinsky, M. and Khain (2000) A. Formation of small-scale droplet concentration inhomogeneity in a turbulent flow as seen from experiments with an isotropic turbulence model, *Proc. 13th Int. Conf. on Clouds and Precipitation*

[11] Kostinski, A. and Shaw (2001) R. Scale-dependent droplet clustering in turbulent clouds, *J. Fluid Mech.*, **434**, 389–398.

[12] A. Aliseda, A. Cartellier, F. Hainaux and J.C. Lasheras (2002) Effect of preferential concentration on the settling velocity of heavy particles in homogeneous isotropic turbulence, *J. Fluid Mech.*, **468**, 77-105, see also this volume.

[13] Falkovich, G., Gawedzki, K. and Vergassola (2001) M. Particles and fields in fluid turbulence. *Rev. Mod. Phys.,* **73**.

[14] V. Klyatskin and D. Gurarie (1999) Coherent phenomena in stochastic dynamical systems, *Sov. Ph. Usp.*, **169,** 171.

[15] T. Elperin, N. Kleeorin, and I. Rogachevskii (1996) Self-excitation of fluctuations of inertial particle concentration in turbulent fluid flow, *Phys.Rev.Lett.,* **77**}, 5373.

[16] R. Shaw and S. Oncley (2001) Acceleration intermittency and enhanced collision kernels in turbulent clouds, *Atmos Res.,* **59-60**, 77–87.

[17] C. Jeffery (2001) Investigating the small-scale structure of clouds using the d-correlated closure: effect of particle inertia, condensation/evaporation and intermittency, *Atmos Res.,* **59-60**, 199–215.

NUMERICAL SIMULATION OF THE COHERENT STRUCTURES IN A HOMOGENEOUS SEDIMENTING SUSPENSION

EVGENY S. ASMOLOV
Central Aero-Hydrodynamic Institute, Zhukovsky, Moscow region, 140180, Russia
aes@an.aerocentr.msk.su

1. Introduction

Long-range hydrodynamic interactions of the particles sedimenting in the fluid under gravity force with small Reynolds numbers lead to the large-scale velocity disturbances. Behaviour of the disturbances of the fluid velocity due to fluctuations of particle concentration is described commonly on the basis of solution of the steady Stokes equations. A stability problem for a dilute homogeneous suspension of spherical particles was considered by Koch & Shaqfeh (1989). The orientation of a particle pair are affected by a fluid shear field induced by particle density disturbances. The coupling between the pairs' velocities and their relative position was suggested as the possible mechanism of the instability. However the linear-stability analysis based on the steady Stokes equations indicated that the suspension is neutrally stable. Nevertheless Segre *et al.* (1997) in their recent experiments observed the structures in a sedimenting suspension which prove that the instability still occurs.

Even though the large-scale flow is a superposition of disturbances induced by many particles it has a lengthscale other than particle radius, $\lambda \gg a$. The different terms in the Navier-Stokes equations should be estimated using λ rather than a. Thus one should take into account unsteady terms to describe the large-scale disturbances. Asmolov (2001) reconsidered the stability problem on the basis of unsteady solution for the large-scale waves of fluid velocity and particle concentration. The unsteady effects cause the suspension with random particle distribution to become unstable. In the present work the unsteady evolution of sedimenting dilute suspension is also studied using the numerical simulation of particles' motion in the periodic cell. The large-scale disturbances of the fluid velocity are sought as the unsteady Fourier series with the wave vector commensurate with the cell size. The conversion of the energy of the basic flow into the disturbance energy is confirmed by the simulations.

2. Linear stability of a dilute suspension

Consider sedimentation of a homogeneous suspension of identical particles. The particle inertia and Brownian motion are neglected. There are two main dimensionless parameters, mean particle volume fraction and particle Reynolds number. Both parameters are small, $c_0 \ll 1$, $\mathrm{Re}_s = \frac{aV_s}{\nu} \ll 1$. The fluctuation lengthscale is assumed to be small compared with the container size, but large compared to the particle radius, $L \gg \lambda \gg a$. The particle effect on the large-scale flow can be

A. Gyr and W. Kinzelbach (eds.), Sedimentation and Sediment Transport, 159–164.

approximated by point-force singularities (Saffman 1973):

$$\frac{\partial \mathbf{u}'}{\partial t} + (\mathbf{u}' \cdot \boldsymbol{\nabla})\,\mathbf{u}' + \frac{\boldsymbol{\nabla} p'}{\rho} - \nu \nabla^2 \mathbf{u}' = \sum_k \frac{\mathbf{F}_k}{\rho} \delta\left(\mathbf{x}' - \mathbf{x}'_k\right). \tag{1}$$

The fluid flow can be decomposed into several parts having different length-scales, $\mathbf{u}' = -U_s\mathbf{e}_3 + \varepsilon \mathbf{u}'_l(\mathbf{x}'/\lambda) + \mathbf{u}'_f(\mathbf{x}'/l)$. Here the first term is the homogeneous upward flow due to the choice of coordinate system which translates with the Stokes settling velocity U_s. The z axis of the frame of reference is directed downwards, $\mathbf{e}_3$ is a unit vector along it. The velocity $\mathbf{u}'_l$ characterizes the space-averaged large-scale motion of the mixture as a whole, and $\mathbf{u}'_f$ involves the fine-grained fluctuations with the lengthscale of order of the mean interparticle distance l or less. The approach based on the decomposition of the flow into several parts with different time- and lengthscales is widely used to study coherent structures in turbulent flows (Hussain 1983). These structures are large-scale quasi-deterministic flows and always originate from instability of some kind. The magnitude of the large-scale disturbances is assumed to be small, $\varepsilon \ll 1$. Fine-grained disturbances are finite in the small vicinity of each particle only. As a result it can be shown that the Reynolds stress due to the coupling between the large-scale and the fine-grained flows can be neglected. Thus the space-averaged dimensionless equations governing the large-scale flow can be linearized and reduced to the unsteady Oseen equations,

$$\boldsymbol{\nabla} \cdot \mathbf{u}_l = 0, \quad \frac{\partial \mathbf{u}_l}{\partial t} - \frac{\partial \mathbf{u}_l}{\partial z} + \boldsymbol{\nabla} p_l - \nabla^2 \mathbf{u}_l = \frac{9}{2\mathrm{Re}_s^2} c\mathbf{e}_3. \tag{2}$$

Here the space coordinates and the time are scaled by Oseen length- and timescales, $l_O = \nu/U_s \gg a$ and $t_O = \nu/U_s^2$. The averaged particle effect on the fluid flow is equivalent to the inhomogeneous volume forces due to momentum released into the fluid because of the particle motion. These volume forces are proportional to the concentration. Even though the fine-grained flow does not enter directly into (2) the microstructure of the suspension still affects the evolution of the large-scale flow via the momentum-source term in the rhs of the second equation (2) which depends on particles' positions and pairwise interactions.

To investigate the linear stability one can search the solution of (2) in the form of travelling wave, $c = c_0 + \varepsilon c^* \exp\left[\mathrm{i}\left(\omega t + \mathbf{k} \cdot \mathbf{x}\right)\right]$, $\mathbf{u}_l = \mathbf{u}^* \exp\left[\mathrm{i}\left(\omega t + \mathbf{k} \cdot \mathbf{x}\right)\right]$. The instability mechanism is due to microstructure of the suspension. Disturbances of particle density induce a fluid shear field. A pair of neighboring particles settles with a velocity which depends on their relative position. In turn, the pair orientations are affected by a fluid shear, so that the pairs seek some preferred orientation in this field. Such a pairs may translate toward the regions of increased particle density, and this results in further growth of density disturbances. The phenomenon is described in terms of pair probability density function P which is governed by the conservation equation,

$$\frac{\partial P}{\partial t} + \boldsymbol{\nabla}_x \cdot (\mathbf{U} P) + \boldsymbol{\nabla}_r \cdot (\dot{\mathbf{r}} P) = 0. \tag{3}$$

P is also sought in the form of small oscillating disturbance to the basic-flow function which is assumed to be random, $P_0(\mathbf{r}_{12}) = n_0^2$ as $r_{12} \geq 2a$. The amplitudes of probability function and concentration and mean particle number density n_0 are related by $c^* = \frac{c_0}{2n_0}\int P^* d\mathbf{r}$. Two velocities enter (3): the velocity of pair's centre of mass and the relative velocity of the two particles. The lengthscale for the flow induced by two close spheres is a particle radius a and is much smaller than λ. For this reason the previous solutions of the steady Stokes equations can still be applied to describe the pair motion:

$$U = -\mathbf{e}_3 + \varepsilon \mathbf{u}_l + \mathbf{v}_t(\mathbf{s}), \qquad \dot{\mathbf{r}} = \varepsilon \mathbf{R}_s(\mathbf{s}) \cdot \mathbf{E}_l(\mathbf{x}), \qquad \mathbf{s} = \mathbf{r}_{12}/a. \tag{4}$$

Here $\mathbf{E}_l = \frac{1}{2}\left[\nabla \mathbf{u}_l + (\nabla \mathbf{u}_l)^+\right]$ is a rate-of-strain tensor. The tensors $\mathbf{v}_t$ (Batchelor 1972) and $\mathbf{R}_s$ (Batchelor & Green 1972) depend on a vector separating the particle centres, $\mathbf{r}_{12} = \mathbf{x}_1 - \mathbf{x}_2$, and decay rapidly with s.

Thus we obtain the complete system for the amplitudes of the waves. The dispersion relation can be derived similarly to Koch & Shaqfeh (1989):

$$\frac{\gamma}{k^2 + \mathrm{i}(\omega - k_3)} \int W \frac{\mathbf{s} \cdot \mathbf{k}}{(k_3 - \mathbf{k} \cdot \mathbf{v}_t - \omega)\, s^2} \left\{ \mathbf{s} \cdot \left(\mathbf{I} - \frac{\mathbf{kk}}{k^2} \right) \cdot \mathbf{e}_3 \right\} d\mathbf{s} = 1, \tag{5}$$

where $\gamma = \frac{27c_0^2}{16\pi \mathrm{Re}_s^2}$ and function W is expressed in terms of $\mathbf{R}_s$. The only difference of the equation obtained is the imaginary term in the denominator. As a result the eigen values are also complex, and some of them correspond to unstable waves. The cause of instability is a phase change between the fluid velocity and concentration waves which is due to extra inertia terms taken into account in the governing equations.

The equation (5) is solved numerically. The three eigenvalues were found instead of two values in Koch & Shaqfeh (1989) work. The two eigenmodes are unstable and related by $\mathrm{i}\omega_1(k, \pi - \theta) = \mathrm{i}\omega_2(k, \theta)$, i.e. the imaginary parts of the two waves with the angles between the gravity and the wavevector, $\theta \equiv \arccos(k_3/k)$, and $\pi - \theta$ are the same while the real ones are opposite. The third mode is stable. Figure 1 (a) presents the complex frequency of a first-mode horizontal wave as a function of the wavenumber. The imaginary part is negative, i.e. the waves are unstable for all k. It has a minimum which corresponds to the most unstable wave. Figure 1 (b) shows the dependence of the complex frequency on the angle between the gravity and the wavevector. The waves propagating at small angles are stable. The most unstable waves are nearly horizontal.

3. Numerical simulation of sedimenting suspension

The effect of the fluid inertia studied by the linear-stability theory can be considered by means of simulations as well. The approach based on the Fourier analysis of the large-scale fluid flow can be extended to simulate the particles' motion in this field. The fluid flow in a cell with the periodic boundary conditions can be presented as the Fourier series, $\mathbf{u}_l(\mathbf{x}) = \sum_{\mathbf{k} \neq 0} \mathbf{u}^*(\mathbf{k}) \exp(\mathrm{i}\mathbf{k} \cdot \mathbf{x})$, with the wave vectors commensurate with the box sizes. The equations (1) give for the Fourier coefficients

$$\mathbf{u}^*(\mathbf{k}) = \tilde{u}(\mathbf{k}) \left(\mathbf{I} - \frac{\mathbf{kk}}{k^2} \right) \cdot \mathbf{e}_3,$$

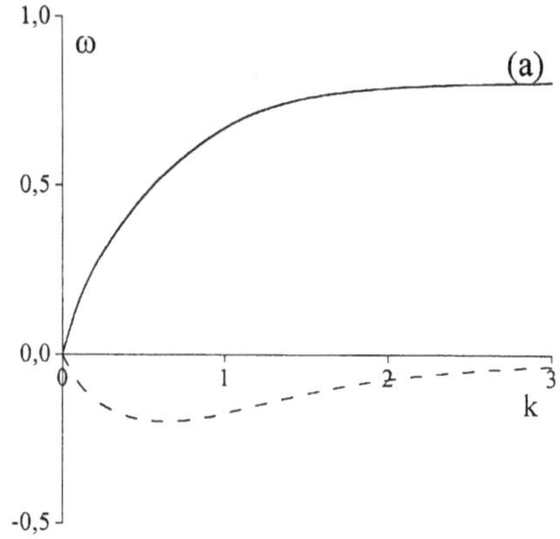

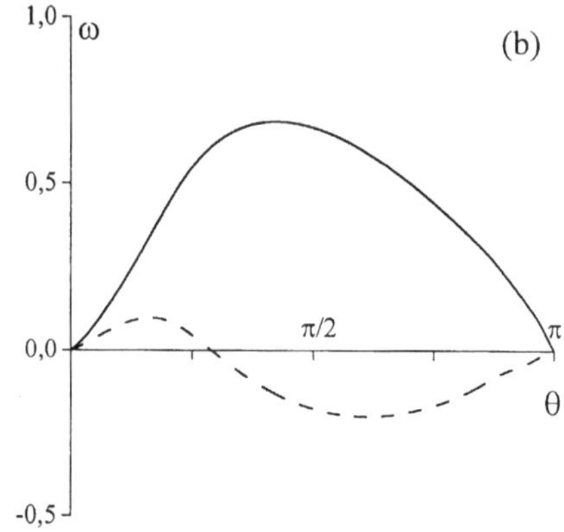

Figure 1: The complex frequency of the first mode at $\gamma = 1$ as a function (a) of k at $\theta = \pi/2$ (horizontal waves) and (b) of θ at $k = 1$. Solid and dashed lines are the real and imaginary parts respectively.

$$\frac{\partial u}{\partial t} - \left(ik_3 - k^2\right) u = \frac{3\mathrm{Re}_s}{4\pi^2 V} \sum_n \exp\left(-i\mathbf{k} \cdot \mathbf{x}_n\right), \tag{6}$$

where V is the cell volume. Equations of particles' motion are similar to (4):

$$\frac{\mathrm{d}\mathbf{x}_n}{\mathrm{d}t} = \mathbf{u}_l\left(\mathbf{x}_n\right) + \sum_i \left[\mathbf{v}_t\left(\mathbf{r}_{ni}\right) + \mathbf{R}_s\left(\mathbf{r}_{ni}\right) \cdot \mathbf{E}_l\left(\mathbf{x}_n\right)\right]. \tag{7}$$

Thus the equations governing the fluid flow are transformed to the system of ordinary differential equations which are integrated over t numerically together with (7) using Runge-Kutta routine.

The linear-stability analysis treats the growth of different modes separately. The lhs of the system (6) are also linear. However there is non-linear coupling between all the harmonics because of the hydrodynamic interactions of close particles and their migration. For this reason the exponential growth of some harmonics cannot be obtained in the simulations. Another difficulty is that the cell sizes in the simulations correspond to the short wavelengths for which the linear analysis predicts the very small growth rate. The instability mechanism exhibits as a tendency of pairs to drift to the regions of enhanced particle concentration. As the energy of the basic flow converts into the disturbance energy the time-averaged wave energy $\overline{u_k^2}$ for some harmonics should be greater than the ensemble-averaged values $\langle u_k^2 \rangle$ for random particle distribution.

The calculations are fulfilled for $c_0 = 5 \cdot 10^{-3}$, $\mathrm{Re}_s = 3 \cdot 10^{-3}$, the number of particles in the cubic cell, $\lambda = V^{1/3}$, is $N = 1000$. All harmonics $k_j = \frac{2\pi m_j}{\lambda}$ with the integers $|m_j| \leq 2$ are taken into account (total number of harmonics is 150). The results of calculations of the ratios $\overline{u_k^2}/\langle u_k^2 \rangle$ vs dimensionless time are shown in Figure 2 for the horizontal and oblique waves propagating at a $\theta = 45°$. The energy of the horizontal waves (see Figure 2 (a)) changes very slowly during a long time. It always remains larger than the random value for one of the wave

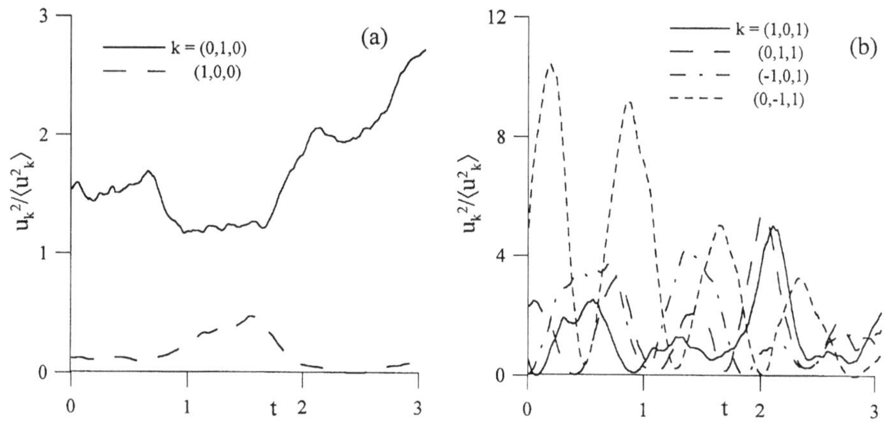

Figure 2: The wave energy as a function of dimensionless time for (a) horizontal, $\mathbf{k} = \frac{2\pi}{\lambda} \times (0,1,0)\,,(1,0,0)$, and (b) oblique, $\mathbf{k} = \frac{2\pi}{\lambda} \times (1,0,1)\,,(0,1,1)\,,(-1,0,1)\,,(0,-1,1)\,,$ waves.

while for another it is always less. Nevertheless the time averaged total energy of the two equivalent waves is close to the random value.

Quite another behaviour is obtained for the oblique waves propagating at a 45° angle to gravity (see Figure 2 (b)). The energy of some harmonics grows to very large values than decays to very small. The coupling between the harmonics can also be seen. As one harmonics decays the another grows. The total time-averaged energy of these four equivalent waves is approximately two times greater than the random value. This means that the mechanism of energy transfer from the basic motion to the disturbances takes place indeed. The calculations for another harmonics show that their energies are close to the random ones.

Figure 3 shows the fluid velocity field for a random initial particle distribution ($t = 0$) across the whole cell width and that calculated at $t = 0.72$. The horizontal wavevectors induce the vertical velocity disturbances only and such a disturbances are dominant for random distribution (see Figure 3 (a)). After some time the amplitudes of oblique waves exceed the random values. This results in the growth of the horizontal components of fluid velocity, and the vortical structures become more pronounced (see Figure 3 (b)).

3. Conclusion

A stability problem for a dilute sedimenting suspension (Koch & Shaqfeh 1989) is reconsidered. The unsteady effects cause the suspension to become unstable to particle concentration disturbances. The unsteady evolution of large-scale fluid flow is studied also using the numerical simulation of particles' motion in the periodic cell. Simulations demonstrate an instability of suspension.

The work was supported by Russian Foundation for Fundamental Research (Grant No. 02-01-00149).

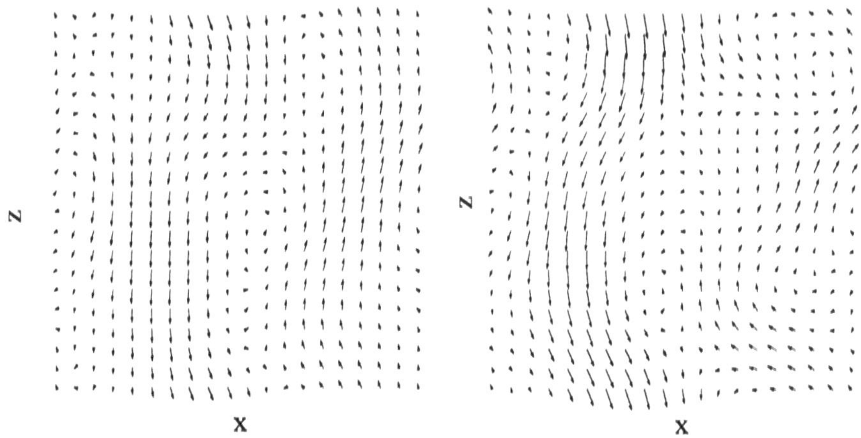

Figure 3: Fluid velocity fields at $y = \lambda/2$ for (a) $t = 0$ and random initial particle distribution in the cell and (b) for $t = 0.72$.

References

Asmolov, E. S. 2001 Instability of a suspension of sedimenting spherical particles. *Submitted to Eur. J. Mech.B/Fluids.*.

Batchelor, G. K. 1972 Sedimentation in a dilute dispersion of spheres. *J. Fluid Mech.* **52**, 245–268.

Batchelor, G. K. & Green, J. T. 1972 The hydrodynamic interactions of two small freely-moving spheres in a linear flow field. *J. Fluid Mech.* **56**, 375–400.

Hussain, A. K. M. F. 1983 Coherent structures - reality and myth. *Physics of Fluids* **26**, 2816–2850.

Koch, D. L. & Shaqfeh, E. S. G. 1989 The instability of a dispersion of sedimenting spheroids. *J. Fluid Mech.* **209**, 521–542.

Saffman, P. G. 1973 On the settling speed of free and fixed suspensions. *Stud. Appl. Math.* **52**, 115–127.

Segre, P. N., Herbolzheimer, E. & Chaikin, P. M. 1997 Long ranged correlations in sedimentation. *Phys. Rev. Lett.* **79**, 2574–2577.

SIMULATION OF PARTICULATE FLOW USING PARTICLE METHODS A P³M ALGORITHM FOR CHARGED PARTICULATES

J. H. WALTHER, S. KERN AND P. KOUMOUTSAKOS
Inst. of Computational Science, ETHZ, 8092 Zürich, Switzerland, walther@inf.ethz.ch

We consider particulate incompressible flows with a two-way coupling. The simulation of these flows is facilitated using vortex particle methods [1] as it tracks adaptively the vorticity generated by the two-way coupling [2]. In addition, a key aspect of the present work is the simulation of charged particles, using efficient P^3M methods to account for the singular interactions.

Assuming an incompressible fluid with constant kinematic viscosity (ν) the fluid motion may be described by the vorticity transport equation

$$\frac{D\boldsymbol{\omega}}{Dt} = (\boldsymbol{\omega} \cdot \nabla)\boldsymbol{v} + \nu\nabla^2\boldsymbol{\omega} + \boldsymbol{\phi}, \tag{1}$$

where $\boldsymbol{v}$ is the velocity, $\boldsymbol{\omega} = \nabla \times \boldsymbol{v}$ the fluid vorticity, and $D/Dt = \partial/\partial t + \boldsymbol{v} \cdot \nabla$ the material derivative. The vorticity source term ($\boldsymbol{\phi}$) is given by

$$\boldsymbol{\phi} = \frac{1}{\rho}\nabla \times \boldsymbol{f}, \tag{2}$$

where $\boldsymbol{f}$ the body force per unit volume, and ρ the fluid density. The fluid velocity is recovered from the vorticity through a Poisson equation with the aid of a solenoidal vector potential, $\boldsymbol{\Psi}$ (such that $\boldsymbol{v} = \nabla \times \boldsymbol{\Psi}$)

$$\nabla^2\boldsymbol{\Psi} = -\boldsymbol{\omega}. \tag{3}$$

The motion of the (N_p) solid particles is governed by Newton's equation of motion:

$$\begin{aligned} \frac{\mathrm{d}\boldsymbol{x}_p}{\mathrm{d}t} &= \boldsymbol{u}_p, \\ \rho_p vol_p \frac{\mathrm{d}\boldsymbol{u}_p}{\mathrm{d}t} &= \boldsymbol{f}_p, \end{aligned} \tag{4}$$

where $vol_p = \pi d_p^3/6$ is the volume of the particles, $\boldsymbol{f}_p$ is the total force acting on the particle, $\boldsymbol{u}_p$ the particle velocity, and ρ_p the particle density. In the present study the total force is comprised of viscous drag forces $\boldsymbol{f}_d = \frac{\pi}{8}\rho C_d d_p^2(\boldsymbol{v}(\boldsymbol{x}_p) - \boldsymbol{u}_p)|\boldsymbol{v}(\boldsymbol{x}_p) - \boldsymbol{u}_p|$, gravity $\boldsymbol{f}_g = vol_p(\rho_p - \rho)\boldsymbol{g}$, and electrostatic forces $\boldsymbol{f}_E = q\boldsymbol{E}(\boldsymbol{x}_p)$, and Re_p is the particle Reynolds number ($|\boldsymbol{u}_p - \boldsymbol{v}(\boldsymbol{x}_p)|d_p\nu^{-1}$), $C_d = C_d(\mathrm{Re}_p)$ is the drag coefficient [2], $\boldsymbol{E}$ is the electrostatic field induced by the presence of the charged particles, and $\boldsymbol{g}$ is the acceleration due to gravity. Notice, the vorticity source term (Eq. (2)) contains contributions from the fluid forces only, here $\boldsymbol{f} = \boldsymbol{f}_d$. The "proactive" method of Sundaram & Collins [3] is employed to handle particle collisions.

The vorticity field is discretized using Lagrangian vortex particles,

$$\boldsymbol{\omega}(\boldsymbol{x}) = \sum_p \eta_\sigma(\boldsymbol{x} - \boldsymbol{x}_p)\boldsymbol{\alpha}, \tag{5}$$

where $\boldsymbol{\alpha}$ is the strength of the vortex elements, and $\eta_\sigma(\boldsymbol{x}) = \sigma^{-3}\eta(|\boldsymbol{x}|/\sigma)$ is a smooth approximation to the Dirac delta function. The vorticity transport equation is solved using a fractional step algorithm [4] with the two-way coupling implemented through the vorticity source term in the viscous sub-step of the algorithm [2]. The N-body problem is

A. Gyr and W. Kinzelbach (eds.), Sedimentation and Sediment Transport, 165–168.

solved using the CIC [5] or Vortex-In-Cell (VIC) algorithm [6] with a second order finite-difference discretisation of the governing Poisson equation (3). The mesh vorticity field is constructed from the particles using the projection

$$\boldsymbol{\omega}(\boldsymbol{x}_m) = \frac{1}{h^3} \sum_p W(\boldsymbol{x}_m - \boldsymbol{x}_p)\, \boldsymbol{\alpha}_p. \tag{6}$$

where $\boldsymbol{x}_m$ denotes the mesh points, and h the mesh spacing, and $W(x)$ is the interpolation kernel. After the solution of the Poisson equation, the mesh velocity is computed from the streamfunction and interpolated back to the vortex particles using

$$\boldsymbol{v}_p = \sum_m W(\boldsymbol{x}_p - \boldsymbol{x}_m)\, \boldsymbol{v}_m. \tag{7}$$

The moment conserving M'4 formula proposed by Monaghan [7] is used for W in both Eqs. (6) and (7).

A similar procedure is employed for the Poisson equation relating the electrostatic potential (ϕ) and the charge density ρ, $\nabla^2\phi = -\rho/\epsilon_0$, where ϵ_0 is the permittivity in vacuum. However, the CIC algorithm is incapable of resolving the $1/r$ singularity of the electrostatic force due to the smoothing incurred during the projection steps Eqs. (6, 7) and the anisotropy introduced by the finite-differences. The proposed influence matrix P^3M algorithm [8] annihilates the CIC resolved field for particles that are close and adds specifically the exact $1/r$ force field in the particle-particle correction step of the algorithm. Thus the mesh resolved electrostatic field $\boldsymbol{E}_m$ induced by a mesh resolved charge density ρ_m is computed for the neighbouring grid cells during the projection step (6), thus

$$\boldsymbol{E}_m = \mathrm{C}\rho_m, \tag{8}$$

where C is the influence matrix. The matrix is formed during the initialisation step of the algorithm by placing a test particle of unit charge at different position within a mesh cell and solving the field equation (3) using the Poisson solver employed in the CIC step of the algorithm. The size of the matrix is determined by the number of neighbouring cell (L) included in the correction and the width of the interpolation kernel (M). The electrostatic field given by (8) is interpolated to the particles and subtracted to cancel the subsequent CIC contribution. Thus, the resulting electrostatic interactions include contributions from distant particles only, the nearby contributions being annihilated through (8). The subsequent particle-particle correction term involves the exact $1/r$ contribution.

To demonstrate the capabilities of the proposed algorithm we present preliminary studies [9] of a spherical suspension of randomly charged particles falling due to gravity in an initially quiescent flow. In previous studies [2], the evolution of the neutral system was found to resemble that of a two immisible fluids with an elevated density and viscosity of the suspension. During the descent, the neutral suspension deforms cf. Fig. 1, and depending on the initial vorticity field, is entrapped by the dynamics of the flow forming a vortex ring cf. Fig. 2a,b. The neutral system starting from rest is unstable and induces a bifucation on the suspension of particles which in turn creates new vortex rings in a cascade like fashion cf. Fig. 2c,d. For weakly charged particles ($f_E \approx 0.01 f_d$) the suspension descents undisturbed (Fig. 3a,b), whereas strongly charged particulates ($f_E \approx f_d$) result in chain formation as demonstrated in Fig. 3c. A detailed parameter study is currently being considered to quantify the influence of the charge magnitude and charge distribution.

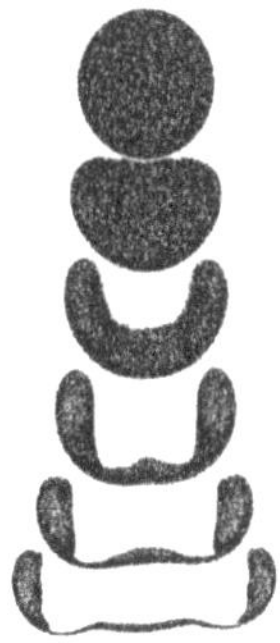

Figure 1: Cross section of neutrally charged suspension falling in an initially quiscent flow [2]. The snapshots are takes at a non-dimensional time of: a) $tU/d_p = 0.00$; b) $tU/d_p = 7.18$; c) $tU/d_p = 10.77$; d) $tU/d_p = 14.36$; e) $tU/d_p = 17.94$; f) $tU/d_p = 21.53$, where U is the Hadamard and Rybczyński velocity of the corresponding immisible drop [10]. The total number of solid particle is 10^5.

References

[1] G.-H. Cottet and P. Koumoutsakos. *Vortex Methods – Theory and Practice.* Cambridge University Press, New York, 2000.

[2] J. H. Walther and P. Koumoutsakos. Three-dimensional particle methods for particle laden flows with two-way coupling. *J. Comp. Phys.*, 167:39–71, 2001.

[3] Shivshankar Sundaram and Lance R. Collins. Numerical considerations in simulating a turbulent suspension of finite-volume particles. *J. Comp. Phys.*, 124:337–350, 1996.

[4] Alexandre Joel Chorin. Numerical study of slightly viscous flow. *J. Fluid Mech.*, 57:785–796, 1973. Part 4.

[5] Charles K. Birdsall and Dieter Fuss. Clouds-in-clouds, clouds-in-cells physics for many-body plasma simulation. *J. Comp. Phys.*, 3:494–511, 1969.

[6] J. P. Christiansen. Numerical simulation of hydrodynamics by the method of point vortices. *J. Comp. Phys.*, 13:363–379, 1973.

[7] J. J. Monaghan. Particle methods for hydrodynamics. *Comp. Phys. Repts.*, 3:71–123, 1985.

[8] J. H. Walther. An influence matrix particle-particle particle-mesh algorithm with exact particle-particle correction. *J. Comp. Phys.*, page accepted, 2002.

[9] Stefan Kern. Simulations of falling drops containing dispersed particles. semester project, ETH, July 2001.

[10] G. K. Batchelor. *An Introduction To Fluid Dynamics.* Cambridge University Press, 1 edition, 1967.

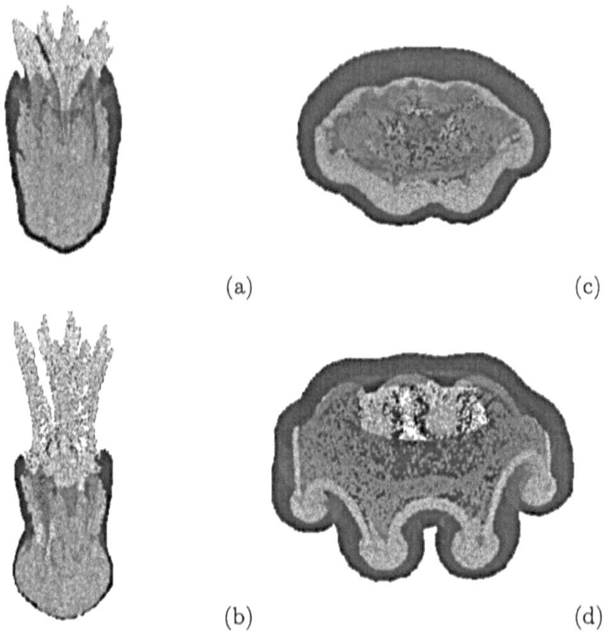

Figure 2: Snapshot (perspective view) from simulations of a neutrally charged falling blob of particles (from [2]). The solid particles shown as white spheres, create vorticity (in red) during their descent which forms a vortex ring. For suspension falling in a non-zero initial vorticity field (a,b), generated by allowing, for a short time, only a one-way viscous coupling, the particle suspension is entrapped by the initial vorticity field which dominates the dynamics of the flow. In the second study the suspension is inserted in a fluid with no initial vorticity field (c,d), and the falling of the suspension imparts vorticity on the fluid elements forming a vortex ring. The vortex ring becomes unstable and new vortex rings are formed along the circumference of the ring in a cascade like fashion.

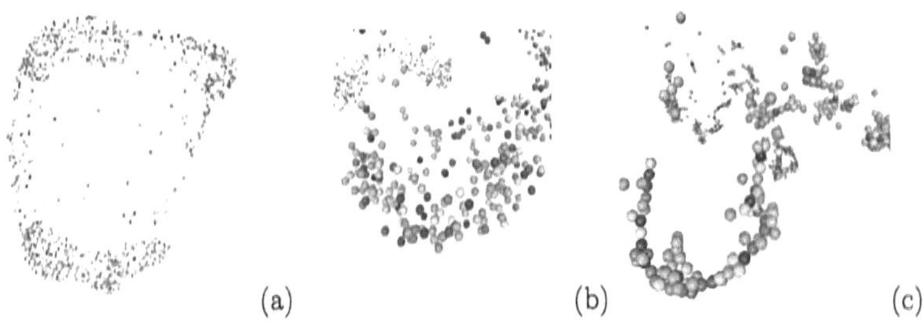

Figure 3: Snapshots of a weakly charged blob (a: topview) (b: close up) and strongly charged blob (c). The particles are coloured according to the charge (red: positive charge, blue: negative charge; green: neutral).

VI

Self Organization in Sediment Transport

LDV, PIV AND LES INVESTIGATION OF FLOW OVER A FIXED DUNE

R. BALACHANDAR[1], C. POLATEL[2], B-S. HYUN[3], K. YU[4], C-L. LIN[5], W. YUE[6] & V.C. PATEL[7]
IIHR – Hydroscience & Engineering[8], The University of Iowa, Iowa City, IA 52242-1585, USA
Contact: v-c-patel@uiowa.edu

1. Introduction

Investigations of flow over sand dunes are too numerous to list here. However, Table 1 gives a summary of some recent experimental studies of flow over fixed dune-like disturbances in open channels. All of them have used LDV to explore various facets of the flow. The present investigation was motivated by the need for a more complete picture of the flow than is available from these previous studies.

We seek to answer a number of questions by combining the complementary capabilities of LDV, PIV and LES. For example, in the treatment of dunes as bed roughness, it is necessary to have knowledge of the friction and pressure distributions on the dune. This information cannot be readily obtained, if at all, by experimental means. Similarly, for advances in modeling of sediment transport at a fundamental level, it is necessary to have information on the so-called coherent structures. Quantitative information about these structures is also difficult to obtain in the laboratory. Numerical modeling may claim to address these and other issues in a more comprehensive way, but it limited by *a priori* assumptions and computer resources. For example, CFD models based on the Reynolds averaged equations are limited by assumptions of turbulence models and fail to provide any information on the coherent structures responsible for bulk of the sediment transport. While LES has the potential to resolve these structures, it also suffers from limitations of sub-grid scale models and inadequate grid resolution. Experimental data required to improve and validate numerical models are best obtained by LDV and PIV, in spite of their limitations.

2. Experimental Arrangement

The experiments are designed to take advantage of the LDV and PIV systems available at IIHR. A train of two-dimensional fixed dunes (some made of Plexi-glass) is attached to the bottom of a 10-m long, 0.61-m wide, re-circulating open channel flume. The dunes are identical in shape and size, and geometrically similar to those used in the experiments of Mierlo & de Ruiter (1988) and in the subsequent numerical studies of Yoon & Patel (1996) and others. LDV and PIV measurements are carried out in the flow over

[1] Visiting Professor, University of Saskatchewan, Saskatoon, Canada
[2] Graduate Research Assistant
[3] Visiting Associate Professor, Korea Maritime University, Pusan, Korea
[4] Graduate Research Assistant
[5] Assistant Professor, Mechanical & Industrial Engineering
[6] Graduate Research Assistant
[7] Director and Professor
[8] Formerly Iowa Institute of Hydraulic Research

A. Gyr and W. Kinzelbach (eds.), Sedimentation and Sediment Transport, 171–178.

the 17th and 18th dunes where the flow is space-periodic, i.e., profiles of mean velocity and turbulence parameters at identical locations on adjacent dunes are identical in shape.

The Reynolds number based on the dune height h and the maximum velocity U_o, which occurs close to the free surface, namely, $Re = U_oh/v$, is held constant nearly constant at 1.0×10^4, but the flow depth d over the dune crest is varied over the range $3<d/h<8$. The wavelength of the dunes is $\lambda/h = 20$, as in a number of previous experiments (Table 1). The measurements are made in the mid-plane of the flume, where the flow in nominally two-dimensional. The sketch in Figure 1 shows the overall flow field and coordinates.

A two-component fiber optic LDV system of conventional design is used. At each measurement point, 15,000 samples are obtained and a standard procedure is used to determine the averages. For the PIV, mirrors and lenses are used to generate a 3-mm thick vertical light sheet from a 5W Laser. The field-of-view, 120 mm x 80 mm in area, is illuminated from the bottom of the flume and imaged with a 640 x 480 pixel CCD camera. The light sheet is translated in the streamwise direction to capture overlapping areas of the flow. Commercial image-analysis software based on a gray-level cross-correlation technique is used to obtain the velocity vectors. A total of 1,000 frames are used to compute the mean and turbulence quantities. Other details of the LDV and PIV systems and data processing are reported in Hyun et al. (2002) along with estimates of measurement uncertainties.

3. Numerical Modeling with LES

The three-dimensional, unsteady, incompressible, filtered continuity and Navier-Stokes equations are solved in conjunction with a dynamic sub-grid-scale eddy-viscosity model. The finite-volume LES numerical code of Cui et al. (2000) provided the starting point. They have conducted extensive simulations in closed channel flows with sinusoidal wavy walls and walls fitted with transverse ribs. The present calculation for flow over a dune is a relatively simple extension of that work because the free surface is not explicitly treated. Instead, calculations are made for a closed channel of twice the depth, with dune shaped top and bottom walls. The simulation domain is one dune wavelength ($\lambda = 20h$) long and 0.625λ (= 12.5h) wide. The flow is driven by a mean streamwise pressure gradient that matches the Reynolds number of the experiments. No-slip conditions are applied at the top and bottom walls, periodic conditions are applied at the inlet and outlet, and also along the sides of the solution domain at each instant. A non uniform grid with 80 x 128 x 48 points in the (x,y,z) directions is used. The first grid point near the walls is set at two wall units. Convergence is claimed at each time step when the sum of the pressure and viscous resistance is in balance with the imposed pressure difference between the duct inlet and outlet. After the flow reaches "steady-state", the solution is continued for about five large-eddy turnover times ($2d/u_\tau$, where u_τ is the friction velocity) to accumulate the statistics. A typical simulation takes about 300 hours of CPU time on a Dell Dimension 8200 (1.8G CPU) computer.

4.1 TIME-AVERAGED FLOW

Figure 2 shows the mean streamlines constructed from PIV and LES for $d/h = 7$ in a small region around flow separation at the dune crest (plotted to the same scale). There is considerable similarity between the two sets of results although the center of the separation eddy predicted by LES is further upstream than measured by PIV. In neither case there is a well-defined separation streamline that originates from the crest and ends on the bed, making it difficult to precisely identify a mean reattachment point by this means. Visual placement of this point is deceptive because the near-wall streamlines in the separation region, in both cases, suggest the possibility of flow converging into the plane of symmetry.

The LES results are processed to determine the variation of the local (time-averaged) pressure and friction coefficients, C_p and C_f, respectively, defined in the usual way, along the dune. These are plotted in Figure 3. Also shown are the r.m.s. of the fluctuations, $\sqrt{c_p'^2}$ and $\sqrt{c_f'^2}$, obtained from LES. It should be noted that these are *only the resolved-scale contributions*. While the mean pressure can be measured by pressure taps on the wall, there is no reliable way to measure the friction coefficient in a flow with pressure gradients and separation. The mean velocity profiles measured by LDV were analyzed by Clauser plots (U/U_o vs log $U_o y/\nu$) to look for semi-log regions from which a friction coefficient could be inferred. Such a region was found only at a measurement station just upstream of the crest. The only experimental data point obtained in this manner is found to be in good agreement with the LES results, as shown in Figure 3a. The remainder of the figure shows that the mean reattachment point, identified by $C_f = 0$, is at $x/h = 4.5$, quite different from that might be inferred from a visual inspection of the streamlines. A detailed analysis of the near-wall LDV data also indicate that the mean reattachment point is close to $x/h = 4.5$, in agreement with LES. The mean pressure distribution is similar in shape to that predicted by Reynolds-averaged models (Yoon & Patel, 1996). The pressure and friction distributions together provide the information necessary to predict the components of bed resistance. Information on fluctuations in the pressure and friction coefficients, provided only by LES, is entirely new and may be useful in developing models that seek to connect local flow properties with sediment movement at the bed. It is interesting to note the similarity in the distributions of the pressure and friction fluctuations over most of the dune. This is perhaps an indication of a high level of organization in the overlaying flow (the vortices described below). Both quantities show large peaks near the mean reattachment point ($x/h = 4.5$) and a gradual decrease toward the dune crest.

Figure 4 shows the profiles, at a station just upstream of reattachment, of mean velocity (U) in the x-direction, the r.m.s. velocity fluctuations ($\sqrt{\overline{u'^2}}$, $\sqrt{\overline{v'^2}}$) in the (x,y) directions, and the Reynolds shear stress ($-\overline{u'v'}$), all normalized by the maximum velocity U_o. The measurements by LDV and PIV are in agreement considering the different values of d/h in the experiments and the expected uncertainties in the data. The LES results for the longitudinal fluctuations and the shear stress are lower that the experiments due, in part, to the fact that they are only the resolved-scale contributions, the effect being the largest in correlations involving the x-component. The profiles show the large peaks in the turbulence parameters associated with the shear layer emanating from the separation at the crest. The much smaller peaks farther out are the remnants of the shear layer generated by the previous dune.

4.2 INSTANTANEOUS FLOW

PIV and LES offer a very graphic view of the instantaneous structures in the flow, as shown in Figure 5. A series of instantaneous velocity fields constructed from PIV and LES are plotted to nearly the same scale. For PIV, these are successive data sets separated by the measurement time interval of 1/30 sec. The LES data, obtained with much smaller time steps, were processed to plot the velocity field at the same time interval of 1/30 sec. It is important to note that these PIV and LES results were selected from a large amount of data to show certain flow features that appear to repeat at random intervals. Although there is no obvious periodicity, similar sequence of events can be identified in the PIV as well as LES data.

The most obvious flow feature is the formation of vortices just downstream of the crest below the shear layer, and their growth as they travel downstream. There is no evidence of instability-induced vortices and their pairing as described by Muller & Gyr (1986). If present, they are not resolved by either PIV or LES. The PIV shows better-defined vortices than LES for reasons yet to be explored. One reason for this difference could be that the PIV data come from an open channel flow while the LES is carried out for a closed channel and does not contain any explicit effect of the free surface. The difference in the flow depth in the two cases is not thought to be a significant factor but only LES that resolves the true free surface can provide more conclusive evidence.

Although there is no periodicity in the flow, it should be possible to determine length and time scales of these vortices by further analysis of the PIV and LES results for a proper quantitative validation of LES. Instantaneous velocity fields measured by PIV and computed by LES in turbulent flow require much further analysis using different transformations to reveal different flow features (see, for example, Adrian et al., 2000). At a qualitative level it is seen that, in the presence of a movable sand bed, sediment thrown into the shear layer at the crest would be brought into the "mean reattachment" zone by the vortices to be transported, intermittently, to both sides of reattachment. Both PIV and LES data on the ramp of the dune (downstream of the field of view shown in Figure 5) indicate weakening and diffusion, and some times breakup and reformation, of the vortices, generating an intermittent flow reminiscent of sweep and ejection events in canonical wall-bounded turbulent flows. These events are displayed much more graphically in movies.

5. Concluding Remarks

This is a very brief account of a comprehensive investigation that has led to a wealth of data in this idealized case of a fixed, two-dimensional dune. More detailed description of the experiments, simulations, and analysis of results will be presented in subsequent publications. The LES code is currently being extended for a more direct treatment of the free surface using the level-set approach. This study provides a starting point for future investigations with a movable bed and sediment transport, and interaction of the bed with the free surface.

This study was initiated during a period when the first and third authors spent their sabbatical leaves at IIHR. The authors acknowledge the cooperation and assistance of Dr. Marian Muste in the conduct of this study.

References

Adrian, R.J., Christensen, K.T. & Liu, Z-C., 2000, Analysis and interpretation of instantaneous turbulent velocity fields, Experiments in Fluids, Vol. 29, pp. 275-290.

Bennett, S. J. & Best, J. L., 1995, Mean flow and turbulence structure over fixed two-dimensional dunes: implications for sediment transport and bedform stability, Sedimentology, Vol. 42, pp. 491-513.

Cui, J., Patel, V.C. & Lin, C-L., 2000, Large-eddy simulation of turbulent flow over rough surfaces, Iowa Inst. Hydr. Res., The University of Iowa, IIHR Report 413.

Hyun, B-S., Balachandar, R., Yu, K. & Patel, V.C., 2002, Relative assessment of LDV and PIV to measure mean velocity and turbulence in water flow, Submitted to Experiments in Fluids.

Lyn, D. A., 1993, Turbulence measurements in open channel flows over artificial bed forms, J. Hydraulic Engineering, ASCE, Vol. 119, pp. 306-326.

McLean, S. R., Nelson, J. M. & Wolfe, S. R., 1994, Turbulence structure over two-dimensional bed forms: Implications for sediment transport, J. Geophysical Research, Vol. 99, No. C6, pp. 12729-12747.

Mierlo, M. C. L. M. & de Ruiter, J. C., 1988, Turbulence measurements over artificial dunes, Report Q789, Delft Hydraulics Laboratory, Delft, The Netherlands.

Muller, A. & Gyr, A., 1986, On the vortex formation in the mixing layer behind dunes, J. Hydraulic Research, Vol. 24, No. 5, pp. 359-375.

Nelson, J. M. & Smith, J. D., 1989, Mechanics of flow over ripples and dunes, J. Geophysical Research, Vol. 94, No. C6, pp. 8146-8162.

Nelson, J. M., McLean, S. R. & Wolfe, S. R., 1993, Mean flow and turbulence fields over two dimensional bed forms, Water Resources Research, Vol. 29, No. 12, pp. 3935-3953.

Smith, B. T. and Ettema. R., 1995, Ice cover influence on flow and bedload transport in dune-bed channels, Iowa Inst. Hydr. Res., The University of Iowa, IIHR Report 374.

Yoon, J. Y. & Patel, V. C., 1996, Numerical model of turbulent flow over sand dune, J. Hydraulic Engineering, ASCE, Vol. 122, No. 1, pp. 10-18.

Weiberg, P. and Nelson, J. M., 1992, Unidirectional flow over asymmetric and symmetric ripples, J. Geophysical Research, Vol. 97, No. C8, pp. 12745-12761.

Table 1: Recent LDV measurements in flow over dunes

Author (s)	Dune Arrangement and Surface Conditions	Wavelength, λ/h	Flow Depth, d/h
Meirlo & de Ruiter, 1988	33 dunes, measurements on 17th dune Sand plastered on concrete	20	3.65, 2.63
Nelson & Smith, 1989	5 dunes, measurements on 4th dune Concrete	20	5
Wiberg & Nelson, 1992	20 ripples, measurements after 15th ripple	8	6, 11
Lyn, 1993	Long train, measurements on 60th dune Sand plastered on wood	12.5	~ 5
Nelson et al., 1993	5 dunes, measurements on 4th dune Concrete dunes	20	~ 5
McLean et al., 1994	20 dunes, measurements on 16th dune Cement dunes	20	~ 4, 5, 14
Bennett & Best, 1995	7 dunes, measurements on 5th dune Concrete	15.75	~ 2.5
Smith & Ettema, 1995	Train of sand dunes at 17 m from the flume entrance	23.6	3.5
Present, 2002	22 dunes, measurements on 17th dune Smooth surface	20	3 - 8

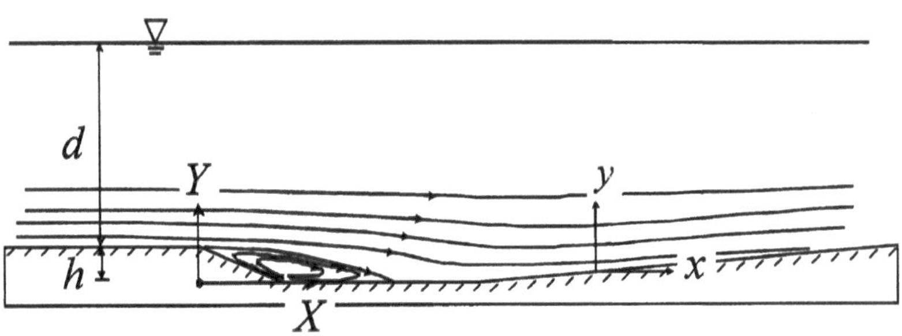

Fig. 1. Schematic and coordinate system

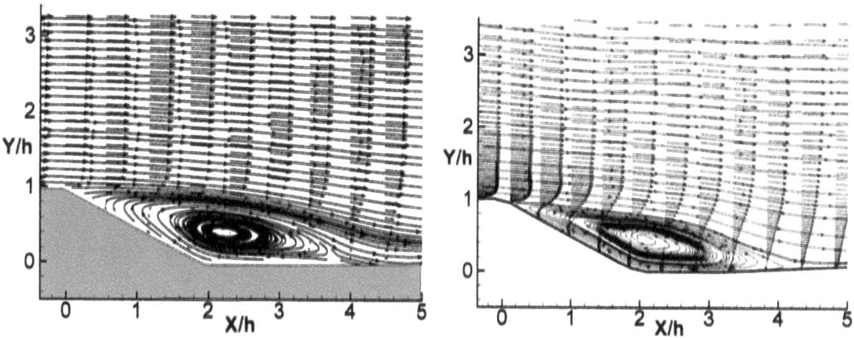

(a) PIV, d/h=7 (b) LES, d/h=6

Fig. 2 Velocity vectors and streamlines

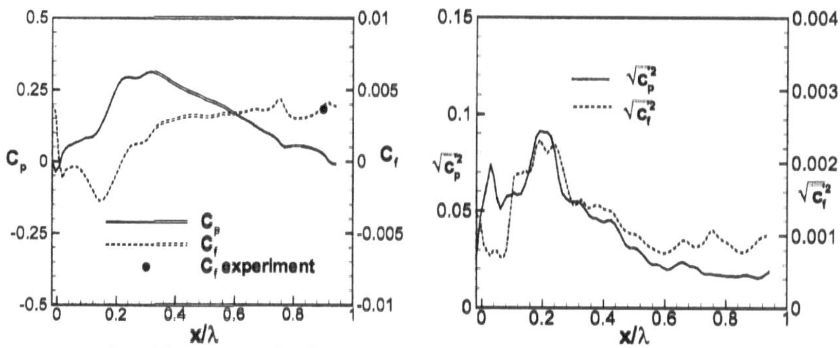

Fig. 3 Time-averaged and r.m.s. pressure and friction coefficients from LES

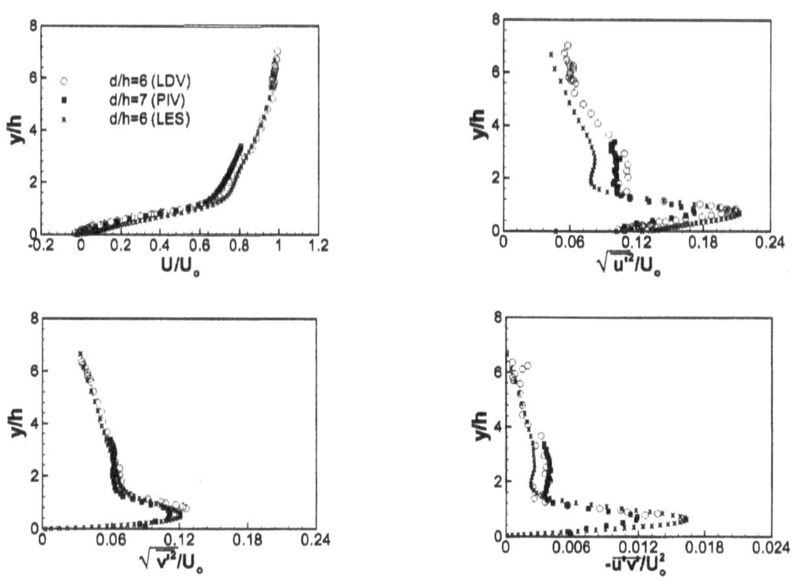

Fig. 4 Comparison of LDV, PIV and LES results at x/h=4

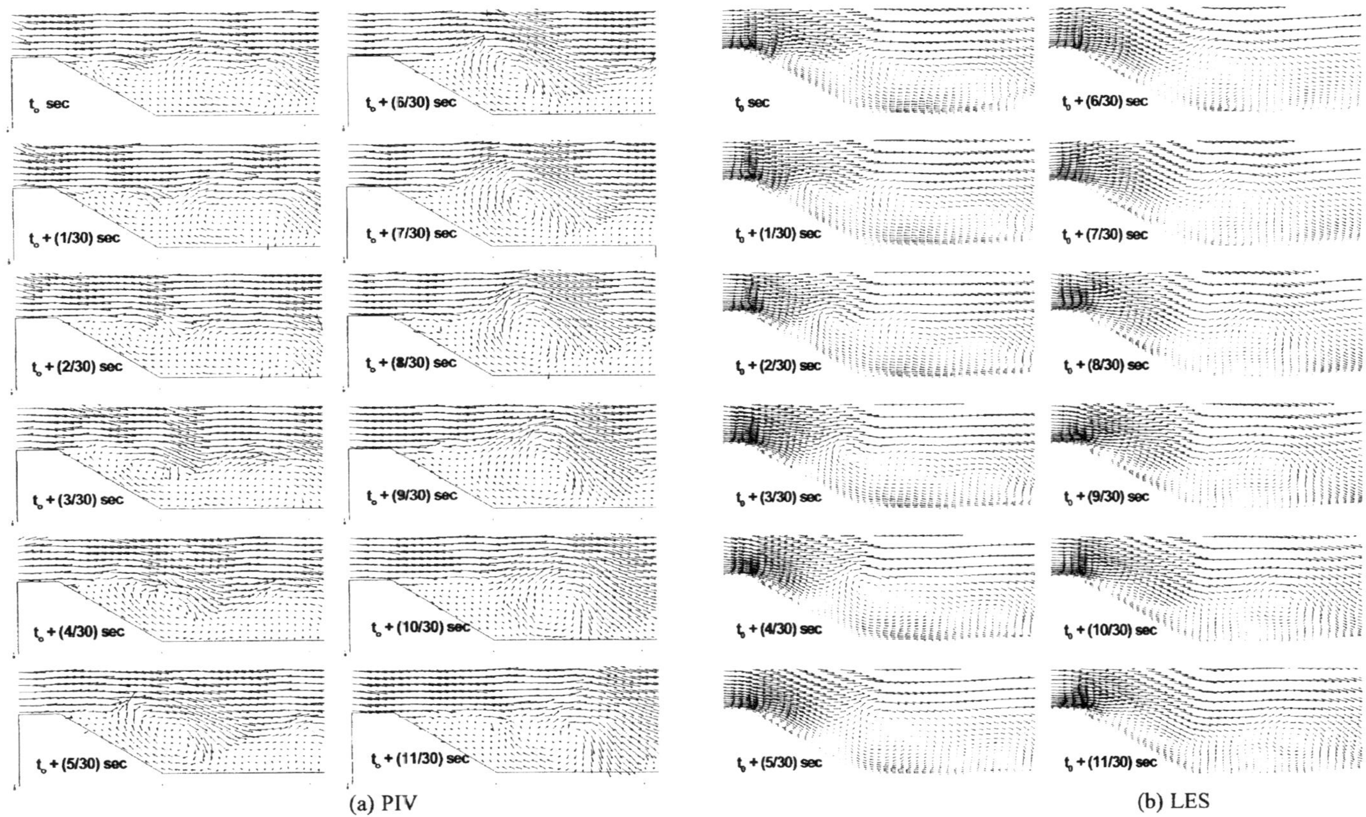

Fig. 5 Time sequence of instantaneous velocity field

SEDIMENT CONTINUITY FOR RIVERS WITH NON-UNIFORM SEDIMENT, DUNES, AND BED LOAD TRANSPORT

ASTRID BLOM[1], JAN S. RIBBERINK[2], GARY PARKER[3]
[1] *Research conducted at Department of Civil Engineering, University of Twente, The Netherlands. Presently at WL Delft Hydraulics, P.O. Box 177, 2600 MH Delft, The Netherlands, e-mail astridblom@freeler.nl*
[2] *Department of Civil Engineering, University of Twente, The Netherlands*
[3] *St. Anthony Falls Laboratory, University of Minnesota, USA*

1. Introduction

Models describing the interaction among grain size-selective sediment transport, the vertical sorting profile within the bed, and bed level changes constitute a critical component of morphological model systems for rivers with non-uniform sediment. This interaction is described in terms of sediment continuity models. Hirano (1971) was the first to develop such a sediment continuity model. In the Hirano active layer model, the bed is divided into a homogeneous top layer, i.e. the active layer, and an inactive substrate. Only sediment in the active layer interacts with the flow and participates in the transport process. Sediment fluxes between the active layer and the substrate occur only when there is a change in the average bed level. In the last few decades, a number of variants to the Hirano active layer model have been proposed in order to overcome several shortcomings and obtain a better description of vertical sediment sorting mechanisms within the bed.
Within the present research, two sets of flume experiments have been conducted to obtain a better insight in these sediment sorting mechanisms (Blom et al., in press). Migrating bedforms such as dunes are mostly characterised by a downward coarsening trend, which results from the avalanching of grains down the lee face. When conditions of partial transport prevail and a significant amount of coarse material does not or barely takes part in the transport process, coarse sediment gathers below the migrating bedforms, as shown in Figure 1. Another mechanism that contributes to the formation of such a coarse bed layer is the winnowing of fines from the trough surface and subsurface.
The Hirano active layer model and its variants do not account for these vertical sorting processes. Moreover, in most sediment continuity models, vertical sediment fluxes only occur when there is a change in the average bed level, whereas flume

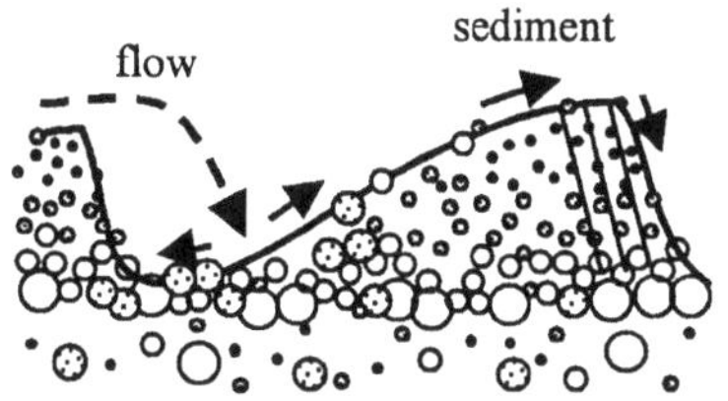

Fig. 1 Downward coarsening of bedform material and the presence of a coarse bed layer below migrating bed forms

experiments have shown that vertical sorting fluxes also occur through the migration of dunes and the variability in trough elevations (Crickmore and Lean, 1962, Ribberink, 1987, Blom et al., in press). Another problem of sediment continuity

A. Gyr and W. Kinzelbach (eds.), Sedimentation and Sediment Transport, 179–182.

models with discrete bed layers is that in certain situations, the set of equations may become elliptic and thus unsolvable (Ribberink, 1987). Apparently, in such circumstances, the bed layer models are not able to describe the sorting mechanisms acting within the bed.
The present paper proposes a new sediment continuity model aimed at conditions where dunes dominate. It is based on 1) the Parker-Paola-Leclair framework for sediment continuity, 2) the Einstein step length theory, 3) a newly-developed lee sorting function, and 4) a newly-developed method to account for the variability in bedform trough elevations. The derivation is too extensive to be included in this paper; details can be found in Blom (200-).

2. The new sediment continuity model

The authors have developed a new sediment continuity model based on the depth-continuous framework for sediment continuity presented by Parker et al. (2000). In this framework, parameters such as the volume fraction content of size fractions and the grain size-specific entrainment and deposition rates vary continuously over depth, but are horizontally averaged over a series of bedforms.
On the stoss face of a bedform, we compute simultaneous entrainment and deposition fluxes, whereas on the lee face entrainment is neglected (Figure 2). We apply the step length formulation first introduced by Einstein (1950) to the stoss face. The Einstein step length is defined as the average distance covered by a particle from the moment it is eroded until it is deposited again. This tells us that the deposition rate of a certain size fraction at a certain point on the stoss face equals the entrainment rate of this size fraction one step length upstream of this point.
A newly-developed lee sorting function describes the grain size-selective deposition rate down the avalanche lee face of a bed form, based on 1) the difference between the specific grain size and the geometric mean grain size, 2) the standard deviation of the mixture, and 3) the bed shear stress.
Another component of the sediment continuity model is a method to account for the variability in bedform trough elevations. A series of bedforms is characterised by its probability density function (PDF) of trough elevations, and all bedform geometrics are assumed to be related to the specific trough elevation. To determine the grain

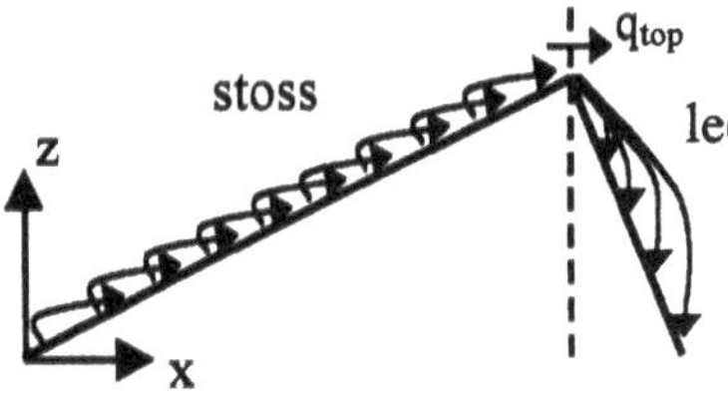

Fig. 2 Schematisation of deposition and entrainment fluxes over, for simplicity, a triangular dune

size- and depth-specific entrainment and deposition rates averaged over the series of bedforms, the entrainment and deposition rates are averaged over all trough elevations while weighed by the PDF of trough elevations.
For equilibrium conditions ($\partial/\partial t=0$), the full sediment continuity model is reduced to an *equilibrium sorting model*, which analytically solves the equilibrium sorting profile from the given bed load transport composition and the PDF of trough

elevations. For non-equilibrium conditions, the full sediment continuity model is reduced to a *sorting evolution model*, which numerically solves the time evolution of both the vertical sorting profile and the bed load transport composition from the initial sorting profile, the given bed load transport rate, and the PDF of trough elevations.

3. Results

The equilibrium sorting model and the sorting evolution model have been calibrated and verified using experiments by Blom et al. (in press) and by Ribberink (1987). Measured PDFs of trough elevations were used as input. Figure 3 shows the measured and predicted sorting profiles for the equilibrium stage of experiment A2. The sediment mixture consisted of three size fractions (0.68mm, 2.1mm, and 5.7mm). It illustrates that the downward coarsening trend is predicted well by the equilibrium sorting model. Another step in the verification procedure is the comparison of the time-evolution of the vertical sorting profile as predicted by the sorting evolution model to the measured time evolution for experiment B2 by Blom et al. (in press), see Figure 4. The predictions agree reasonably well with the data.

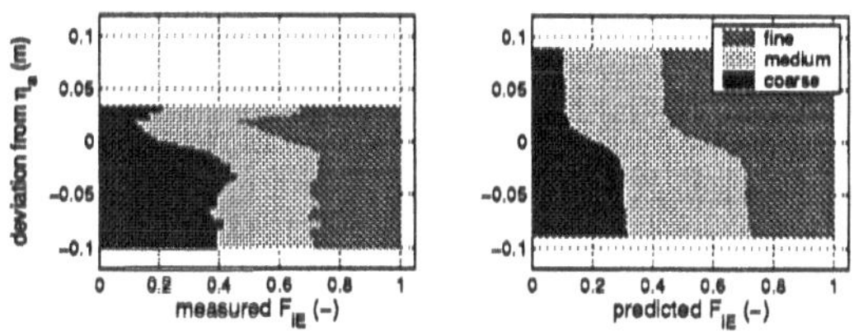

Fig. 3 Measured sorting profile in equilibrium stage of experiment A2, and equilibrium sorting profile as predicted by the equilibrium sorting model. F_{iE} denotes the volume fraction content of size fraction i at bed elevation z. η_a denotes the average bed level.

4. Conclusions

The sediment continuity model introduced in this paper seems promising. It is the first sediment continuity model that incorporates sediment sorting fluxes through 1) bedform migration, 2) the variability in bedform trough elevations, and 3) changes in average bed level. A submodel for the time evolution of the PDF of trough elevations needs to be developed in order to use the model as a predictive tool.

References

Blom, A. (200-) - A sediment continuity model for rivers with non-uniform sediment and bed forms, Ph.D. thesis, University of Twente, to be published early 2003.

Blom, A., Ribberink, J. S. and H. J. de Vriend (in press) – Vertical sorting in bed forms – flume experiments with a natural and a tri-modal sediment mixture. Water Resources Res. In press.

Crickmore, M. J. and G. H. Lean (1962) - The measurement of sand transport by means of radioactive tracers, Proc. Roy. Soc. London A., Vol.266, pp.402-421
Einstein, H. A. (1950) - The bed-load function for sediment transportation in open channel flows, US Dept. Agric., Soil Conserv. Serv., T.B. No.1026.
Hirano, M. (1971) – River bed degradation with armoring. Trans. of J.S.C.E., Vol.3, pp.194-195.
Parker, G., Paola C., and S. Leclair (2000) - Probabilistic Exner sediment continuity equation for mixtures with no active layer. J. Hydraul. Eng., ASCE, Vol.126, No.11, pp.818-826.
Ribberink, J. S. (1987) - Mathematical modelling of one-dimensional morphological changes in rivers with non-uniform sediment, Ph.D. thesis, Delft University, the Netherlands.

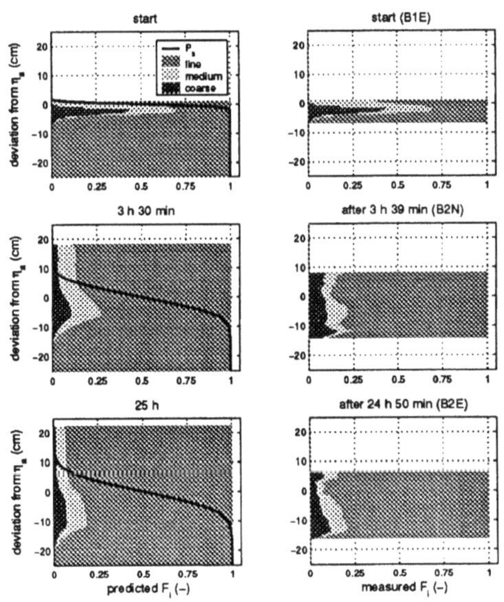

Fig. 4 Measured sorting profiles in experiment B2, and sorting profiles as predicted by the sorting evolution model.

THE SELF-ORGANIZATION OF RIPPLES TOWARDS TWO-DIMENSIONAL FORMS

ALBERT GYR
Inst. of Hydromechanics and Water Resources Management, Federal Institute of Technology, CH-8093 Zurich, gyr@ihw.baug.ethz.ch

1. Statement of the problem

Ripples are two-dimensional bed forms of dune-like shape which develop from a flat bed under well defined conditions. Two-dimensional ripples are the result of an interaction of the erodible bed with a three-dimensional turbulent flow field, $\underline{u} = \underline{\bar{u}} + \underline{u}'$. The question is then: Why do ripples become two-dimensional and periodic? One may suggest that due to the flow separation on the crest of the ripples, a strong two-dimensional component is superimposed onto the three-dimensional turbulent channel flow. Nevertheless this explanation is too simple, since it implies that two-dimensional bed forms are already present. In case of an infinitesimal two-dimensional disturbance, one can analyze the stability of the bed by introducing a sediment transport equation. This approach is problematic for two reasons. The primary periodicity is unknown, and the turbulent flow is represented in the transport equation by the mean wall shear stress, $\tau_W \approx \overline{u'v'}\big|_W$, only. Moreover, film sequences show that the early sediment transport produces a three dimensional bed configuration. It is thus not clear how three-dimensional preforms evolve into two-dimensional ripples. This self-organization needs to be explained. A first hint is the fact that the sand grains have to be smaller than $k^+>12.5$, with $k^+= ku_\tau/\nu$ and $u_\tau=$ sqrt(τ_ω/ρ) for ripples to form. This implies that the separation on individual roughness elements are strong enough to disturb the self organization process. In other words, one doesn't need a hydraulically smooth wall, with $k^+<5$ where the grains are embedded in the viscous sublayer. The roughness is thus small enough that the flow is still dominated by viscous effects. This is especially important for the topology of the separation.

2. Preforms

When the water discharge in an open channel with a flat bed of fine sand is gradually increased from zero, arrow-shaped preforms start to develop. In case the discharge is increased fast, erosion-dells are forming instead. Flow separation occurs on both of these preforms. The separation field is responsible for a propagation and reproduction of these disturbances in the flow direction. Due to this process the disturbances propagate along narrow streamwise trails.
The main information about a flow field is contained in its topology, and especially in its vortex skeleton, see Figure 2. The topology of the approach flow

A. Gyr and W. Kinzelbach (eds.), Sedimentation and Sediment Transport, 183–186.

of the bed form can be described as an open separation bubble. Along the sides of the body longitudinal vortices in the flow direction are present. The flow around the entire bed form is viscous dominated, see Figure 1.

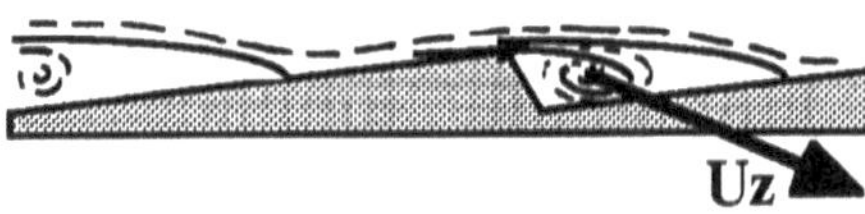

Figure 1: The essentialy viscous separation in the axial center plane generates an open buble

The essential feature of this configuration is that a part of the boundary layer flow is incorporated into the separation zones, and needs to be discharged laterally The corresponding vortices are thus helical.

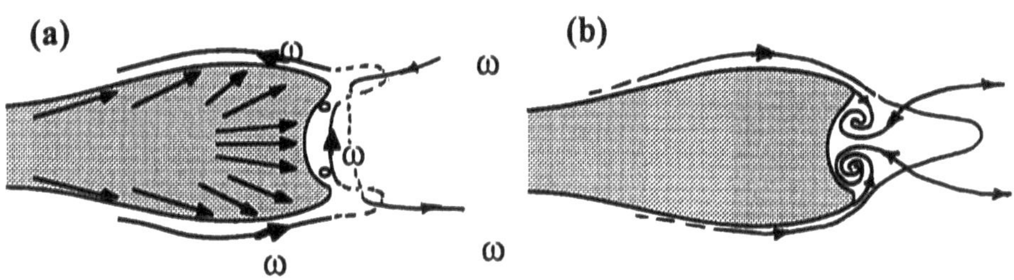

Figure 2: The vortex skeleton in an top-view (a), together with the topological picture of the wall-streamlines (b) .

In the wall near zone behind the vertical ridges vertical separation-vortices are produced as shown in the picture of the topology of the wall streamlines. The flow in the frontal- as well as in the lateral- vortices is directed towards the vertical ridge on the back side of the sediment bodies. This implies that the helical flow must have a stagnation point in this region, and a vortex break down would have to occur. What happens in these areas is very difficult to conceptualize.However, the interaction of the vortex tube elements can be described by the law of Biot- Savart. Observations show that the separation vortices axis are bent due to these interacting processes.
An equilibrium of the vorticity is established between the front- and side-vortices. This equilibrium is responsible for the width between the preforms which reproduce in stripes by the sediment transport of the flow.

3. Self organization into two-dimensional ripples

The formation of two-dimensional ripple forms must be the result of a disturbance of the afore mentioned equilibrium of vorticity. The most natural disturbance is due to the interaction of flows around adjacent sediment elevations. This in fact is the process which is observed.

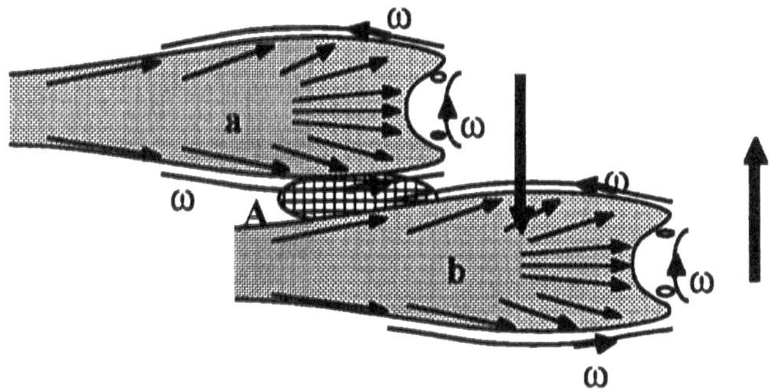

Figure 3: A sche-matic of the self organization due to a annihilation of vorticity. The lateral vortices are organizing the preforms towards two-dimensional ripples.

Two adjacent sediment elevations interact by weakening the two longitudinal separation vortices between them in area A, Figure 3. This as a direct result of the new adjusting asymmetry of the flow fields over the bed forms, and of the even more important annihilation of the vorticity by vorticity diffusion from the adjacent vortex of opposite sign. The lateral vortices force out in the direction of the neighboring sediment elevations which due to the sediment transport start to merge into a two-dimensional ripple form. The wavelength of the new two-dimensional fronts correspond to the length of a single sediment elevation, which means it is given by mean transport intermittency of the "large" structure of the turbulent flow, $\lambda \approx T_B U$.

4. Quantitative results

Most of the experimental data are taken from Schmid (1985). For the evaluation of the flow parameters near a single bedform the cast of a naturally developed bedform was used. The flow was thus examined without the sediment transport which is essential for the formation of the preforms in fine sand. Nevertheless, the flow field could be measured with sufficient accuracy to balance the mass and circulation flux into and out of the separation using their preservation laws. By neglecting the mass flux through the separatrix and the annihilation of vorticity in the boundary layer of the recirculation zone, the inflow can be estimated by the flow trough $\hat{\delta}_0^+$, see Figure 4, by integration over all the seven measured lateral profiles. The discharge is estimated by the fluid transported into the helical front vortex passing the area α. In the caracteristic sample $\hat{\delta}_0^+ \approx 8$ and $Q^+ \approx 2600$. This value can be compared with the content of the separation bubble $\forall^+ \approx 15 \cdot 10^4$. This means that the exchange of the total fluid in the separation takes about 60 s. For an estimate of the circulation it was assumed that the vortices are circular. The result is $\Gamma^+ \approx 550$, 110, 550 for the crest, the vertical and the longitudinal vortices, which shows the suggested equilibrium. Figure 5 shows a typical formation of a two-dimensional bed form.

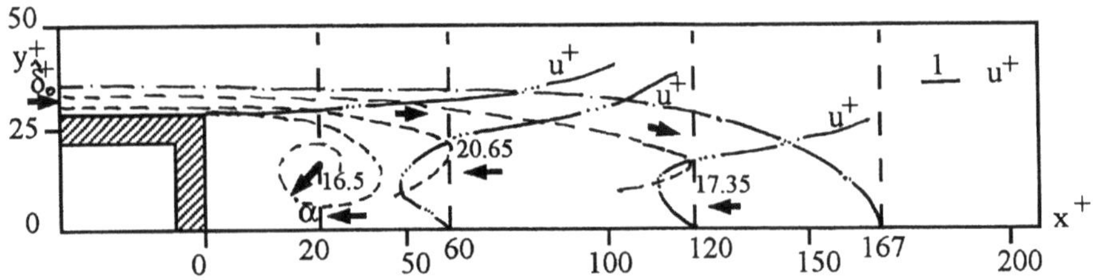

Figure 4: The velocity profile in the axial center plane.

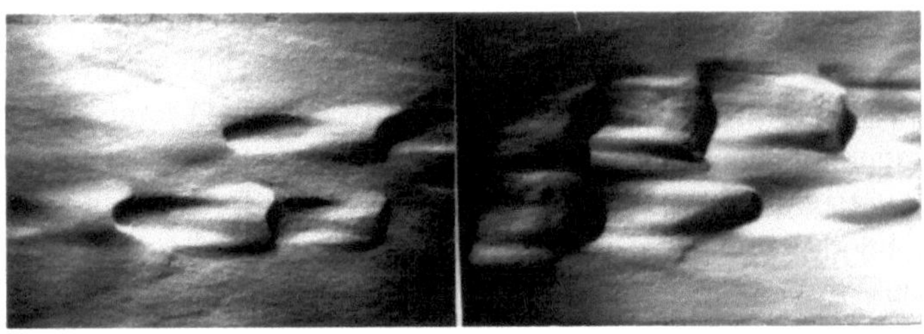

Figure 5: View of a self organizing ripple. The time interval between subsequent pictures is 3 min.

5. Summary

Ripples are created by sediment transport of fine sand. They develop from three-dimensional disturbances of the flat bed due to essentially viscous separations. The self organisation is the result of the interaction of flows over adjacent preforms, in which vorticity of part of the longitudinal vorticity between them is annihilated. Conclusion: When bed annihilation forms occur due to sediment transport, models for the sediment transport must include separation mechanisms to be physically relevant.

6. References

Schmid, A. Wandnahe turbulente Bewegungsabläufe und ihre Bedeutung für die Riffelbildung. Diss. ETHZ Nr.7697, 1985.

AN INTEGRO-DIFFERENTIAL MODEL FOR THE DYNAMICS OF AEOLIAN SAND RIPPLES

HEZI YIZHAQ,[1] NEIL J. BALMFORTH,[2] ANTONELLO PROVENZALE [1,3]
[1]*I.S.I Foundation, V.le. Settimio Severo 65, I-10133 Torino, Italy. E-mail address: hezi@andromeda.isi.it,* [2] *Dept. of Applied Mathematics and Statistics, UCSC, Santa Cruz, CA 95064, USA. E-mail address: njb@cse.ucsc.edu,* [3] *CNR-ISAC, Corso Fiume 4, I-10333 Torino, Italy. E-mail address: a.provenzale@isac.cnr.it*

1. Introduction

Aeolian ripples are commonly found in sand deserts, often atop dunes and on sandy beaches. Ripple wavelengths range from a few centimeters to tens of meters, and ripple amplitudes range from a few millimeters to a maximum of a few centimeters (Sharp, 1963; Wilson, 1972). A classic reference to ripples and dunes is the book "The Physics of Blown Sand and Desert Dunes" by R.A. Bagnold (1941). A theoretical model for the dynamics of aeolian ripples has been proposed by Anderson (1987), and it is used here as a starting point for our own formulation. Recently, there has been a resurgence of interest in developing mathematical models of the dynamics of aeolian ripples, a fact that has led to a series of papers based on different types of phenomenological and mathematical assumptions (Nishimori and Ouchi, 1993; Werner and Gillespie, 1993; Hoyle and Woods, 1997; Terzidis et al., 1998; Prigozhin, 1999; Valance and Rioual, 1999; Cshaók et al, 2000; Makse, 2000; Balmforth et al., 2001)

The physical mechanism responsible for the formation of sand ripples is thought to be the action of the wind on loose sand. When the wind strength is above some threshold, grains are displaced by the direct action of the wind, and are lifted into the air. Even for strong winds, however, sand grains are too heavy to stay suspended and return to the ground. During their flight, the grains reach a velocity that is approximately that of the wind, and upon their impact with the surface, they impart their energy and momentum to the sand, and eject other grains. For sufficiently large wind velocities, the bombardment by sand grains accelerated by the wind generates a cascade process, and an entire population of saltating grains hopping on the sand surface emerges. During strong winds, the layer of saltating grains can reach a thickness of more than one meter.

Experimental results (Anderson, 1987) indicate that the bombardment process generates two populations of moving grains: grains that are ejected with large energy form one component of the population; these grains reach higher elevations and are directly accelerated by the wind. The high-energy grains hit the surface elsewhere and generate a cascade process; these grains form the "saltating" population. The second population consists of grains that are ejected with low energy, and stay close to the sand surface. These crowling grains form what is called the "reptating" population. Low-energy grains typically jump to a distance of a few hundred times the average grain size (Andreotti et al. 2002), and continue to roll on the sand surface after landing. The exchange flux between the reptating and the saltating populations is thought to be small (Anderson, 1987).

A. Gyr and W. Kinzelbach (eds.), Sedimentation and Sediment Transport, 187–194.

Here we adopt the view of ripple formation that was mathematically formalized by Anderson (1987, 1990). According to this interpretation, the only role of saltating grains is to bring energy into the system, extracting it from the wind that blows above the sand surface. In this view, ripple formation is entirely due to spatial changes in the reptation flux. Linear stability analysis of a model built on these premises shows that the wavelength of the fastest-growing perturbation is determined by the reptation length and not by the saltation length as previously suggested by Bagnold. The fastest-growing wavelength turns out to be roughly six times the mean reptation length.

However, there are two problems with Anderson's model: The first is that large wavenumbers remain linearly unstable in his original model. Secondly, Anderson's approach is purely linear and thus it is valid only for small ripples and short times.

We overcome the first problem by introducing a standard correction to the reptation flux, that makes it slightly smaller on the windward (stoss) slope and slightly larger on the lee slope. This correction to the reptation flux stabilizes short wavelengths and shifts the maximum of the reptation flux closer to the ripple crest, stopping excessive growth of the model ripples (Wilson, 1972). We then consider the full nonlinear version of the modified Anderson model and numerically solve the integro-differential equation that describes the dynamics of one-dimensional aeolian ripples. We also mention a weakly nonlinear expansion of the integro-differential model, along the lines previously discussed by Balmforth et al. (2001).

The rest of this paper is organized as follows: In section 2 we derive the model, in section 3 we discuss the results of linear stability analysis and in section 4 we discuss some numerical results. Finally, in section 5 we give a summary and some conclusions.

2. The integro-differential model

Following Anderson (1987) we build an heuristic model of sand transport based on the Exner equation,

$$(1-\lambda_p)\rho_p \frac{\partial \zeta}{\partial t} = -\frac{\partial Q}{\partial x} , \tag{1}$$

where $\zeta(x,t)$ is the local height of the sand surface at point x and at time t, ρ_p is the density of a sand grain, λ_p is the porosity of the bed (typically taken as 0.35), and $Q(x,t)$ is the sand flux which includes both the saltation and the reptation fluxes. This equation shows that erosion $(\partial\zeta/\partial t < 0)$occurs in regions where the sediments flux is increasing $(\partial Q/\partial x > 0)$, deposition $(\partial\zeta/\partial t > 0)$ occurs where the flux is decreasing $(\partial Q/\partial x < 0)$, and there is no change in the surface height where the transport rate is constant $(\partial Q/\partial x = 0)$.

Saltating grains are accelerated by the wind to speeds that are close to the wind speed. These grains follow a ballistic path to the next impact site. If their arced trajectory is long compared to the size of the bed undulations, it seems plausible that the angle at which the grains descend back to the bed is dictated largely by the wind speed and that the flux of saltating grains is fairly uniform. Hence we assume that, for ripple dynamics, the flux of saltating grains, Q_s, can be taken as constant, and we do not consider it further in the Exner equation ($\partial Q_s / \partial x = 0$). However, we should keep in mind that saltating grains are the driving force of the reptation flux, and that for other aeolian bedforms, such as dunes, this approximation is not valid.

The reptation flux at a given point x and time t, Q_r, is obtained by a sum on all the grains that are passing by that point at that time. This is given by all the grains that have been ejected upwind and are still on their reptation flight. If all the grains had the same reptation length $\bar{a}$, the flux at x would be proportional to the total number of grains that have been ejected between $x-\bar{a}$ and x (note that this introduces a basic non-locality in the dynamics of aeolian ripples). In reality, the grains have a distribution of reptation lengths, and this should be taken into account in the expression for the flux. To derive an explicit expression for the reptation flux, we follow Anderson (1987) and write

$$Q_r^0(x) = m_p n_p \int_{-\infty}^{\infty} d\alpha\, p(\alpha) \int_{x-\alpha}^{x} N_{im}(x')dx' , \qquad (2)$$

where m_p is the mass of each particle, n_p is the average number of reptating grains ejected by the impact of one saltating grain, and $p(\alpha)$ is the probability distribution of reptation lengths. The value of n_p depends on the impact velocity and on the diameter of the grains (Forrest & Haff, 1992); in our model we assume n_p to be constant.[1] The probability distribution $p(\alpha)$ has to be determined experimentally; it is quantified by the "splash function" introduced by Ungar and Haff (1987).

Because the saltation flux is uniform, and the fixed angle ϕ at which the grains descend back to the ground is assumed to be constant, the number density of impacting grains changes only because of variations in the bed slope. Based on geometrical considerations, we obtain

$$N_{im}(x) = N_{im}^0 \left(1 + \frac{\tan\theta}{\tan\phi}\right)\cos\theta = N_{im}^0 \frac{1+\zeta_x \cot\phi}{\sqrt{1+\zeta_x{}^2}} , \qquad (3)$$

where N_{im}^0 is the number density of impacting grains on a flat surface.

Expression (2) for the reptation flux leads to a situation where linear instability is not confined to a finite range of wavenumbers (Anderson, 1990). However, there are at least two physical effects that modify the expression of the reptation flux, and lead to a regularization of the system.

Following Prigozhin (1999), it can be shown that (assuming for simplicity that the bed slope is constant) the mean reptation length should be corrected as $\bar{a}$ = $\bar{a}_0(1-\cot\gamma\zeta_x)$, where $\bar{a}_0$ is the mean reptation length on a flat surface and γ is the ejection angle of reptating grains (see Figure 1). This correction leads to a mean reptation length that is shorter on the windward slope and longer on the leeward slope of the bedform.

---

[1] Andreotti et al. (2002) argue that n_p should be smaller for a positive slope along the wind direction than for a negative slope.

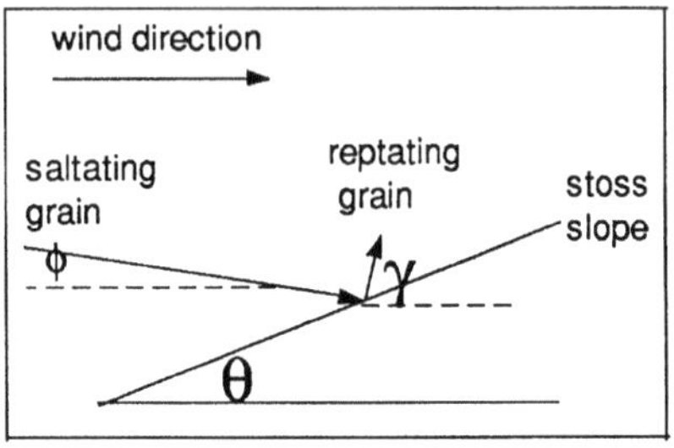

Figure 1: Schematic view of a typical collision. All angles are referred to the horizontal surface. A saltating grain approaches the inclined bed at angle ϕ. A reptating grain is ejected at an angle γ. The inclination of the bed is θ ($\theta > 0$ on the windward slope). The saltating grain ricochets (not shown in the figure) and continues the cascade.

A second effect is that part of the energy of the saltating grain is imparted to individual surface grains, which vibrate rapidly, and as a result can roll or slide down the slope (Whitehouse & Hardisty, 1988). Whitehouse and Hardisty termed this mechanism an "impact-induced gravity flow," which acts to reduce the flux on the windward slope and to increase it on the lee face. Their empirical estimate of this correction gives:

$$Q_{slope} = Q_{flat}\left(1 - \frac{\tan\theta_r}{\tan\theta_r + \tan\theta}\right) \approx Q_{flat}\left(1 - \frac{1}{\tan\theta_r}\tan\theta\right). \quad (4)$$

where Q_{flat} is the flux on a flat surface, Q_{slope} is the flux that includes the correction due to the gravity flow, θ_r is the angle of repose (usually taken to be 33^0), and θ is the bed inclination (positive on the windward face).

Consideration of these two effects leads to the following expression for the reptation flux:

$$Q_r(x) = (1 - \mu\zeta_x)Q_r^0(x), \quad (5)$$

where the parameter μ heuristically includes both the Prighozin and the Whitehouse & Hardisty corrections discussed above. The value of μ should be determined empirically.

Inserting expression (5) into the Exner equation (1) leads to the following equation

$$\zeta_t = Q_0\left\{\mu \int_{-\infty}^{\infty} d\alpha p(\alpha) \int_{x-\alpha}^{x} F(x')dx' + (1-\mu\zeta_x)\int_{-\infty}^{\infty} d\alpha p(\alpha)[F(x-\alpha)-F(x)]\right\}, \quad (6)$$

where $Q_0 = m_p n_p N_{im}^0 \cot\phi / (\rho_p(1-\lambda_p))$ and $F(x) = (\tan\phi + \zeta_x)/\sqrt{1+\zeta_x^2}$. Note that equation (6) breaks down when the lee slope of the ripples exceeds the impact angle ϕ of the saltating grains. In such conditions, the ratio N_{im}/N_{im}^0 becomes unphysically negative. Thus, we add the restriction F(x) = 0 if the bed slope exceeds the impact angle ϕ. This condition introduces a local shadowing effect and it is limited only to the lee slope, in contrast to a "complete shadowing" (Sharp, 1963) where the shadow zone can be extended to the windward slope.

3. Linear stability analysis

We assume an infinitesimal sinusoidal perturbation on an otherwise flat granular bed, given by

$$\zeta(x,t) = \zeta_0 \exp(ik(x-ct)) \tag{7}$$

where ζ_0 is the amplitude of the perturbation. Inserting (7) into (6), using the linear approximation $F(x) \approx \tan\phi + \zeta_x$ and discarding all nonlinear terms we obtain the dispersion relation (Anderson, 1987; Balmforth et al., 2001)

$$\frac{c}{Q_0} = 1 - \hat{p}(k) - ik\mu\bar{a}\tan\phi \tag{8}$$

where $\hat{p}(k)$ is the Fourier transform of $p(\alpha)$ and $\bar{a} = \int_{-\infty}^{\infty} p(\alpha)d\alpha$ is the mean reptation length. If we choose for p(α)an exponential distribution

$$p(\alpha) = \begin{matrix} \chi e^{-\chi\alpha} & \alpha > 0 \\ 0 & \alpha \le 0 \end{matrix} \tag{9}$$

where χ is a parameter, we find that

$$\frac{c_r}{Q_0} = \frac{k^2}{k^2+\chi^2}, \quad \frac{c_i}{Q_0} = \frac{k\chi}{k^2+\chi^2} - \mu k\bar{a}\tan\phi \tag{10}$$

where c_r is the real part of c, which determines the drift speed of the ripples, and c_i is the imaginary part that determines the growth rate of the perturbations. Instability occurs for $c_i \ge 0$. Figure 2 shows the dispersion curves for different values of μ, for φ=10° and for $\bar{a} = 1$ cm. The slope-dependent correction of the flux stabilizes the shorter wavelengths. Moreover, it is clear that the destabilizing effects must now overcome the stabilizing effect, introduced by including the flux correction, in order to form ripples. Note, also, that low impact angles favor destabilization of the bed.

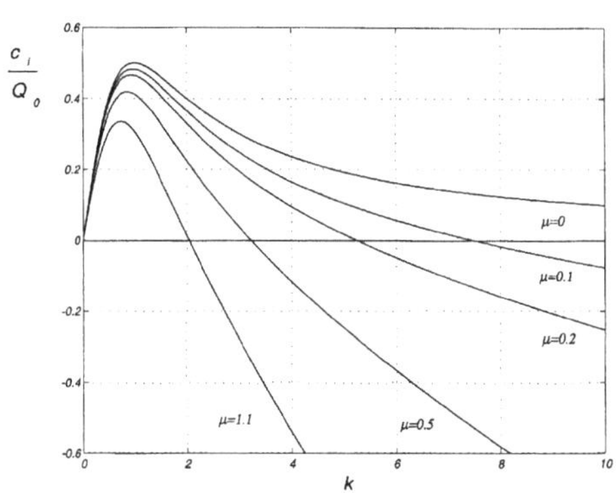

Fig. 2: Dispersion curves for the linearized version of model (6) for different values of μ and for the exponential probability distribution (9). Note that for μ=0, the original Anderson model is recovered. Higher values of μ result in a narrower range of unstable wavenumbers. Other parameter values φ=10° are and $\bar{a} = 1$ cm.

Careful experimental work is needed in order to measure the value of m and give it a plausible value. For example, for $\overline{a}=1$ cm and ϕ=10°, the maximum wavelength $\lambda_{max}=2\pi/k_{max}$ increases from $4\overline{a}$ for μ=0.1 to $8\overline{a}$ for μ=1.1.

4. Numerical results

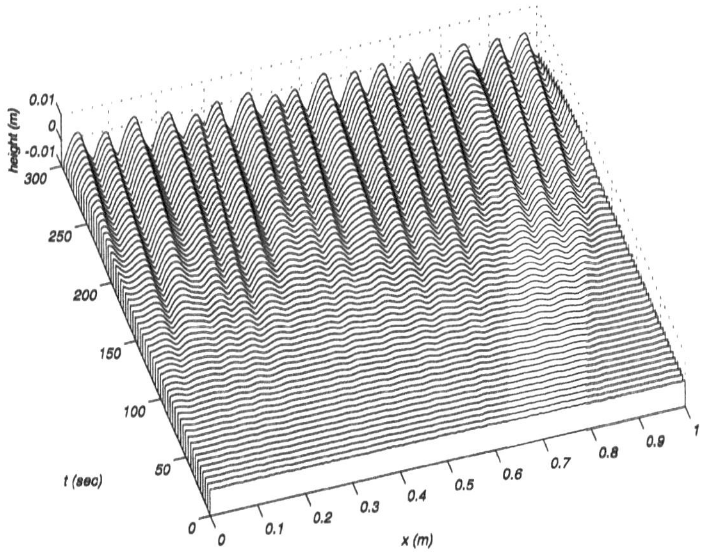

Fig. 3: Time evolution of the ripple pattern for μ = 0.9 and uniform impact angle ϕ = 10°. Ripples grow from initial small random perturbations and their mean wavelength increases because of merging between different ripples. The asymmetry in the ripple profile starts to develop as the lee-slope angle exceeds the impact angle of saltating grains and a shadow zone appears. The drift velocity is approximately 1.2 cm/s, in good agreement with field measurements (Sharp, 1963).

In this section we show some of the results obtained from numerically solving the integro-differential model (6) for realistic parameter values. The saltation flux is $N_{im}^{0}=10^7$ impacts m^{-2} s^{-1}, which according to Anderson (1990) is the saturated flux for a wind shear velocity of 0.5 m/s and for fine sand with grain diameter of 0.25 mm. For the probability distribution we choose χ=1.651, which gives $\overline{a}=1$ cm. This value is close to the typical value of 8 mm reported for example by Andreotti et al. (2002). We solved eq. (6) numerically, assuming periodic boundary conditions and using an implicit finite-difference scheme on a spatial domain with size L = 1 m, grid spacing Δx=1 mm and time step Δt=0.001 seconds. Figure 3 shows the time evolution of the ripple pattern that emerges from random initial conditions. The initial evolution of the slightly perturbed bed surface is dictated by linear theory, and the wavelength of the fastest linearly growing wave dominates in the early stages of the evolution. Ripples begin as small amplitude bedforms or mottles on the surface (see fig. 2 in McEwan et al., 1992). The growing ripples remain almost symmetrical until the lee-slope angle exceeds the saltation impact angle. The asymmetry continues do develop as the ripples grow and the downwind slopes become steeper. Further development of the ripple pattern is due to a coarsening process, which leads to an increase of the ripple wavelength. Data from field observations (Werner & Gillespie, 1993) and from wind tunnel experiments (Anderson, 1990) are consistent with an asymptotic power-law increase of the mean ripple wavelength $\Lambda(t)$. Weakly nonlinear heuristic models of the dynamics of sand ripples (Csahók et al., 2000; Makse, 2000) predict a similar behavior, $\Lambda(t) \propto t^{\beta}$ with $0.27 < \beta < 0.5$. We find that the full model (6) also leads to approximate power-law growth of the average ripple wavelength, with $\beta \approx 0.35$, at intermediate times. At later times, the integro-differential model (6) is consistent with a saturation of the ripple wavelength to a constant value (even though we cannot exclude a slow logarithmic growth of the wavelength).

5. Discussion and conclusions

In this work we have discussed an integro-differential model for the dynamics of aeolian sand ripples. The model is based on an extension of the classic approach by Anderson (1987). The present model relies upon the introduction of a physically-motivated flux correction, due to which the reptation flux decreases on the windward slope and increases on the lee slope. This type of correction is similar in shape to analogous terms introduced in the study of bedforms under water (e.g., Richards, 1980). Numerical solution of the integro-differential equation with realistic parameter values reveals the formation of ripples that are very close in size and shape to natural desert ripples. It remains for future experiments to determine the value of the phenomenological parameter μ, which defines the flux sensitivity to the bed slope.

In a previous work, a long-wave weakly nonlinear approximation of the integro-differential equation was introduced (Balmforth et al. 2001). In such a purely local approach, instability of the flat bed emerges due to the presence of a negative viscosity term, $-\zeta_{xx}$, which is stabilized at small wavelength by a hyperviscous term, $-\zeta_{xxxx}$. The growth of the ripple amplitude is saturated by the presence of quadratic and cubic nonlinear terms. A comparison between the results of the full integro-differential model and its weakly nonlinear expansion will be given in a future work, as well as an extension of the model to the case of two-dimensional ripples.

Finally, we note that in the present model we have assumed the saltation flux to be uniform, and the ripple wavelength depends solely on the reptation flux. Large enough undulations of the bed will break this ideal picture of a uniform saltation flux. In such cases, the saltation flux can depend on the bed topography, similarly to what has been discussed here for the reptation flux. Additionally, the feedback between the variations of the bed topography and the shear stress of the wind cannot be discarded for large bedforms (Richards, 1980). This may explain the formation of large-scale ripples, whose dominant wavelength may then depend on the mean saltation length as suggested by Ellwood et al. (1975).

References

Anderson, R.S. (1987) A theoretical model for impact ripples. *Sedimentology*, **34**, 943-956.
Anderson, R.S. (1990) Eolian ripples as example of self-organization in geomorphological systems. *Earth Sciences*, **29**, 77-96.

Andreotti, B., Claudin, P., & Douady, S., (2002). *The European Physical Journal B*, in press
Bagnold, R.A (1941) *The Physics of Blown Sand and Dunes*. Methuen, London, 265 pp.
Balmforth, N.J., Provenzale, A., & Whitehead, J. (2001) The language of pattern and form. In *Geomorphological Fluid Mechanics*, N.J. Balmforth and A. Provenzale Eds., Springer.
Cshaók, Z., Misbah, C., Rioual, F., & Valance, A. (2000) Dynamics of aeolian sand ripples. *The European Physical Journal E*, **3**, 71-86.
Ellwood, J.M., Evans, P.D. & Wilson, I.G (1975) Small scale aeolian bedforms. *Journal of Sedimentary and Petrology*, **45,** 2, 554-561.
Forrest, S.B., & Haff, P.K (1992) Mechanics of wind ripple stratigraphy. *Science*, **255**, 1240-1243.
Hardisty, J. & Whitehouse, R.J.S. (1988) Evidence for a new sand transport process from experiments on Saharan dunes. *Nature*, **332**, 532-534.
Hoyle, R., & Woods, A. (1997) Analytical model of propagating sand ripples. *Phys. Rev. E*, **56**, 6861.
Makse, H.A (2000) Grain segregation mechanism in aeolian sand ripples. *The European Physical Journal E*, **1**, 127-135.
McEwan, B.B., Willetts & Rice, A. (1992) The grain/bed collision in sand transport by wind. *Sedimentology*, **39**, 971-981.
McLean, S. R (1990) The stability of ripples and dunes. *Earth-Science Reviews*, **29**, 131-144.
Nishimori, H., & Ouchi, N. (1993) Formation of ripple patterns and dunes by wind-blown sand. *Phys. Rev. Lett.*, **71**, 197.
Prigozhin, L. (1999) Nonlinear dynamics of aeolian sand ripples. *Physical Review E, **60**, 729-733.*
Sharp, R.P (1963) Wind ripples. *J. Geology*, **71**, 617-636.
Richards, K.J (1980) The formation of ripples and dunes on an erodible bed. *J Fluid Mech.* **99**, 597-618.
Terzidis, O., Claudin, P., & Bouchaud, J.P (1998) A model for ripple instabilities in granular media. *The European Physical Journal B*, **5**, 245-249.
Ungar, J., & Haff, P.K (1987) Steady state saltation in air. *Sedimentology*, **34**, 289-299.
Valance, A., & Rioual, F. (1999) A nonlinear model for aeolian sand ripples. *The European Physical Journal B*, **10**, 543-548.
Werner, B.T, & Gillespie, D.T (1993) Fundamentally discrete stochastic model for wind ripples. *Physical Review letters*, **71,** 19, 3230-3233.
Wilson, I.G, (1972) Aeolian bedforms-their development and origins. *Sdimentology*, **19**, 173-210.

SIMULATION OF SEDIMENTATION AND MIXING IN DEEPLY-SUBMERGED GRAVITY CURRENTS

F. NECKER, C. HÄRTEL, L. KLEISER
ETH Zürich, Institute of Fluid Dynamics, CH-8092 Zürich, Switzerland

E. MEIBURG
UCSB, Dept. Mech. and Env. Engineering, Santa Barbara, CA 93106, USA

1 Introduction

Gravity currents form when a heavier fluid intrudes into a body of lighter fluid under the influence of gravitational forces. Particulate gravity currents are a special class of such flows with the difference in density being caused by a differential loading with suspended particles. Typical examples of particulate gravity currents in geophysics are turbidity currents in oceans or lakes, which can lead to substantial erosion and transport of sediment (cf. Kneller & Buckee, 2000).

In the present study we address some fundamental aspects of particulate gravity currents by means of three-dimensional (3D) simulations of flows in a plane channel. A principle sketch of the initial configuration is shown in Figure 1. We consider deeply-submerged flows here, where the height of the suspension layer is only one tenth the height of the layer of ambient fluid. The channel is of dimensionless size $L_1{*}L_2{*}L_3{=}72{*}2{*}20$, the size of the suspension reservoir is $L_1^s{*}L_2^s{*}L_3^s{=}6{*}2{*}2$ (half the reservoir height is employed for non-dimensionalization). For this flow, we will examine the sedimentation of particles and the mixing of ambient and interstitial fluid.

A detailed description of the numerical approach used is given in Necker *et al.* (2002) and Härtel *et al.* (2000). The simulations are based on the Navier-Stokes equations augmented by a source term, that accounts for the two-way coupling between fluid and particles. The particles are treated in an Eulerian way by a transport equation for the particle-number density. This approach is most efficient in the case of dilute suspensions of small particles, which we consider. An important parameter is the ratio of buoyancy forces to viscous forces, which can be quantified by a Grashof number $Gr{=}g'(L_3^s/2)^3/\nu^2$. Here, ν is the kinematic viscosity of the fluid, and g' is the reduced gravity. The flow also depends on the dimensionless settling speed u_s of the particles (normalized by $\sqrt{g' L_3^s/2}$). We have set $Gr{=}1.25 \cdot 10^6$ and $u_s{=}0.02$ in the present study, both values being representative for the conditions in more recent experimental studies of particulate gravity currents (see Bonnecaze *et al.*, 1993; de Rooij *et al.*, 1999).

2 Flow structure

An impression of the flow evolution is gained from Figure 2, where isolines of the spanwise-averaged particle concentration field are shown for the lower part of the channel. A characteristic feature of the deeply-submerged flow is that the

A. Gyr and W. Kinzelbach (eds.), Sedimentation and Sediment Transport, 195–200.

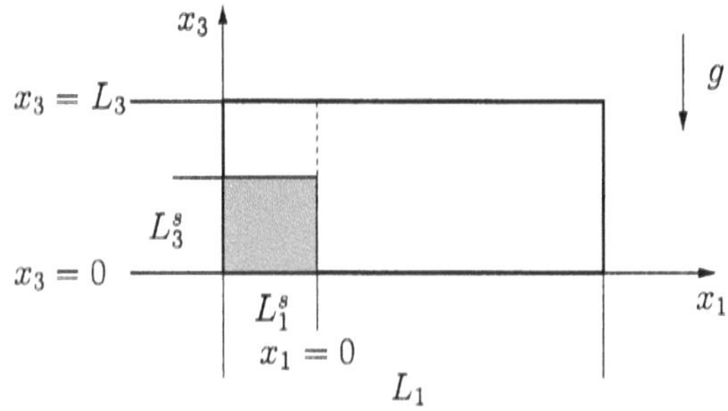

Figure 1: *Principle sketch of the initial set-up in a plane channel (length L_1, width L_2, height L_3. A subvolume of the channel (length L_1^s, width $L_2^s = L_2$, height L_3^s) is filled with particle-laden fluid. Gravity acts vertically.*

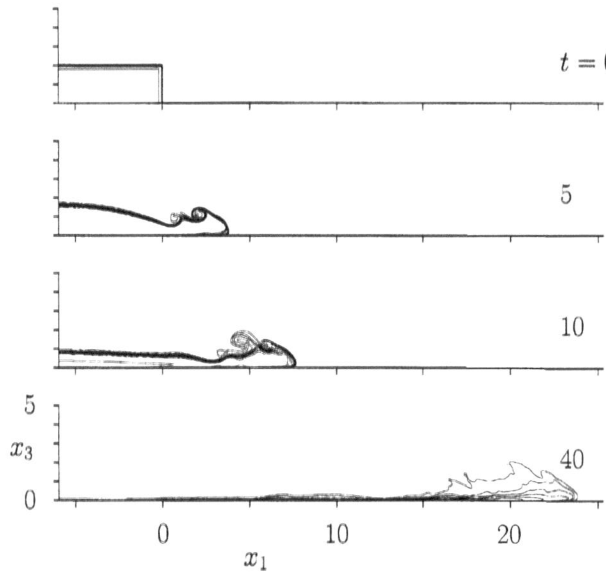

Figure 2: *Isolines of spanwise-averaged particle concentration.*

fluid motion is fundamentally different in the reservoir region and around the front propagating to the right. Behind the head of the front intense vortices develop, but the concentration field in the rear part of the flow remains smooth. This difference is also seen from Figure 3, where an isosurface of concentration is shown at time t=10. Moreover, the isosurface reveals the presence of the typical lobe-and-cleft instability at the leading edge.

The concentration fields in Figure 2 indicate that the front propagates at constant speed during the early flow stages. To make this more evident, the temporal evolution of the front position x_f is plotted in the left graph of Figure 4, together with the time history of the mass of suspended material. The front

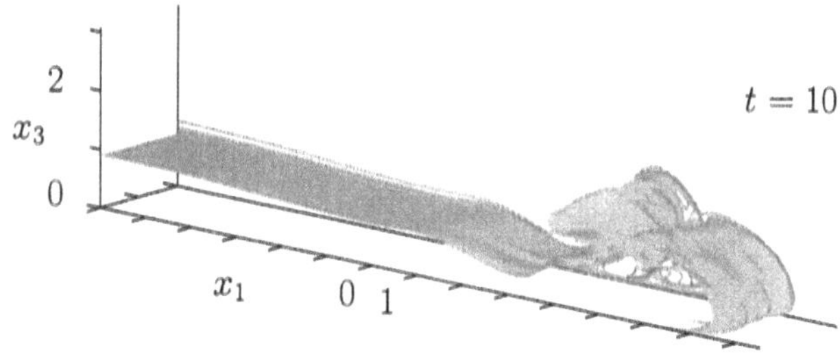

Figure 3: *Isosurface of particle concentration at t*=10. *Isovalue set to* 0.25.

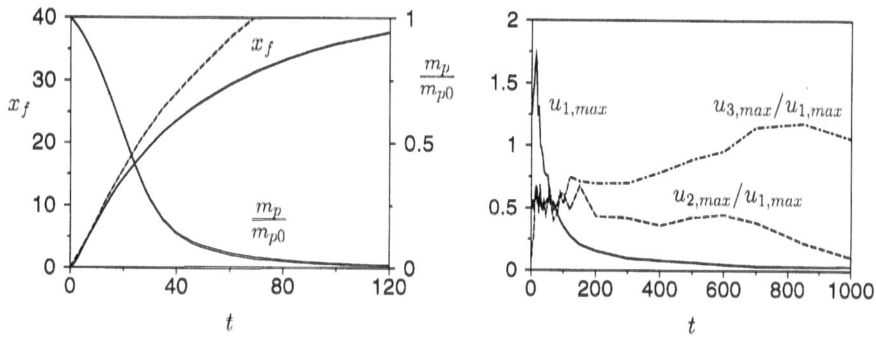

Figure 4: *Left: Front position $x_f(t)$ and mass $m_p(t)$ of suspended particles (normalized with the initial mass m_{p0}). Solid lines: Results for the flow shown in Figure 2. Dashed line: 2D result for density-driven gravity current. Right: Temporal evolution of maximum streamwise velocity $u_{1,max}$ (solid line). The dashed and dot-dashed lines denote, respectively, the maximum spanwise and vertical velocity $u_{2,max}$ and $u_{3,max}$, normalized by $u_{1,max}$.*

speed $u_f = dx_f/dt$ is constant initially, taking a value of $u_f \approx 0.78$. This is clearly smaller than the maximum streamwise velocity which is found within the front (cf. the right graph of Figure 4). After about t=20, the front speed continuously decays. Note that this is also the time when the speed of the particulate front starts deviating more strongly from the speed of a density-driven gravity current. Eventually, the front comes to rest at a run-out length of about x_f=45.

3 Sedimentation and Mixing

The time history of the total mass m_p of suspended material, given in Figure 4, shows a rapid decrease of m_p during the first 30 time units when approximately 75% of all particles settle. Subsequently, the loss of particles decreases, leaving a few percent of material in suspension for rather long times. In order to examine the particle sedimentation at the bottom quantitatively, we evaluated the instantaneous sedimentation rate $\dot{m}_s$, and the result is depicted in Figure 5.

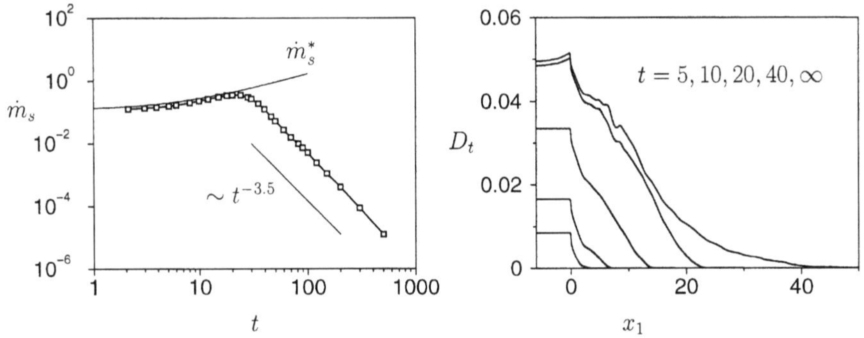

Figure 5: *Left: Sedimentation rate* $\dot{m}_s$ *at the bottom wall as function of time. Estimate* $\dot{m}_s^*$ *(see text) calculated with a front speed of* u_f*=0.78. Right: Non-dimensional particle deposit* D_t *as function of* x_1 *for five different times. The curve for* t*=∞ gives the final distribution after all particles have settled.*

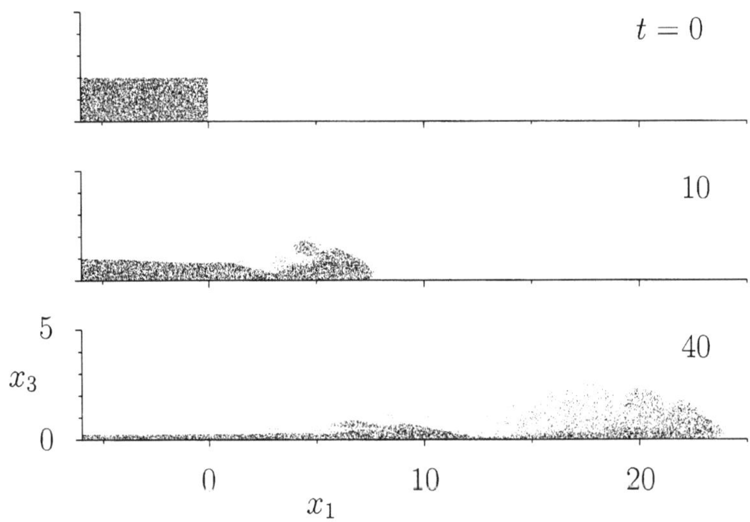

Figure 6: *Dispersion of interstitial fluid visualized by markers (early times).*

The settling of particles produces a sediment layer at the bottom of the tank, which is non-uniform in streamwise direction. This is illustrated in the right graph of Figure 5, where a mean deposit D_t per unit span (cf. Necker *et al.*, 2002) is plotted. In the initial reservoir (x_1<0), the sediment layer grows almost linearly during the first 20 time units. After all particles have settled (t→∞), the largest values in D_t are encountered in this region. This is markedly different from what is found with lock-release flows, where a distinct maximum develops downstream of the initial lock (see Bonnecaze *et al.*, 1993; Necker *et al.*, 2002).

The second point we wish to examine is the mixing of interstitial fluid and ambient fluid. In lock-release flows, a thorough mixing across the full channel height

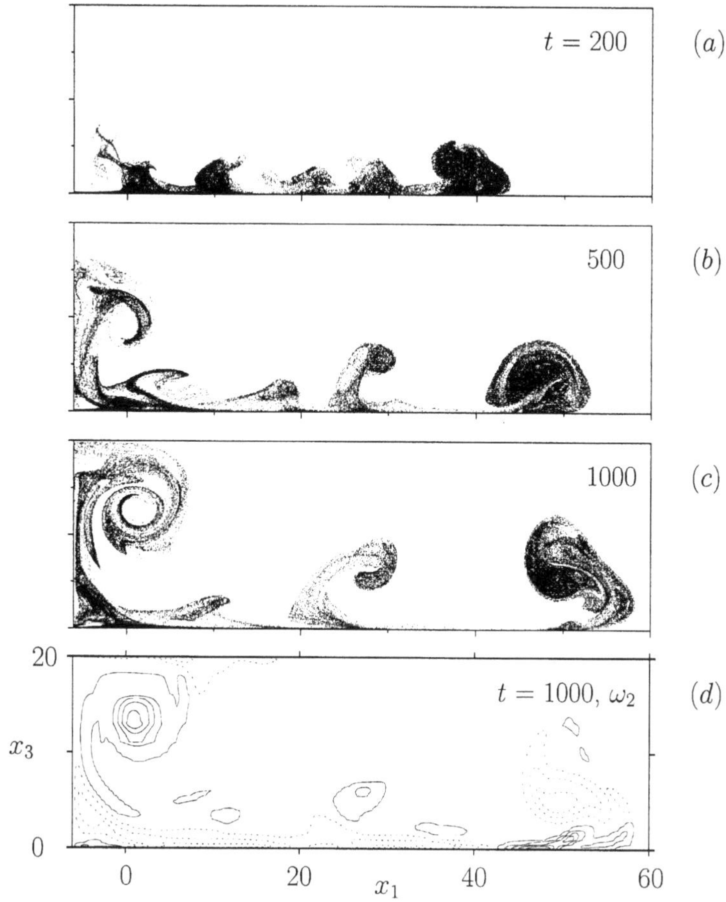

Figure 7: *(a)-(c) Dispersion of interstitial fluid visualized by passive markers (late times). (d) spanwise-averaged vorticity component* ω_2 *at time* $t = 1000$.

occurs after complete sedimentation, which can be attributed to the persistence of large-scale vortices (Necker *et al.*, 2002). To visualize the mixing, passive markers were added to the suspension and then tracked along their pathlines during the flow. The marker distribution is shown in a side view in Figures 6 and 7 for early and late times, respectively. At early times, the marker field resembles the shape of the gravity current. The pronounced head of the current, as well as the almost undisturbed flow structure in the rear part, are clearly recognized from the marker distribution. After t=40, when most particles have settled, the fluid motion slowly ceases. However, weak vortices remain which continue mixing interstitial fluid and ambient fluid. At t=200, markers are already distributed over a layer of thickness 3-4 times the height of the initial reservoir; at t=1000, markers extend over the full channel height. At this late stage, the vortices in the (x_1,x_3)-plane have grown to a length scale much larger than the channel width, leading to essentially 2D motion. This is visible from Figure 4, where the time history is depicted of the maximum vertical velocity and the maximum spanwise velocity (both normalized with the maximum streamwise velocity, which slowly decays with time). The structure of the resulting vortices is visualized in Figure 7d by isolines of the spanwise-averaged vorticity. How strongly the mixing in the final stage is affected by the fact that a narrow channel width promotes large-scale 2D motion, can only be clarified by future simulations of flows in a much wider computational domain.

References

BONNECAZE, R. T., HUPPERT, H. E. & LISTER, J. R. 1993 Particle-driven gravity currents, *J. Fluid Mech.* **250**, 339-369.

DE ROOIJ, F. & DALZIEL, S. B. 1998 Time-resolved measurements of the deposition under turbidity currents, in: *Proc. Conf. Sediment Transport and Deposition by Particulate Gravity Currents*, Leeds, 7-9 September 1998.

HÄRTEL, C., MEIBURG, E. & NECKER, F. 2000 Analysis and direct numerical simulation of the flow at a gravity-current head. Part 1: Flow topology and front speed for slip and no-slip boundaries, *J. Fluid Mech.* **418**, 189-212.

KNELLER, B. & BUCKEE, C. 2000 The structure and fluid mechanics of turbidity currents: a review of some recent studies and their geological implications, *Sedimentology* **47**, 62-94.

NECKER, F., HÄRTEL, C., KLEISER, L. & MEIBURG, E. 2002 High-resolution simulations of particle-driven gravity currents, *Int. J. Multiphase Flow* **28**, 279-300.

VII

Problems related to Field Measurements and Inputs needed for Numerical Models

RIVER BRAIDING IN RELATION TO UNSTEADY WATER AND SEDIMENT TRANSPORT

HELMUT M. HABERSACK
Department for Water Management, Hydrology and Hydraulic Engineering, Universitaet fuer Bodenkultur Vienna, Muthgasse 18, 1190 Vienna, Austria
email: haber@edv2.boku.ac.at

1. Introduction

In Europe nearly all larger rivers were channelised over the last 100 years, beginning mostly at the second half of the 19th century (Vischer, 1986). Thereby especially Alpine rivers experienced a remarkable reduction of river bed width, combined with shortening of flow length (Habersack & Schneider, 2000). Longterm negative consequence is river bed degradation, following increased shear stresses and prohibition of side erosion and bedload input from the catchments. This leads also to a change in river morphology from former often braided sections to single thread channels. In order to compensate these effects over the past ten years a significant increase in channel widening can be observed (Habersack et al., 2000). Increases in effective bed channel widths cause changes in river morphology, eventually beginning of braiding (if sediment input is sufficient). For planning purposes (e.g. river bank protection) the prediction of river morphology is essential, therefore knowledge of involved processes is necessary.

2. Processes involved in river braiding

Field data and physical models lead to an at least qualitative knowledge of braided river morphology covering the general characteristics of the bed topography, channel network form, mechanisms of braiding initiation, processes of channel migration and avulsion, sedimentary processes of bar formation and growth, morphology and development of confluence, unsteadiness and non-uniformity of bed material transport rate (Ashmore, 2001). There exists a characteristic length scale related to total discharge of the river, where stability theory of bars etc. can lead to predictive power. Braiding intensity is empirically related to stream power and bed material size. Concerning sediment transfer and change in channel morphology Carson and Griffiths (1987) argued, that, in braided rivers, transfer is primarily from lateral erosion of banks along the downstream margins of lateral and medial bars, through confluence zones, and onto the bar at the downstream bifurcation. As up to now there are only few field data available some results given below are valuable for understanding the key processes.
Another process is the extreme variation of bedload transport for a given discharge. According to Hoey (2001) confluence are extremely important for braided river morphology, showing local maxima in shear stress and transport rate.

3. Existing modeling concepts and deficits

In principal there exist two types of modeling concepts concerning stream braiding (Paola, 2001):

A. Gyr and W. Kinzelbach (eds.), Sedimentation and Sediment Transport, 203–206.

• traditional approach of modeling using conventional PDE techniques
• simplified model approaches, reproducing the main behavior of braided rivers by dealing with a few crucial aspects

Modeling of fully developed braiding using conventional PDE techniques has been hindered by the difficulty of solving a nonlinear system of equations on a complex, constantly deforming domain. A variety of processes are not described adequately concerning sediment transport and interaction with river morphology. On the other hand simplified cellular models can reproduce many of the main features of braiding but their parameters are difficult to constrain. Paola (2001) suggests that the two modeling approaches should be integrated to improve simulation of the dynamics of braided rivers. One of the main deficits in modeling braided morphology is the lack of understanding of various processes, like side erosion in combination with unsteady sediment transport, interaction between bedforms (e.g. gravel sheets) and transport of individual particles, as well as morphological consequences. Only the understanding of these processes can lead to e.g. an extension of the now existing one dimensional bedload transport formulas (Habersack & Laronne, 2001) and a following integration into numerical models. Field measurements highlight parts of these processes and results can be used for model calibration and validation.

4. Results from field measurements and monitoring

Between November 1998 and July 1999 the braided Waimakariri River in New Zealand was studied by means of radio-tracking gravel particles during floods as well as aerial surveys of the planform development. At the study reach the Waimakariri River is about 1000 m wide, has a slope of 0.0038 m m^{-1}, an average discharge of 116 m^3 s^{-1}, a one hundred years flood of 4300 m^3 s^{-1} and a subsurface mean grain-diameter of 24 mm. The radio tracer technique consists of a transmitter, that is implanted into natural or artificial pebbles, an antenna system receiving the radio signals along the river side and thereby locating flow paths of the stones. The signals finally reach a receiver that is connected to a notebook, where the received information can be stored and manipulated with a special software (Habersack, 2001). During the study period a variety of different floods could be observed. Floods like in Fig. 1a cause no overtopping of individual braids (are called freshes in New Zealand and peak at about several hundred m^3 s^{-1}) whereas Figure 1b shows a flood with a return period of several years and bankful discharge (Figure 1c).
Freshes are dominated by the process of side erosion and individual grain transport (Figure 2). As these events occur several times per year quantitatively these small floods cause a significant contribution to the overall sediment transport budget. The observed side erosion reached a value of about 30 m, mainly caused by high shear stresses in the outer curve, followed by bottom erosion of the banks and bank collapses due to gravity effects. Larger floods lead to a frequent breakdown of the "organized" braided morphology, where bar migration and development/modification of diffluence/confluence structures dominate (Ashmore, 2001). Bedload transport is characterized by large variations which are associated with the development and migration of the morphology. In numerical models bedload discharge formulas are incorporated, like the one by Einstein, 1937. One of his assumptions was that a mean step length (average jump) of a particle would be about 100 times the grain diameter. Measurements in New Zealand showed that there is an inverse relationship between relative roughness and step length, scaled by the particle diameter (Habersack, 2001).

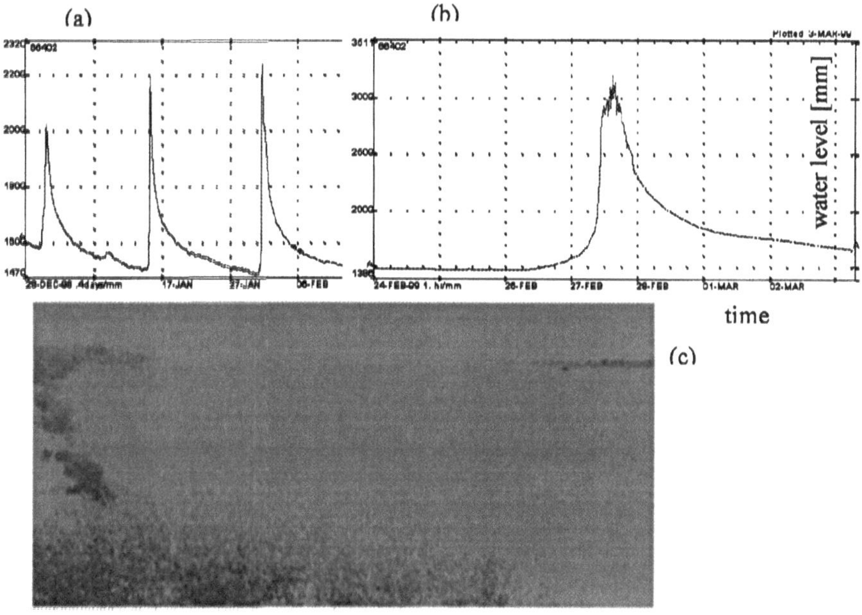

Fig. 1: Monitored floods at the Waimakariri River; freshes (a) and major flood (b, c)

Whereas Einstein assumed an exponential distribution for step lengths and rest periods we found that the combination of the distribution of step lengths and rest periods leads to the so-called gamma exponential model, which is mainly used in order to calculate the spatial probability density through time

$$f_t(x) = \lambda_1 e^{-(\lambda_1 x + \lambda_2 t)} \sum_{n=1}^{\infty} \frac{(\lambda_1 x)^{n \cdot r - 1}}{\Gamma(nr)} \frac{(\lambda_2 t)^n}{n!}$$

where λ_2, which acts as a scale parameter, = the number of rest periods per time unit [1/minute], λ_1 (the inverse of the mean step length) = scale parameter [1/meter], $\Gamma(\,)$= gamma function, and r = shape parameter of $\Gamma(\,)$, x = displacement of the particle from the origin, t = time taken during this displacement, and n = number of single step lengths and rest periods.

It could be shown, that as long as floods stay within the banks of individual braids (floods in Figure 1a) the theory of single stream may be adopted. Larger floods (Figure 1 b) cause additional effects, like growth and transport of large scale gravel sheets which necessitate further process laws to be established. These erosion and deposition zones could be identified by the tracer measurements, which allows the study of characteristic distances between these features (for the Waimakariri River the distances had a length of about 200 m).

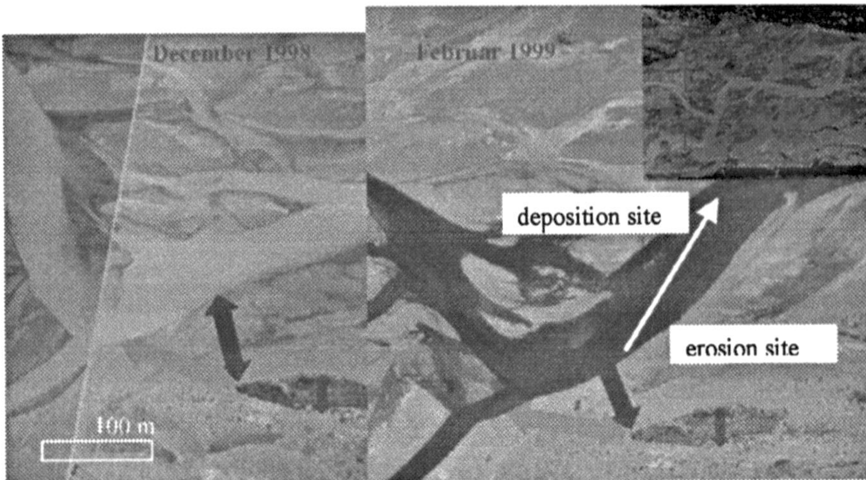

Fig. 2: Side erosion caused by small floods (freshes) and mean transport length during an event

5 Conclusion

This paper discusses the complex processes of river braiding in relation to unsteady water and sediment transport. Existing numerical braiding models can only reproduce the main features of braided systems. The prediction of the detailed morphology that will develop for a given flood needs a better understanding of these processes, leading to modified transport formulas. Field measurements using a radio-tracking measurement system in New Zealand gave insight into the interrelations between river braiding and sediment transport. Thereby essential assumptions of stochastic bedload formulas could be verified or improved.

6. References

Ashmore, P. (2000): Braiding phenomena: statics and kinetics, Gravel-Bed Rivers V (ed. M. P. Mosley), New Zealand, 95-114.

Carson, M. A.& Griffith, G. A. (2001): Bedload transport in gravel channels, Journal of Hydrology, 73, 315-334.

Einstein, H. A. (1937): The bedload transport as probability problem, Mitteilung der Versuchsanstalt fuer Wasserbau an der ETH Zürich, pp. 110.

Habersack, H., Koch, M., Nachtnebel, H.P. (2000): Flussaufweitung in Österreich - Entwicklung, Stand, Ausblick. Österr. Wasser- und Abfallwirtschaft, 7/8, 143-153.

Habersack, H.M., Schneider, J. (2000): Ableitung und Analyse flussmorphologisch relevanter Parameter von historischen Karten. Wasser & Boden, 52, 6, 55-59.

Habersack, H.M. (2001): Radio-tracking gravel particles in a large braided river in New Zealand: a field test of the stochastic theory of bed load transport proposed by Einstein, J. Hydrological Processes, 15, 3, 377-391.

Habersack, H.M., Laronne, J.B. (2001): Bedload texture in an Alpine gravel bed river. Water Resources Research, 37, 12, 3359-3370.

Hoey, T., Cudden, J. & Shvidchenko, A. (2001): The consequences of unsteady sediment transport in braided rivers, Gravel-Bed Rivers V (ed. M. P. Mosley), New Zealand, 121-142.

Paola, C. (2001): Modeling stream braiding over a range of scales, Gravel-Bed Rivers V (ed. M. P. Mosley), New Zealand, 11-46.

Vischer, D. (1986): Schweizerische Flusskorrektionen im 18. und 19. Jahrhundert, VAW-Mitteilung 84.

ON THE VALIDITY AND LIMITS OF CONTINUOUS-PHASE MODELLING OF SEDIMENT TRANSPORT IN ESTUARIES AND COASTAL ZONES

ERIK A. TOORMAN
Hydraulics Laboratory, Civil Engineering Department, Katholieke Universiteit Leuven
Kasteelpark Arenberg 40, B-3001 Leuven, Belgium
erik.toorman@bwk.kuleuven.ac.be

1. Introduction

Numerical models for sediment transport (ST) for large-scale engineering applications are based on a continuous-phase approach, i.e. they consider the fluid-particle mixture as a continuous, incompressible phase with varying density. In addition to the hydrodynamics of the suspension, the mass balance equation for the sediment particles is solved. This approach assumes that the particles follow the average fluid motion, except for the effect of gravity, which introduces the settling flux in the sediment transport equation. This assumption seems to be justifiable for volumetric concentrations up to 1%, above which particle-particle interactions become important (Villaret & Davies, 1995). Moreover, measurements of velocity lags between the two phases (e.g. Rashidi *et al.*, 1990; Best *et al.*, 1997; Muste & Patel, 1997) indicate that the relative difference in the case of sediment particles usually is below 5% of the fluid velocity, which is considerably smaller than the accuracy achieved (or desired) in sediment transport simulations. Hence, these effects need not to be simulated in detail.

2. Engineering Sediment Transport Models versus Two-phase Models

Because of the wide spectrum of spatial and time scales in coastal problems (e.g. turbulence versus tides), the developments in ST models are directly coupled to advancements in computing performance. Over the past decade 3D ST models have been developed for large-scale engineering purposes. 2DV(ertical) models have been used already for a longer time, mainly as research tool. These models require one or the other turbulence closure. The majority of the engineering models still use algebraic mixing-length models. The latest generations have moved to the two-equation k-ε model. Both turbulence closures are applied to determine the vertical turbulent diffusion only, since the horizontal grid scale usually is too large (of the order of 100 times the vertical grid size). The introduction of a LES-type turbulence model for the horizontal mixing is one of the latest improvements.

Initially, these models neglected most of the turbulence modulation effects due to suspended particles, since concentrations in the field are often small. The only effect accounted for most often is the damping of turbulence due to buoyancy effects (i.e. stratification). Typically, empirical closure relationships for the turbulent Schmidt number are taken from atmospheric science (e.g. the Munk-Anderson damping functions). Recently, efforts are being undertaken to investigate in greater depth the validity of these models.

Research, funded by the European Commission under the MAST3 programme, has shed new light on several aspects (Toorman *et al.*, 2002). An important shortcoming has been

A. Gyr and W. Kinzelbach (eds.), Sedimentation and Sediment Transport, 207–210.

revealed in the estimation procedure of the bed shear stress, due to inconsistent application of the damping functions to the bed boundary conditions, resulting in an important overestimation of the bed shear stress when saturation conditions are reached. A consistent methodology has been developed, which in addition succeeds in predicting the right order of magnitude in drag reduction in the case of cohesive sediment, as observed in highly concentrated rivers and in laboratory experiments (Toorman *et al.*, 2002; Toorman, 2002). Other remarkable findings have been made regarding the conditions for saturation (Toorman, 2002).

A different approach is adopted by physicists. Many of them are using much more complicated models based on one or the other two-phase approach (e.g. Besnard & Harlow, 1988). However, often these models are restricted to small scale problems of idealized fluid-particle systems (even more in the case of DNS and LES models; e.g. Boivin *et al.*, 1998). Considering the complexity of the characterization of real sediments in nature, and in particular cohesive sediments, where the particles actually are flocs, the detailed physical models require much empiricism or tuning to get the results right. But even for the ideal homogeneously sized spherical particles, there seems to be not one common approach for the modelling of the turbulence modulation and solutions can only be obtained for a simplified set of equations with semi-empirical closures for the various correlations of the turbulent fluctuations in the field variables (e.g. Elghobashi & Abou-Arab, 1983).

3. The Problem of Scales

An important limitation of the engineering models is the fact that they run on very coarse vertical grids and cannot account for the processes in the benthic boundary layer. This is a major gap in the continuous-phase approach. More precisely, low-Reynolds modelling approaches, which allow bridging the gap, are not adapted to particle-laden flows and very few to rough bottoms. These models even cannot be applied to large-scale problems with uneven bottoms, where topographic roughness (including bed forms) is much larger than the viscous sublayer thickness, but much smaller than the horizontal grid size. There are conflicting interests. Validation of models against laboratory data is useful, but does not give any guarantee that the performance will be correct to flows over complex geometries and with natural sediments.

Figure 1, showing experimental flume data, illustrates the problem. In this example, the actual high-concentrated benthic layer is ±2 cm thick. The importance of interparticle collisions in this layer is clear from the Reynolds stress distribution. This thickness is about two orders of magnitude smaller than typical vertical grid spacing used in 3D coastal modelling. The typical horizontal grid spacing is two more orders larger. Hence, the (often badly known) details of the bottom roughness cannot be captured. Instead, an equivalent bed roughness for a flat bottom between two subsequent grid points has to be defined, without resolving the flow field in detail in the lowest centimeters.

In practice, the following procedure is proposed. Simulations have to be carried out on realistic small scale problems (typically of the order of 10 m depth and similar order width)

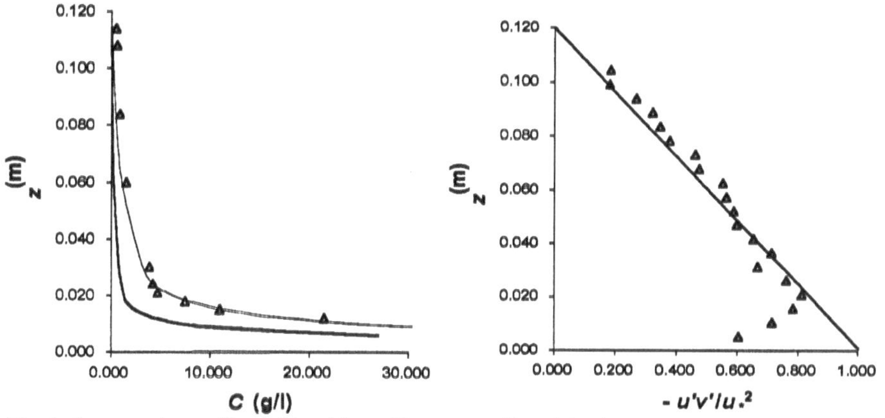

Fig. 1: Concentration profiles (left) and Reynolds stress profiles (right) for suspension flow with sand (d_{50} = 0.135 mm) measured in a laboratory flume by Cellino (1998). Run Q40S003 (u_* = 0.029 m/s): △ = data, curve = Rouse profile fit (above the reference level 0.020 m the Rouse number Z = 1, below Z = 2.4). Run Q75S03 (u_* = 0.055 m/s): ① = data, curve = Rouse profile fit (Z = 1 above the reference level 0.024 m, Z = 1.8 below).

realistic small scale problems (typically of the order of 10 m depth and same order width) for a variety of real and/or sediment-modulated apparent bed roughnesses with both a detailed model and a coarse grid model. A parameterization for the effective bed roughness then has to be formulated based on intercomparison of coarse model simulation results with numerical and/or experimental data. Some suitable data is already available in the literature (e.g. for flow over rippled sand beds).

The study is carried out with the author's 2DV research code FENST-2D, employing the continuous-phase approach, which comprises various low-*Re* models and a hybrid two-layer model. The latter approach, which allows coarser vertical grid spacing, seems promising, as it may allow the simulation of the thickening of the viscous sublayer under specific flow condition to heights above the vertical grid size.

Evidently, the major problems occur near the bed where the highest concentrations occur and particle-particle interactions can become important. The continuous-phase approach then requires an additional, semi-empirical stress closure for this mechanism.

4. Accuracy and Complexity versus Uncertainty

A probabilistic approach would become necessary to demonstrate that the uncertainty on the model parameters may increase too much with regard to the reduction in model uncertainties by increasing the model complexity (i.e. improving the physics), such that the overall simulation uncertainty increases anyhow. Indeed, considering in particular the high uncertainty on the composition and grain size distribution of suspended and bed surface particles and the erosion strength and the effective roughness of the bed surface, one can wonder how large the uncertainties will be on many of the model parameters, which often have been determined empirically. This would provide the best justification to keep using the traditional, computationally much cheaper continuous-phase approach for engineering problems. In practice, more attention should be given to sensitivity analysis.

Another issue to be considered are the intrinsic errors of the numerical techniques implemented in the code. Recent intercomparisons between various engineering codes has shown that they can produce very different results for the same test case (e.g. Violeau et al., 2002). Numerical diffusion, necessary to stabilize the solution in order to obtain a robust code, is believed to be one of the major causes.

5. Conclusion

The continuous-phase approach is not valid in high-concentration regions, usually restricted to the thin benthic layer. This layer cannot be modelled explicitly due to discretization generated scale problems. A semi-empirical approach combined with a two-layer turbulence model is suggested. The problem then also requires a parameterization of the effective bottom roughness. The validity of the continuous-phase approach should be evaluated further by intercomparisons with detailed two-phase models on well-defined test cases, preferably laboratory experiments.

6. References

Besnard, D.C. & F.H. Harlow Turbulence in multiphase flow. *Int. J. Multiphase Flow*, **14**, 679-699, 1988.

Best J., S. Bennett, J. Bridge & M. Leeder Turbulence modulation and particle velocities over flat sand beds at low transport rates. *J. Hydr. Eng.*, **123**,1118-1129, 1997.

Boivin, M., O. Simonin & K.D. Squires. Direct numerical simulation of turbulence modulation by particles in isotropic turbulence. *J. Fluid Mech.*, **375**, 235-263, 1998.

Cellino, M. Experimental study of suspension flow in open-channels. PhD thesis, Dept. de Génie Civil, Ecole Polytechnique Fédérale de Lausanne, 1998.

Elghobashi, S.E. & T.W. Abou-Arab (). A two-equation turbulence model for two-phase flows. *Phys. Fluids*, **26**, 931-938, 1983.

Muste, M. & V.C. Patel Velocity profiles for particle and liquid in open-channel flow with suspended sediment. *J. Hydr. Eng.*, **123**(9):742-751, 1997.

Toorman, E.A.. Modelling of turbulent flow with cohesive sediment. In: *Proc. in Marine Science*,**5**: *Fine Sediment Dynamics in the Marine Environment* (J.C. Winterwerp & C. Kranenburg), 155-169, Elsevier Science, Amsterdam, 2002.

Toorman, E.A., A.W. Bruens, C. Kranenburg & J.C. Winterwerp. Interaction of suspended cohesive sediment and turbulence. In: *Proc. in Marine Science*,**5**: *Fine Sediment Dynamics in the Marine Environment* (J.C. Winterwerp & C. Kranenburg, eds.), 7-23, Elsevier Science, Amsterdam, 2002.

Rashidi, M., Hetsroni, G. & Banerjee, S. Particle-turbulence interaction in a boundary layer. *Int. J. Multiphase Flow*, **16**, 935-949, 1990.

Villaret, C. & A.G. Davies Modeling sediment-turbulent flow interactions. *Appl. Mech. Rev.*, **49**, 601-609, 1995.

Violeau, D., S. Bourban, C. Cheviet, M. Markofsky, O. Petersen, W. Roberts, J. Spearman, E. Toorman, H.J. Vested & H. Weilbeer Numerical simulation of cohesive sediment transport: intercomparison of several numerical models. In: *Proc. in Marine Science*,.**5**: *Fine Sediment Dynamics in the Marine Environment* (J.C. Winterwerp & C. Kranenburg), 75-89, Elsevier Science, Amsterdam, 2002.

MODELING THE HYDRAULICS AND EROSION PROCESS IN BREACH FORMATION DUE TO OVERTOPPING

P.WANG & R.KAHAWITA
Ecole Polytechnique de Montreal, T.T. Quach & T. Mai Phat, Hydro Quebec
rene.kahawita@polymtl.ca

1. Introduction

Many cases of breach initiation in a dike or embankment dam due to overtopping have been recorded internationally, very frequently resulting in complete failure of the structure. This in turn has led to extensive material and human loss downstream. In fact, overtopping is cited as the more common reason for failure rather than 'piping' in rockfill or earth dams. Overtopping generally occurs due to large inflows into the reservoir that are either caused by excessive rainfall or by the failure of an upstream dam. The resulting volume of water entering the reservoir can well be beyond spillway evacuation capacity. The overtopping failure of two embankments in 1996 on the Chicoutimi river due to excessive rainfall is a recent example in Canada.
The significant increase in global warming will, according to climatologists, result in strong variability and extremes in precipitation patterns. This means that spillways designed on the basis of expected flood return periods ("maximum expected flood") evaluated from hydrological analyses of data from previous years could prove to be dangerously inadequate if a definite shift in precipitation frequencies and intensities manifests itself. It is quite realistic to expect that the frequency of dam failure episodes attributable to climate change is very likely to increase in the future. The increasing number of recently reported failures appear to bear this out.
Simulation of embankment dam breach events and the resulting floods are crucial to characterizing and reducing threats due to potential dam failures. Development of effective emergency action plans requires accurate prediction of inundation levels and the time of flood wave arrival at a given location. If population centers are located well downstream of a dam, details of the breaching process have little effect on the result; travel time, attenuation, and other routing effects predominate. In such cases, several techniques of predicting surge wave propagation times are available. However, in a growing number of cases, the location of population centers near a dam makes accurate prediction of breach parameters (e.g., breach width, depth, rate of development) crucial to the analysis.
This work is focused on the development of a numerical model (together with some preliminary results) to predict breach formation in an embankment dam due to overtopping and forms part of a research program initiated between Hydro-Quebec, the National Sciences and Engineering Research Council of Canada and the Ecole Polytechnique de Montreal.

2. Model Development

The St. Venant equations for unsteady free surface flow in one dimension describe the conservation of mass and momentum:

A. Gyr and W. Kinzelbach (eds.), Sedimentation and Sediment Transport, 211–220.

$$\frac{\partial \zeta}{\partial t}+\frac{1}{b}\frac{\partial Q}{\partial x}=q \qquad (1)$$

$$\frac{\partial Q}{\partial t}+\frac{\partial}{\partial x}\left(Q^2/A\right)+gA\frac{\partial \zeta}{\partial x}+gAS_f=0 \qquad (2)$$

Note that they are presented here in non-conservation form

Here Q is discharge, ζ is water surface elevation above datum, A is cross sectional area, b is lateral width, q is lateral inflow, x is the streamwise coordinate and t is time. S_f is the friction slope evaluated using the Manning Formula.

The sediment continuity equation in one dimension is:

$$\frac{\partial z_b}{\partial t}+\frac{\partial s_b}{\partial x}=q_s \qquad (3)$$

Where s_b is sediment transport rate per unit width, q_s is sediment inflow and z_b is bed elevation

Equations (1) to (3) define the physics of the problem. In addition, a sediment transport relation to evaluate s_b is required. The sediment transport formula used in the computation is a slightly modified version of the expression developed by Smart (1984). In principle however, any transport formula may be chosen for the evaluation of s_b. The main difficulty of course is in the formulation of a suitable formula that incorporates the non-equilibrium, impulsive nature of the erosion.

Large amounts of suspended sediment result in non-Newtonian behavior of the mixture, and so work is currently in progress to incorporate these effects into the formulation.

2.1 NUMERICAL ALGORITHM

During the breaching process, fairly high velocities are likely to be encountered with the flow regimes changing between subcritical, transcritical and then supercritical along the downstream face of the dike. Very frequently, a hydraulic jump occurs at, or very close to the toe of the dike. An essential attribute of the numerical algorithm for the hydraulics must therefore be its ability to continuously handle different flow regimes including automatic shock capturing. There are very few algorithms that have these desirable properties. One such class of schemes is based on Riemann solvers and has been frequently described in the literature. There still remain however, certain difficulties associated with the treatment of the source terms in the governing equations.

Within the context of this project, a one-dimensional Riemann solver using various methods of flux determination has been developed. The solver is based on a two-dimensional finite volume model developed by Tchamen & Kahawita (1998).

This section is devoted to outlining the development of an alternate one-dimensional scheme that is being refined for application to the process of breach formation. Parallel development of a coupled algorithm for the erosion appears promising enough to warrant the development of a numerical model (based on these algorithms) for the complete process. The algorithm is based on a staggered grid stencil and differs from the finite volume schemes mentioned earlier using Riemann solvers which are characteristic based. Both methods provide robust solutions, however the present technique is easier to code and implement. It is loosely based on a technique proposed by Stelling et al. (1998).

The sediment transport module uses a variation of the Lax scheme together with suitable formulae for the volumetric sediment transport. The model as currently developed, has the advantage of being able to treat transitions from sub to supercritical flow and vice-versa including the appearance of shocks which are handled without any special "treatment" of the numerics in critical regions. Discretization of the equations is inspired by the work of Stelling et al. (1998). The numerical algorithm is based on a staggered grid stencil.

As indicated in Figure (1), depths and water surface elevations are defined at 'i' the center of a cell. Velocities are defined at the boundaries i+1/2 and i-1/2 of the cell. The model uses a 'positive depth' formulation for the discretisation of the continuity equation so that subject to certain restrictions, dry bed conditions may be treated. For the same reason, the bottom shear stress is modeled in an implicit manner to avoid numerical difficulties as water depth becomes vanishingly small. Thus equation (4) is an example of the discretization of the momentum equation.

$$\frac{u_{i+1/2}^{n+1}-u_{i+1/2}^{n}}{\Delta t}+\frac{1}{\overline{A}_{i+1/2}}\frac{\left(\overline{Q}_{i+1}u_{i+1/2}-\overline{Q}_{i}u_{i-1/2}\right)}{\Delta x}-\frac{u_{i+1/2}\left(\overline{Q}_{i+1}-\overline{Q}_{i}\right)}{\Delta x}$$

$$+g\frac{\left(\zeta_{i+1}^{n+\theta}-\zeta_{i}^{n+\theta}\right)}{\Delta x}+\frac{n^{2}\left|u_{i+1/2}^{n}\right|u_{i+1/2}^{n+1}}{\left(R_{i}^{n}\right)^{4/3}}=0$$

where

$$\overline{A}_{i+1/2}=\frac{1}{2}\left(A_{i}+A_{i+1}\right)$$

$$\overline{Q}_{i}=\frac{1}{2}\left(Q_{i+1/2}+Q_{i-1/2}\right)$$

$$\overline{Q}_{i+1}=\frac{1}{2}\left(Q_{i+3/2}+Q_{i+1/2}\right)$$

$$\zeta^{n+\theta}=\theta\zeta^{n+1}+(1-\theta)\zeta^{n}$$

The discretisation of the sediment continuity equation (3) follows along similar lines. The sediment transport formula used in the computation is a slightly modified version of the expression developed by Smart (1984).

2.2 INITIAL AND BOUNDARY CONDITIONS

Since the model is applied to breach formation, the following scenario is assumed. Inflow into the reservoir is assumed that causes the water surface elevation to exceed the crest level resulting in overtopping. Small scale laboratory experiments indicate that the rate of increase in water level has an important influence on the failure time of the structure. Presently, the test runs assume that a constant water surface elevation in the reservoir that slightly exceeds the crest elevation is maintained during the complete breaching process. Some runs have also used level pool routing techniques with a given inflow into the reservoir so that a dynamic variation in the water surface elevation occurs as the breach enlarges.

3. Description of Tests

The test cases are divided into two groups, the first group compares the performance of the hydraulic component of the model to some known analytical solutions. Included in this group are the simulation results for a dambreak wave formed by sudden removal of a vertical plane retaining 10 meters of water in a rectangular smooth channel of zero slope. The objectives of these tests is to ascertain whether the algorithms are robust enough to handle sub and supercritical regimes within the same computational domain.

Two cases are studied:

(A) The downstream bed is considered dry

(B) The downstream water depth is 1 meter.

Initial conditions are zero velocity everywhere and constant depths as specified earlier.

Analytical solutions to these classical problems have been known for some time based on the method of characteristics and negative wave theory. Computation of Case (A) has been repeated for a rough, dry bed with a Manning n of 0.04. No analytical solution to this problem is available.

The next series of tests that use the fully coupled model for the hydraulics and breach erosion were undertaken on an 80 meter dam. This test approximately represents the Teton dam (91 m) failure on June 5th 1976 that killed 11 people and rendered 25000 homeless. Property damage was about U$400 million. The reservoir capacity was estimated at $304 * 10^6$ m^3. For purposes of this simulation, a rectangular reservoir with a constant inflow of 500 m^3/sec was assumed. Under these conditions, overtopping of the crest by about 10cm occurs. The breach formed is assumed to be trapezoidal in section with bottom width equal to 0.32 times the depth.

The following additional parameters describing the material properties were used in this simulation:

Dam composed of homogenous material $d_{90} = 0.005$ m, $d_{30} = 0.0006$ m, d_m=0.0021 m and an s = 2.56.

4. Results

The first group of results are presented in figures (2) and (3). Figure (2) compares the water surface profile with the analytical (parabolic) solution. The results are virtually indistinguishable. It is pertinent to mention here that unlike many other schemes that require a very small but non-zero water depth downstream for computational purposes, the downstream bed in these computations used a true zero water depth. For a rough downstream bed, the water surface profile, deviates from the parabolic profile as the bed is approached, with a discontinuous front appearing at the bed. This behavior is physically intuitive and has been observed experimentally. It is pertinent to add that no analytical solution for this case is available.

Figure (3) is the result obtained with a downstream water depth of 1 meter. In this case, a discontinuous surge wave propagates on the surface of the still water downstream as required from analytical considerations. The difference between the numerical and analytical solutions is virtually indistinguishable.
We now turn to the results obtained for the Teton Dam failure. The progressive breach development as well as the water surface elevations are illustrated in Figure (4). The behavior is quite remarkable; progression is quite slow for the first
1.25 hours after overtopping, then incredible amounts of material are removed in the next six to ten minutes! The flow is supercritical almost everywhere; at 1 hour 18 minutes, a hydraulic jump is clearly indicated at the downstream toe of the dam due to material accumulation. A consequence of the large discharge and velocities towards the final stages as the reservoir empties, is the removal of most of the breach material as it is washed downstream.
Breach outflow hydrographs are indicated in figures (5) and (6). Figure (5) indicates the influence of bed roughness and figure (6) the influence of the downstream slope of the dam on the outflow hydrograph. These hydrographs compare favorably with the estimates based on observations [Ray et al. (1978)]. The bed roughness (characterized by the Manning Coefficient) does not appear to influence the magnitude of the peak discharge but does have a significant effect on its timing. This is important when minutes can make a significant difference in early warning procedures designed to save human lives. Therefore use of numerical models as predictive tools will require good estimates of surface roughness. The downstream slope (characterized by 'm' the cotangent of the slope angle), not surprisingly, does influence the timing of the peak outflow discharge as seen in figure (6). A steeper downstream slope (smaller m) results in higher velocities and more intense erosion.

The sediment discharge hydrograph is illustrated in figure (7). It is very sharp with a peak of about 14000 m^3/sec. The peak corresponds to the point of maximum discharge when large amounts of material are eroded. The total amount of material removed during this (admittedly reconstructed!) episode is $1.24 * 10^6$ m^3
A longitudinal view of the breach in plan is shown in figure (8). The computed shape seems to agree with experimental laboratory experiments, notably those of Das (1997) in India who conducted a series of experiments as part of his Ph.D. thesis. He makes the following observations:

- The breach channel once formed, displays a contracting section, a throat and a divergent section
- The breach cross-section during the erosion process is almost trapezoidal in shape in the contracting region, rectangular at the throat and again trapezoidal in the divergent section.

In these simulations, since the section is constrained to remain trapezoidal, a rectangular throat will not be predicted. Nevertheless, the results are generally encouraging and hopefully the model can be improved by trying to account for the different variables in a physically consistent manner.

5. Concluding Remarks

The model described here is far from complete. Much work needs to be done, chief of which is removal of the restriction that every cross-section remains trapezoidal. Some initial work on updating the sectional characteristics on the basis of a local shear stress appears to be promising. The effect of vegetation on the downstream erosion process is being modeled as is the effect of riprap protection. A further avenue that is under active exploration is the incorporation of non-Newtonian fluid behavior into the hydraulic formulation. Finally, the quality of the results obtained is highly dependent on the particular sediment transport formula used in the computations. More accurate formulae applicable to high speed flows on steep slopes need to be developed and validated.

6. References

Das, Pradip Kumar "Breach Erosion Process of Homogeneous Earthfill Dams and Fuse Plugs due to Overtopping Flow" Unpublished Ph.D. Thesis, Department of Civil Engineering, Indian Institute of Technology Kanpur, India. (1997)

Ray,H.A., Kjelstom L.C., Crosthwaite,E.G. and Low,W.H. "The Flood in Southeastern Idaho from the Teton Dam Failure of June 5th 1976" U.S. Geological Survey Open-File Report 77-765, Boise, Idaho. (1978)

Smart, G.M. "Sediment Transport in Steep Channels" Journal of Hydraulic Engineering, ASCE **110**, 267-276, (1984)

Stelling, G.S., Kernkamp, H.W.J. and Laguzzi, M.M. "Delft Flooding System: A powerful tool for inundation assessment based upon a positive flow simulation" Proceedings of the Third International Conference on Hydroinformatics, Babovich & Larsen, Copenhagen, Denmark. (1998)

Tchamen, G.W. and Kahawita, R.A. "Modelling Wetting and Drying effects over Complex Topography" Invited paper for Special Edition of Hydrological Processes, An International Journal, Vol. **12**,.1151-1182 John Wiley (1998)

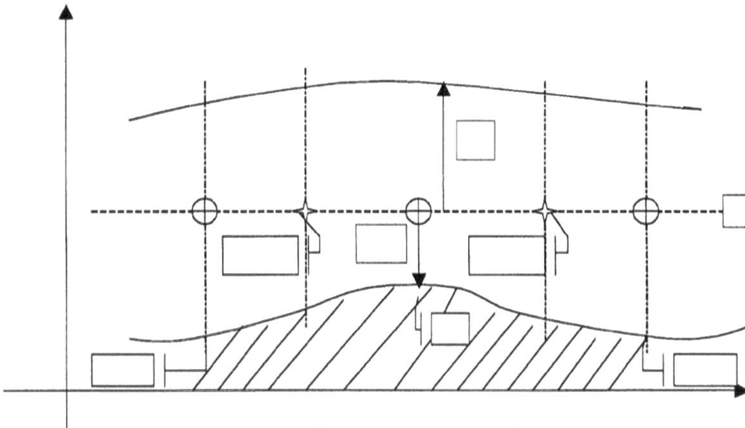

Figure 1 : Definition Sketch

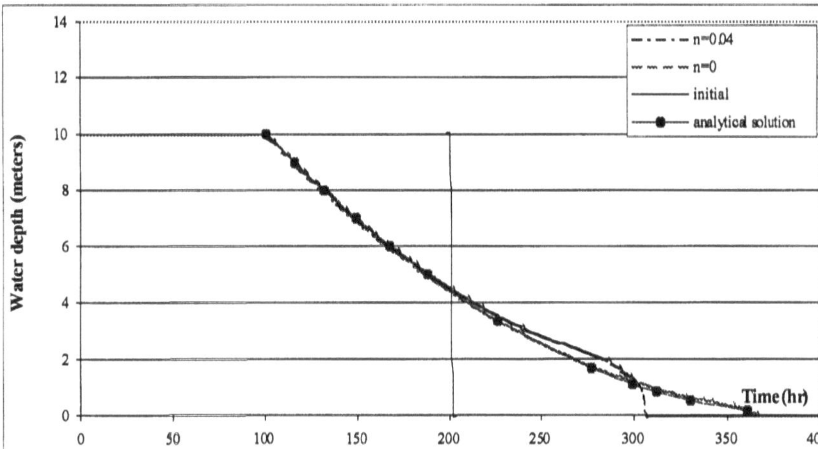

Figure 2 : Comparison of Dambreak Solution (Dry bed downstream)

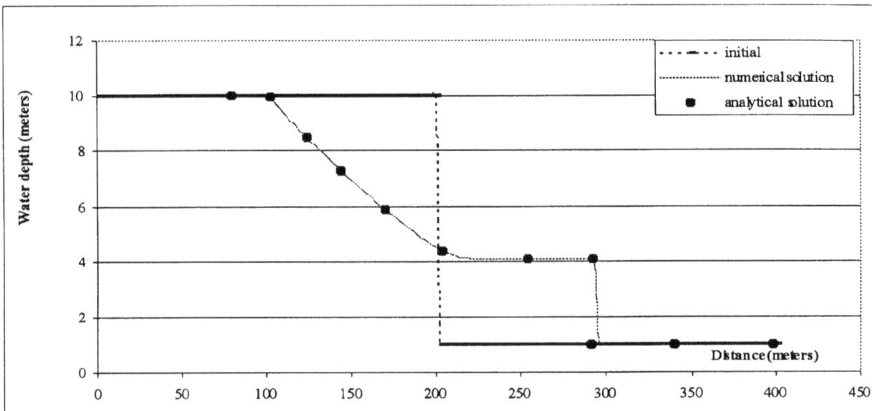

Figure 3 : Comparison of Dambreak Solution (Wet Bed downstream)

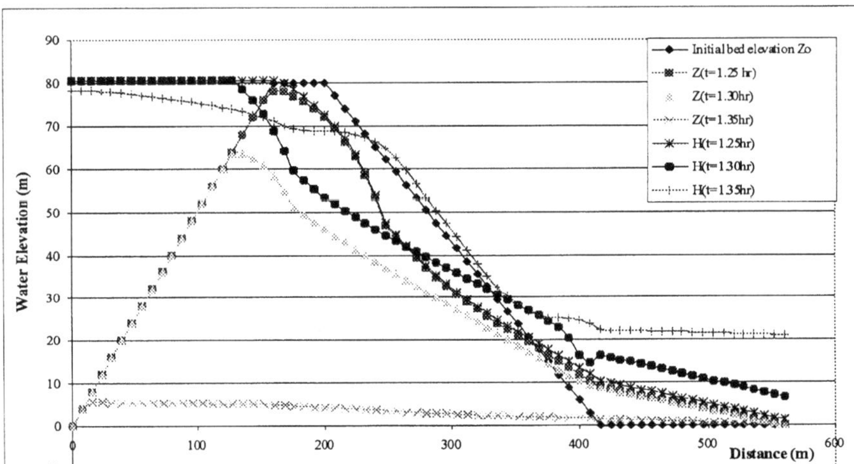

Figure 4: Teton Dam Breach Evolution

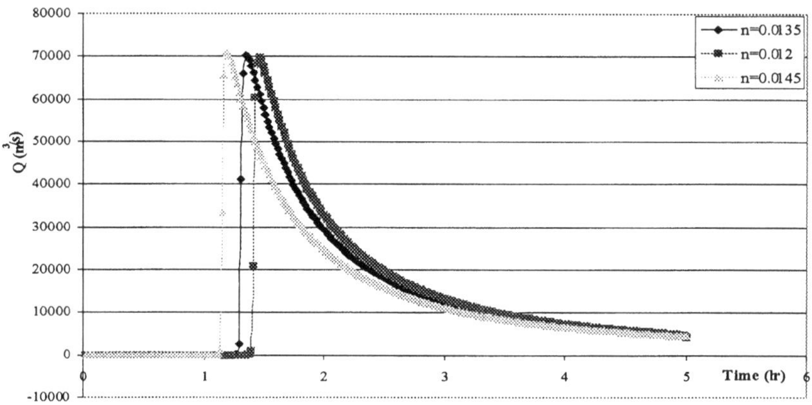

Figure 5 : Influence of Manning Coefficient on Outflow Hydrograph

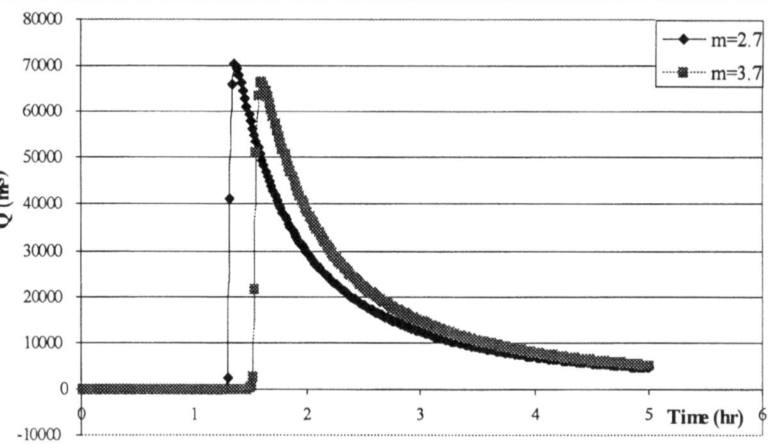

Figure 6 : Influence of Downstream Slope on Outflow Hydrograph

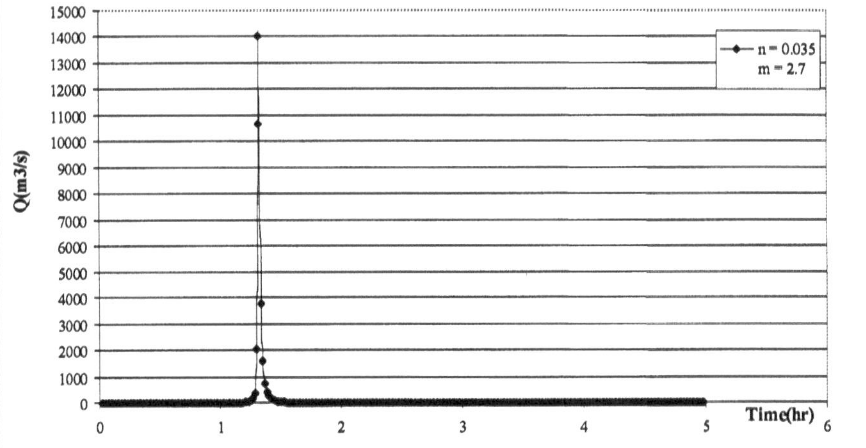

Figure 7 : Sediment Outflow Hydrograph

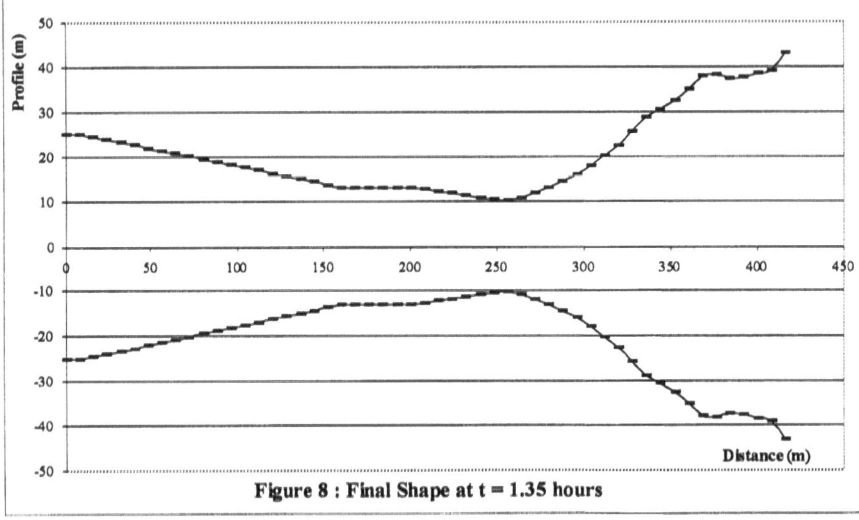

Figure 8 : Final Shape at t = 1.35 hours

GRAIN-SIZE SPECIFIC SUSPENDED SEDIMENT TRANSPORT AND FLOW RESISTANCE IN LARGE SAND-BED RIVERS

SCOTT WRIGHT and GARY PARKER
Saint Anthony Falls Laboratory Mississippi River at 3rd Avenue SE Minneapolis MN 55414 USA wrig0240@tc.umn.edu

1. Introduction

Grain-size specific suspended sediment transport can be computed, given the mean velocity and concentration profiles, by integration of the sediment flux over the flow depth:

$$q_{si} = \int_{\delta}^{h} uC_i dy \tag{1}$$

where u is flow velocity, C_i is volume concentration of grain-size i, δ is the top of the bedload layer, and h is the flow depth. Einstein (1950) used the standard logarithmic velocity profile and Rouse suspended sediment profile to develop the first grain-size specific suspended sediment transport predictor. The Rouse profile requires the near-bed concentration, and Einstein used his bedload predictor to provide this boundary condition. Also, the logarithmic velocity profile requires the near-bed velocity (or roughness height) as a boundary condition. Since Einstein, several advancements have been made in the solution of (1). Most of these have focused on the prediction of the near-bed concentration, for both uniform sediment and mixtures (see Garcia and Parker, 1991, for a review). Other researchers have focused on the roughness height, particularly in the presence of dune bedforms, and density stratification effects (Smith and McLean, 1977, McLean, 1991, 1992). This research follows along these lines, and has been driven by the desire to model large, low-slope, sand-bed rivers. Modifications to existing relations for near-bed concentration and roughness height are proposed which extend their validity to the range of interest. Also, a simplified relation is proposed to account for density stratification effects, which have been shown to be particularly relevant for low-slope, sand-bed rivers (Wright and Parker, in press A).

2. Density stratification

The logarithmic velocity profile and Rouse suspended sediment profile result from an eddy viscosity closure assumption for steady, uniform flow:

$$K_m \frac{du}{dy} = u_*^2 (1-\eta) \tag{2}$$

$$K_s \frac{dC_i}{dy} = -v_{si} C_i \tag{3}$$

where K_m is eddy viscosity, K_s is sediment diffusivity, u_* is shear velocity, $\eta = y/h$ is dimensionless vertical distance from the bed, and v_{si} is settling velocity of grain-size i. A parabolic eddy viscosity and Schmidt number of unity:

A. Gyr and W. Kinzelbach (eds.), Sedimentation and Sediment Transport, 221–227.

$$K_m = K_s = \kappa u_* h(1-\eta) \tag{4}$$

leads to the log-law velocity profile and Rouse equation, where κ is the von Karman constant.

The effects of density stratification, which inhibits vertical mixing in the same manner as stable stratification in the atmosphere, can be accounted for by modifying the parabolic eddy viscosity profile. Smith and McLean (1977) and McLean (1991, 1992) applied an empirical adjustment developed for stable stratification in the atmosphere, and Wright and Parker (in press A) used the Mellor-Yamada turbulence closure, which maintains the buoyancy terms and thus accounts for density stratification. Both approaches result in a dependence of the eddy viscosity on the Richardson number, which varies in the vertical. Wright and Parker (in press B) showed that the typical vertical variation in Richardson number in sand-bed rivers is small enough that it may be assumed constant. For constant Richardson number, the eddy viscosity adjustment is simplified greatly:

$$\frac{K_m}{u_* h} = \alpha\kappa\eta(1-\eta) \tag{5}$$

where α is a constant stratification adjustment which is a function of the Richardson number. Wright and Parker (in press B) further showed that the parameters of primary importance in the Richardson number are the slope and concentration. The methodology is further simplified by using the near-bed concentration as surrogate for the local concentration (an iteration is avoided). Data from several sand-bed rivers that span a wide range of conditions leads to the following simplified stratification adjustment:

$$\alpha = \begin{cases} 1 - 0.06\left(\dfrac{C_{5t}}{S}\right)^{0.77} & \text{for } \dfrac{C_{5t}}{S} \le 10 \\[2ex] 0.67 - 0.0025\left(\dfrac{C_{5t}}{S}\right) & \text{for } \dfrac{C_{5t}}{S} > 10 \end{cases} \tag{6}$$

where S is slope and C_{5t} is the total concentration at 5% of the flow depth, the prediction of which is discussed in the following section. The above formulation assumes a specific gravity of 2.65 for the sediment. The stratification adjusted log-law and Rouse equation then result from (2), (3), (5), and (6):

$$\frac{u}{u_*} = \frac{1}{\alpha\kappa} ln\left(30\frac{y}{k_c}\right) \tag{7}$$

$$\frac{C_i}{C_{ai}} = \left[\frac{1-\eta}{\eta}\frac{\eta_a}{1-\eta_a}\right]^{Z_{Ri}} \tag{8}$$

where k_c is a composite (skin-friction and form drag) roughness height, C_{ai} is the near-bed concentration at point η_a, and

$$Z_{Ri} = \frac{v_{si}}{\alpha\kappa u_*} \tag{9}$$

is the stratification adjusted Rouse number for grain-size i.

3. Near-bed concentration and shear stress partitioning

Several relations exist for predicting the near-bed concentration of an equilibrium suspension, most of which were developed empirically from laboratory data. The relations typically assume a primary dependence on the Shields stress due to skin-friction, and secondary dependence on other parameters determined by dimensional analysis. Garcia and Parker (1991) review several relations and proposed a new relation, and showed that new relation along with those van Rijn (1984) and Smith and McLean (1977) performed the best for a large dataset of flume experiments. The Garcia-Parker relation was further generalized for sediment mixtures using field data from the Niobrara and Rio Grande Rivers, both relatively small, steep, sand-bed rivers.

In order to use a near-bed concentration predictor in the presence of bedforms, the bed shear stress must be partitioned between skin-friction and form drag, because form drag is not applied to entrain sediment. Again, several methodologies exist for accomplishing this. Einstein (1950) computed skin-friction stress by considering a flow with equivalent mean velocity over a flat bed (with known grain roughness), and backing out the stress from the logarithmic velocity law. Smith and McLean (1977) considered the flow structure over a dune bed and used spatial averaging to develop a relation which requires the bedform dimensions to be known. Engelund and Hansen (1967) used the Einstein partitioning, dimensional considerations and similarity, along with a large dataset from flume experiments to develop a direct relation between skin-friction Shields stress and total Shields stress. Note that this also serves as a depth-discharge relation, because it can be used in conjunction with Einstein partitioning to relate composite roughness height to grain roughness height.

Preliminary calculations using the Garcia and Parker (1991) entrainment relation for mixtures (the only one formally generalized) and each of the skin-friction partitioning techniques resulted in gross over-prediction of the discharge-weighted mean concentration for data from large, low-slope sand-bed rivers. This is because the relations, particularly the Garcia-Parker entrainment relation and Engelund-Hansen partitioning, were developed using flume data and field data from small, sand-bed rivers. Here data from several large, low-slope sand-bed rivers have been used to illustrate these deficiencies (see Wright and Parker, in press B, for details of the dataset). Figure 1 shows the measured versus predicted near-bed entrainment (concentration per bed fraction) of grain-size i using the Garcia-Parker relation (and Einstien stress partitioning) for the Rio Grande and Mississippi Rivers. Note the over-prediction of entrainment for the low-slope Mississippi. Figure 2 shows skin-friction Shields stress (τ_{*sk}) (computed using a stratification adjusted Einstein partitioning, Wright and Parker, in press B) versus total Shields stress (τ_*), for several sand-bed rivers, along with the Engelund-Hansen relation. Note the over-prediction of skin-friction stress for the larger sand-bed rivers (Mississippi, Atchafalaya, Red).

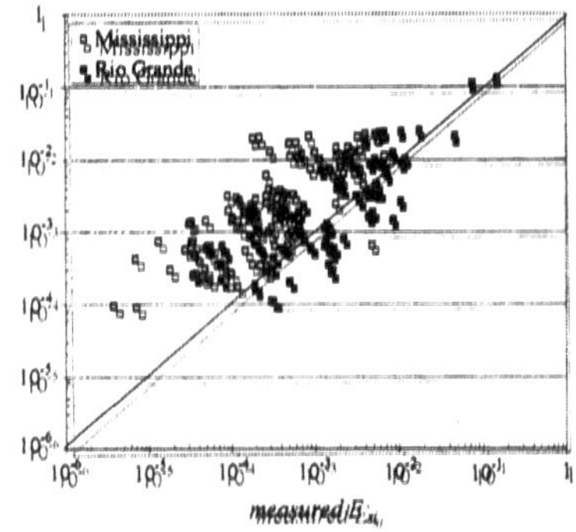

Figure 1 Performance of the Garcia-Parker entrainment relation.

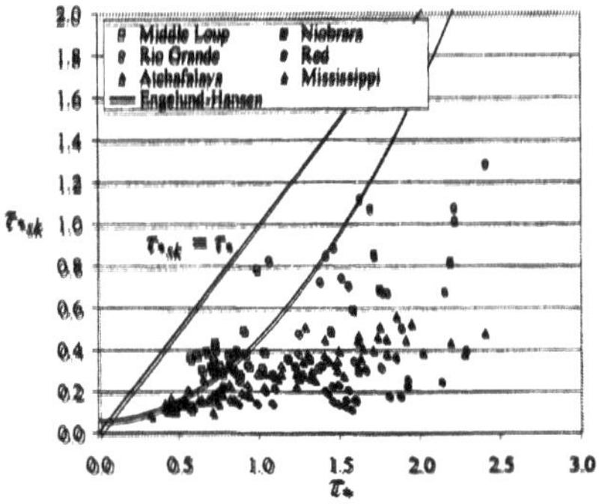

Figure 2 Performance of the Engelund-Hansen stress relation.

In the development of the Garcia-Parker entrainment relation the slope (or equivalently h/D_{50}), which comes out in the dimensional analysis, is dropped because the range in the data is insufficient. The addition of the Mississippi River data allows the inclusion of the slope, and here it is simply folded into the entrainment parameter in a power form. The final relation becomes:

$$E_{si} = \frac{B(\lambda X_i)^5}{1 + \frac{B}{0.3}(\lambda X_i)^5} \tag{10}$$

$$X_i = \left(\frac{u_{*sk}}{v_{si}} Re_{pi}^{0.6}\right) S^{0.07} \left(\frac{D_i}{D_{50}}\right)^b \tag{11}$$

$$b = \begin{cases} 0.2 \ \ for \ \ \frac{D_i}{D_{50}} \le 1 \\ 0.7 \ \ for \ \ \frac{D_i}{D_{50}} > 1 \end{cases} \tag{12}$$

where $E_{si} = C_{5i}/F_{bi}$ denotes concentration (at 5% of depth) per fractional bed content, F_{bi}, of sediment in the ith grain-size range, $\lambda = 1 - 0.28\sigma_\phi$ represents the overall suppression of entrainment due to mixture effects, σ_ϕ is the standard deviation of the bed sediment on the sedimentological ϕ-scale, u_{*sk} is shear velocity due to skin-friction, $Re_{pi} = \sqrt{RgD_i}D_i/\nu$ is particle Reynolds number, ν is kinematic viscosity, D_i is characteristic diameter of the ith grain-size range, D_{50} is median grain diameter of the bed material, and $B = 5.7 \times 10^{-7}$.

Similarly, the Engelund-Hansen relation can be extended to the range of large, low-slope, sand-bed rivers. A primary difference between large rivers and their smaller, steeper brethren is the Froude number (Fr) attained at bankfull discharge. While smaller rivers may approach $Fr \approx 1$, larger rivers tend to realize $Fr \approx 0.2 - 0.3$ even for very high flows (Wright and Parker, in press B). Given that Fr is an important parameter governing bedform stability (Engelund 1970, Smith 1970) and transitions

(Vanoni 1974), it is not surprising that smaller rivers tend to transition to upper regime plane bed at a lower total Shields stress than larger rivers (Julien and Klaasen, 1995). This is illustrated in Figure 2, where skin-friction stress is seen to remain lower at higher total stress for the larger rivers. The effect can be parameterized by using the grouping $\tau_* Fr$ in place of only τ_* in the Engelund-Hansen relation. The best collapse of the data was obtained with the following relation:

$$\tau_{*sk} = 0.05 + 0.7\left(\tau_* Fr^{0.7}\right)^{0.8} \qquad (13)$$

Figures 3 and 4 show the performance of the modified relations for near-bed concentration and skin-friction shear stress respectively. Figure 3 illustrates the collapse of the Mississippi data by including slope in Garcia-Parker relation, and Figure 4 shows the collapse achieved by incorporating the Froude number into the Engelund-Hansen relation.

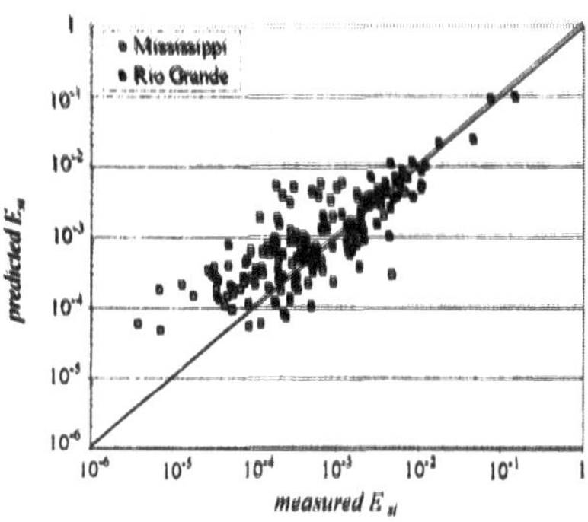

Figure 3 Performance of modified entrainment relation.

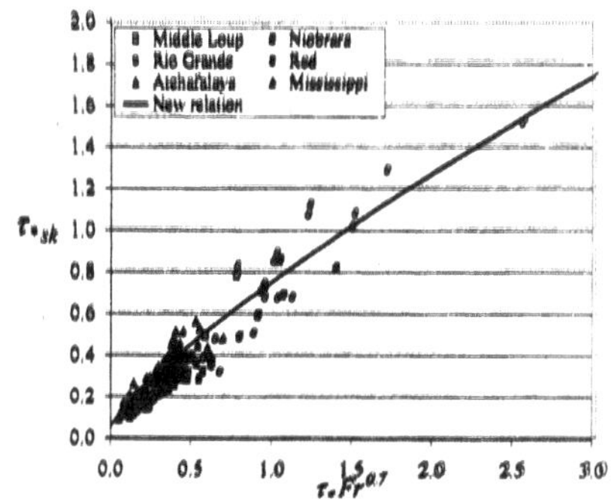

Figure 4 Performance of the modified stress relation.

4. New transport relation

The stratification adjusted profiles, (7) and (8), can be used along with the modified Garcia-Parker and Engelund-Hansen relations to develop a grain-size specific suspended sediment transport predictor. First, an equivalent power-law form of the velocity profile is introduced, which simplifies the integration (primarily for development of a depth-discharge predictor, see Wright and Parker, in press B):

$$\frac{u}{u_*} = \frac{9.70}{\alpha}\left(\frac{y}{k_c}\right)^{1/6} \qquad (14)$$

Substitution of (14) and (8) into (1) yields the following expression for the bed-material suspended sediment transport rate:

$$\frac{q_{si}}{u_* h} = \frac{9.70}{\alpha} C_{5i}\left(\frac{h}{k_c}\right)^{1/6} I \qquad (15)$$

$$I = \int_{\eta_a}^{1} \eta^{1/6} \left(\frac{1-\eta}{\eta} \frac{\eta_a}{1-\eta_a} \right)^{Z_{Ri}} \tag{16}$$

The integral in (16) has no analytical solution and thus must be determined numerically. Since η_a is a constant set equal to 0.05 here, I is a function of Z_{Ri} only. The integral was computed numerically for a range of Z_{Ri} typical of sand suspensions, leading to following analytical approximation:

$$I \cong \begin{cases} 0.679 \exp\{-2.23 Z_{Ri}\} & for\ Z_{Ri} \le 1 \\ 0.073 Z_{Ri}^{-1.44} & for\ Z_{Ri} > 1 \end{cases} \tag{17}$$

Equations (15), (16), and (17) comprise the final relation. Required input includes depth, slope, and bed material grain-size distribution. The stratification adjustment, α, is given by (6). The concentration at 5% of the flow depth of grain-size i is given by (10)-(12). The skin-friction Shields stress (and shear velocity) is given by (13). The composite roughness, k_c, is determined from the Einstein partitioning technique (Wright and Parker, in press B):

$$\frac{k_c}{k_s} = \left(\frac{\tau_*}{\tau_{*sk}} \right)^4 \tag{18}$$

Thus k_c can be computed from (13) and a relation for k_s, such as $k_s = 3D_{90}$ (van Rijn, 1984). Finally, (14) can be integrated over the flow depth to yield a relation between depth and discharge (see Wright and Parker, in press B, for details).

The performance of the new relation (for discharge-weighted mean concentration, C_m) for three large, low-slope sand-bed rivers is shown in Figure 5. For comparison the method was also applied to the same dataset, but without the inclusion of the stratification adjustment, and with the original Garcia-Parker entrainment relation along with the original Engelund-Hansen shear stress partitioning. Without the modifications to the relations presented here, the concentration is grossly over-predicted, because Engelund-Hansen over-predicts skin-friction shear stress and Garcia-Parker over-predicts near-bed concentration for large, low-slope, sand-bed rivers.

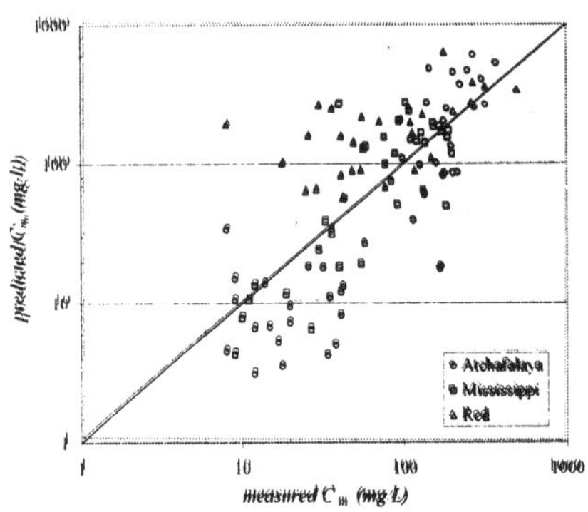

Figure 5 Performance of the new transport relation.

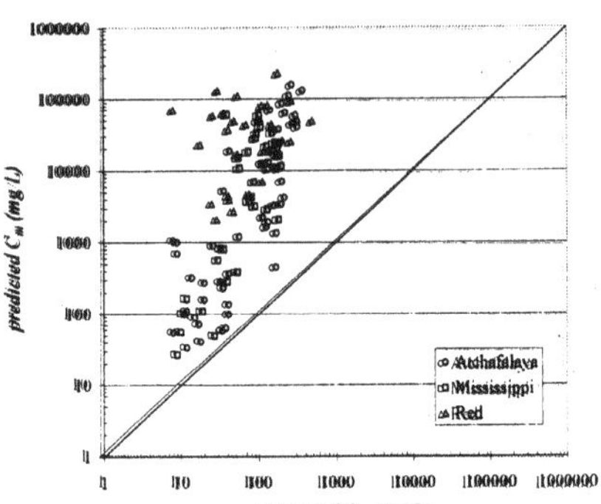

Figure 6 Performance using the non-modified auxiliary relations.

5. Summary

A new relation for the prediction of grain-size specific suspended sediment transport has been presented. The logarithmic velocity profile and Rouse suspended sediment profile were modified to account for the effects of density stratification on turbulent mixing, then integrated over the flow depth to obtain the transport rate. The near-bed concentration boundary condition, on a grain-size specific basis, is obtained from a modified version of the relation of Garcia and Parker (1991), which was shown to over-predict near-bed concentration for conditions typical of large, low-slope, sand-bed rivers. Skin-friction/form drag partitioning is accomplished using Einstein partitioning (modified for stratification effects) in order to develop a modified version of the relation of Engelund and Hansen (1967), which was also extended to the range of conditions of large, sand-bed rivers. The modified Engelund-Hansen relation also provides the near-bed velocity boundary condition (composite roughness height). The new relation was compared to the relation using the original versions of Garcia-Parker and Engelund-Hansen and without stratification effects, and significant improvements were shown. Without modification of the original relations, the discharge-weighted mean concentration may be over-predicted by one or two orders of magnitude for large, low-slope, sand-bed rivers. Though the methodology has been compared only to data from large, sand-bed rivers here, the relations should be of sufficient generality for application over the range of sand-bed rivers. However as always, comparison of the predictions with field data from a particular site is recommended before using the method for design or management purposes.

References

Einstein, H.A., 1950. "The bed load function in open channel flows", United States Department of Agriculture, Technical Bulletin No. 1026, September.

Engelund, F., 1970. "Instability of erodible beds", *Journal of Fluid Mechanics*, 42(2), 225-244.

Engelund, F., and Hansen, E., 1967. *A monograph on sediment transport in alluvial streams.* Technical University of Denmark, 62 pp.

Garcia, M., and Parker, G., 1991. "Entrainment of bed sediment into suspension", *Journal of Hydraulic Engineering*, 117(4), 414-435.

Julien, P.Y., and Klaasen, G.J., 1995. "Sand-dune geometry of large rivers during floods", *Journal of Hydraulic Engineering*, 121(9), 657-663.

McLean, S.R., 1991. "Depth-integrated suspended-load calculations", *Journal of Hydraulic Engineering*, 117(11), 1440-1458.

McLean, S.R., 1992. "On the calculation of suspended load for noncohesive sediments", *Journal of Geophysical Research*, 97(C4), 5759-5770.

Smith, J.D., 1970. "Stability of a sand bed subjected to a shear flow of low Froude number", *Journal of Geophysical Research*, 75(30), 5928-5940.

Smith, J.D., and McLean, S.R., 1977. "Spatially averaged flow over a wavy surface", *Journal of Geophysical Research*, 82(12), 1735-1746.

Vanoni, V.A., 1974. "Factors determining bed forms of alluvial streams", *Journal of the Hydraulics Division*, 100(3), 363-377.

van Rijn, L.C., 1984. "Sediment transport, part II: suspended load transport", *Journal of Hydraulic Engineering*, 110(11), 1613-1641.

Wright, S., and Parker, G., in press A. "Density stratification effects in sand-bed rivers", *Journal of Hydraulic Engineering*.

Wright, S., and Parker, G., in press B. "Flow resistance and suspended load in sand-bed rivers: simplified stratification model", *Journal of Hydraulic Engineering*.

TRANSPORT THRESHOLDS IN GRAVEL-BED RIVERS

PETER C. KLINGEMAN Ph D
Oregon State University, Civil, Construction and Environmental Engineering Department, Corvallis, Oregon USA 97331-2302 peter.klingeman@orst.edu

Abstract

Flow conditions that produce incipient motion of bedload at threshold conditions in coarse-bed rivers are illustrated using data from the Oak Creek, Oregon, bedload research facilities. A dynamically varying transport conditions occur between threshold and full transport. The threshold for initiation of motion is not sharp. A second threshold, usually overlooked, occurs on the recession limb of runoff events and differs not only in "critical" discharge value but also in composition of the bedload and dynamics of the reforming stable bed. Bed reformation and subsequent disturbance relate to hydrograph dynamics and changing turbulence in ways that are not yet well described.

1. Introduction

The exposed stable surface of a gravel-cobble bar at the edge of a river at low flow shows at least seven features about the bar material. These particle features -- size, size gradation, mineralogy, shape, imbrication, hiding, armoring -- are all relevant to the consideration of transport thresholds in gravel-bed rivers.

Specific flow conditions that cause initiation and cessation of bed material motion are significant in channel dynamics, channel morphology, and engineering design. Past studies of the initiation of particle motion gave little attention to cessation of motion. It may have been assumed that the two conditions and thresholds were identical or that the second condition was not important for stable channel design.

2. Transport Thresholds and Incipient Motion Considerations for Bed Material

Most studies of sediment transport thresholds focus on the initiation of particle motion -- the "beginning" or "incipient" or "threshold" condition when a particle first starts to move after a long period of rest or stability. The particle becomes "entrained" in the flow. Once entrained, the particle may continue to move sporadically or steadily. With sporadic stop-and-go particle motion, it is not clear that a motion threshold has actually been passed, whereas with general particle motion (other than brief stops), it may be assumed that the threshold has been passed. The actual threshold for motion is fuzzy, rather than sharp. The precise discharge and moment at which incipient motion occurs in a channel is a subjective determination. Some observers consider it to occur when the first few particles start moving; others believe it only occurs when particles move over a significant portion of the bed surface.

A. Gyr and W. Kinzelbach (eds.), Sedimentation and Sediment Transport, 229–236.

Thus, incipient motion may cover a small range of discharges, depending on one's choice of definition.

While partly a matter of definition, this threshold condition is also "blurred" by the physics of the bed (e.g., relative sizes of bed material present), the recent deposition history for the exposed particles, whether or not nearby particles have been disturbed, and flow turbulence. Incipient motion is the threshold between a stable state and a dynamic state for particles in the bed of a stream. This is generally due to increasing water discharge, such as on the rising limb of a hydrograph during a runoff event caused by storm runoff, snowmelt, or increased releases of water from upstream reservoirs.

This is the first transport threshold passed. A second "threshold" is passed on the falling limb of the same hydrograph event. This is the cessation of the particle's motion and its return to a static, stable condition of inactivity.

3. Prediction of Bed Stability and Incipient Motion

Evaluations of the flow conditions that produce incipient motion of coarse river-bed material date back many decades. Most early results were based on laboratory work with relatively uniform material placed in flumes and involved few test parameters. Field validations often raised questions that could not be addressed in the prediction equations, which tended to retain simplifications or adjustment coefficients.

The main parameters that influence transport thresholds and incipient motion of individual particles are shown in Table 1, grouped by category with some annotations. Additional parameters that influence the overall transport threshold for the stream and the collective incipient motion of the channel bed are shown in Table 2. For point bedload sampling to assess incipient motion and to select a critical discharge, the parameters given in Table 1 would be of greatest influence. For cross-channel sampling, the parameters given in Table 2 could have more influence on the choice of critical discharge for incipient motion.

Many parameters listed in Table 1 are not included in prediction equations. For example, particle shape may determine whether a particle is imbricated in the bed surface and affect the local flow strength needed for particle disturbance; yet it does not appear in prediction equations. Other parameters listed in Table 1 may be considered as involving too much detail. For some, data are not available. For others, there may be data but the manner of incorporation into predictions is unknown. For most parameters, an "averaging" approach may be considered adequate (e.g., mean particle size if the bed is heterogeneous or average particle density if the bed has mixed mineralogical characteristics). Parameters given in Table 2 are morphologic in nature. To date, most have not been included in equations used for engineering design or morphologic description.

With so many parameters involved in incipient motion, many investigators favor a stochastic approach to determining incipient discharge, velocity, or shear stress. The simplest of these may be to make a few measurements of discharge and determine the corresponding values for a parameter such as mean boundary shear stress, at times of limited but differing bedload transport, plot the results, assume linearity, and extrapolate back to zero discharge to get the critical shear stress!

Table 1. Parameters Affecting Transport Thresholds for Individual Particles in Gravel-Bed Rivers

- Sediment Characteristics
  - Representative size or sizes
  - Size gradation
  - Particle density (based on mineralogy)
  - Particle shape (based on mineralogy and transport history)
  - Particle orientation in the bed or bar (ability to withstand flow stresses)
  - Deposition history
- Fluid Properties
  - density (based on type of fluid and temperature)
  - Viscosity (based on temperature)
  - Dissolved solids (affecting electric charge, dispersion/attraction)
  - Suspended solids (potentially affecting fluid behavior and viscosity)
- System Extensive Properties
  - Gravitational attraction
- Flow Characteristics and Flow Structure
  - Flow depth
  - Flow velocity
  - Velocity distributions (velocity profiles, lateral/longitudinal changes)
  - Velocity gradients at boundaries (determining local shear stresses)
  - Boundary layer thickness (relative to particle sizes)
  - Flow patterns (converging/diverging flows; wakes)
  - Flow turbulence (eddies, time-varying conditions)

Table 2. Additional Parameters Affecting General Transport Thresholds for the Channel System in Gravel-Bed Rivers

- Channel Characteristics
  - Transverse shape and compactness
    - (affects depth and hydraulic radius for a given discharge)
  - Longitudinal slope (affecting stream power and tractive force)
  - Presence & distribution of rigid boundary material (e.g., bedrock)
  - Presence & distribution of flexible boundary material (e.g., vegetation)
  - Presence & distribution of anomalous bed material
    - (e.g., logs or large boulders on a gravel bed)

4. Observations of Particle Initiation and Associated Transport Processes

Research at Oak Creek, Oregon (Milhous 1973; Helland-Hansen et al. 1974; Heinecke 1976; Saluja 1982; Mohammadi 1986; Matin 1994) has provided detailed data describing incipient motion and bedload transport processes in a gravel-bed stream. Data were collected using a unique vortex bedload sampler that allowed extraction of the entire bedload. Studies included bedload transport, flushing flows for natural beds and beds placed in an adjacent large flume, particle shape effects, and movement of individual coded tracer particles at both sites. Detailed spatial sampling of surface and sub-surface "layers" of bed material, continual measurement of bedload during runoff events, and analysis of discrete size fractions of the bedload have provided much information regarding bedload transport processes. (The vortex sampler was operated almost continually during rainy-season runoff. Thus, nearly all of the bedload for the year was collected, separated into short-interval samples, and analyzed.) Comparisons were made to relate moving particles of many sizes to the size gradation of the underlying bed material. Oak Creek findings were compared with those at other field sites and general relations were developed (Klingeman & Emmett 1982; Parker et al. 1982; Parker & Klingeman 1982; Diplas 1987; Shih and Komar 1990a, 1990b). The facility was "retired from service" in 2001.

Beds of typical gravel-bed streams such as Oak Creek may be stable each year for all but a few days or a short season when flows may be large enough to exceed incipient-motion values. Incipient motion does not occur everywhere on the bed surface at the same time. Hence, spatial sampling of the bed is helpful for understanding initial particle movement and effects of channel morphology.

Incipient motion does not begin with only the smaller particles in the bed surface. Milhous (1973) showed that particles subject to incipient motion are similar to most sizes present in the bed surface. Small particles rest in depressions of the natural heterogeneous bed surface or are protected (hidden) in the wakes of larger particles. Helland-Hansen et al. (1974) showed that a disturbance-blowout process may occur which allows large and small particles to move together at conditions near incipient motion. Particle vibration is one indication that bed motion is about to begin. Exposed particles respond to the turbulence of passing flow, which causes fluctuating pressure differences and shear stresses that lead to fluctuating lift and drag forces on a very short time scale. As individual particles were dislodged from an otherwise stable-looking bed, flow entered the vacated pocket, altering local water velocities near other particles (and hence lift and drag forces acting on them), which helped mobilize several neighboring particles almost simultaneously to cause a local "burst" or "blowout" of a group of particles before the bed was again stable. Disturbed particles moved along the bed until finding new resting places in weaker currents. As stream discharge increases, such disturbances and blowouts became more frequent. At this time, it may become subjective as to whether the bed is more at rest or more in motion, but a transport threshold is clearly being crossed.

Milhous (1973) showed that the surface layer (armor layer; pavement) of a gravel-bed stream acts as a "reservoir" for fines such as sand. It also controls the transport threshold and associated bedload transport by preventing sand and finer bed material from being entrained in the flow unless armor particles are first moved. When the bed is stable, the flow only "sees" the surface. Therefore, to predict incipient motion, care must be taken to sample only the bed surface to determine the particle size distribution and calculate a "critical" discharge for disturbing the armor layer.

Milhous (1973) described how recession flows restock this reservoir of fines as the armor layer redevelops. Bedload transport continues during recession flows as discharges become progressively smaller than those that initiate particle motion during rising limbs of hydrographs. Large particles become stable and small particles begin to settle into the newly created void spaces. Later, as these fill, the smaller particles continue to be transported over the armor surface until they reach zones of weaker shear stress or until flow has diminished so much that they drop out of transport. This process causes differences to occur in the sizes moving at any given flow as recession continues, with progressive changes among transport rates for sand and small gravel. Sand transport may continue well out into the recession.

Beschta and Jackson (1979) demonstrated the infilling of a clean gravel armor layer by fines in the sand size range, as well as sand flushing behavior for a stable armor layer. Mohammadi (1986) found that flushing flows imposed on a relatively uniform gravel bed (e.g., flume) can clean the void spaces in the top layer quite effectively at discharges smaller than those needed to cause incipient motion of the gravel. Some flushing even occurs below the top layer of the bed, particularly near riffles. But if flows are applied to bed material that is more heterogeneous, flushing dies not appear to be as thorough.

Differences have been found in "critical discharge" values dividing bed stability and incipient motion in rising limbs of runoff hydrographs for several events measured in sequence, even with enough intervening time for full bed stability to be achieved (Milhous 1973). Apparently, differing residual physical effects occur during recessions that can cause a range of "critical" discharges to occur over time. This contributes to difficulties in specifying a single critical value.

Milhous (1973) also showed that bedload and suspended load were still measurable, but variable with time, at discharges below the "critical" discharge for the armor layer. A plot of water discharge versus sediment load for a season or for a series of runoff events produced "scatter" which might be attributed to uncertainty in the value of critical discharge. However, the chronological sequence of sampling shows that other processes are involved: (1) variability in incipient motion conditions from one storm runoff event to another, such that prior runoff can influence incipient motion for the following event; and (2) bedload transport on the recession limb continuing at values below the critical discharge value used for mobilization, at rates differing from transport rates on the rising limb.

Saluja (1982) extended these observations and found that there is no unique bedload relation that includes both the rising limb and falling limb of runoff events. For all discharges spanning rising and falling limbs of the events, the representative median size for all samples was larger on the rising limb than on the falling limb, supporting observations of sustained transport of smaller bedload during recession flows.

Many particles in a natural bed have shapes that allow them to rest in imbricated, hydrodynamically stable orientations and resist flow entrainment (e.g., flat disk-like particles). This should make some shapes less likely to be disturbed than others. In contrast, particles deposited rather abruptly and randomly, such as from a quick drop of discharge, might not take on such resisting orientations. Komar and Li (1986) showed that interactions of individual particles with neighboring particles was important for initiation of motion in Oak Creek. They proposed a pivoting angle between particles for use in stability evaluations and found that imbricated and flat particles had higher pivoting angles and greater stability near threshold conditions. Matin (1994) found that gravel particles of mixed shapes that were not well integrated into the armor layer initiated motion in a manner that is independent of particle shape; there was no evident selectivity for initiation of motion based on particle shape.

Matin (1994) confirmed earlier work that flows smaller than the critical flow for incipient motion of a gravel bed flushed out not only small particles from the bed, but also larger particles. At these low flows and extremely low transport rates, the transport nevertheless appeared to be functionally related to the strength of the flow. This behavior supports a probabilistic view of particle motion due to random turbulence. It furthermore points out the weakness of a concept for a sharp threshold of movement.

Matin (1994) made several conclusions regarding general bedload transport in Oak Creek, based on division of bedload and bed material samples into several size classes and weight distributions of individual particles. There is a tendency to produce an equal mobility of various particle-size fractions (and weights) within the total bedload. A progressive increase occurs in the median sizes of bedload (especially in sizes of the larger particles involved in transport) as water discharge increases. There is a mimicking of bed material composition by bedload as discharge increases, including the high degree of skewness and progressive reduction of frequencies of the smaller size fractions present.

5. The "Equal Mobility" Approach to Analysis of Bedload Transport

One outgrowth of Oak Creek sediment transport research was development of several predictive equations for bedload transport as the data base grew (Milhous 1973; Heinecke 1976; Saluja 1982; Matin 1994). These tended to be site-specific because no data from other streams were included.

Parker merged Oak Creek data with other field data sets for coarse-bed rivers to develop generalized relations (Parker et al. 1982; Parker & Klingeman 1982). These were based on properties of the subarmor bed material and involved non-dimensional analysis and curve fitting techniques. Ten particle-size ranges were used to describe bed material and bedload. An outgrowth of this 1982 work was use of the term "equal mobility" to describe observed behavioral features of bedload transport in gravel-bed streams. According to the equal mobility hypothesis, bed armor regulates entrainment of particles by the stream, resulting in various sizes being approximately equal in mobility, with particle sizes transported at rates proportional to their presence in bed material. For a wide range of sizes present in the bed, bedload thus typically contains all sizes and tends to match the bed material in size gradation. The equal mobility hypothesis has been supported by subsequent studies but remains controversial. The initial Parker 1982 equations relied on coefficients based on Oak Creek data, due to its greater detail than other field data sets. This provided a barrier for some applications if investigators did not collect independent field data to obtain new coefficients.

Shih and Komar (1990a, 1990b) used a different approach to evaluate transport rates in Oak Creek. They found a systematic evolution of the bedload particle-size distributions, demonstrating that the distributions become progressively coarser and more skewed with increasing flow stage. They noted that Parker's 1982 equations made the equal mobility assumption as a first-order approximation to develop bedload relationships for Oak Creek data, yielding reasonable results but not accounting for observed variations in bedload particle-size distributions at different flowrates. They proposed higher-order solutions for predicting transport rates that did not assume equal mobility, as also did Diplas (1987).

Dawdy and co-workers used extensive field measurements in several Oregon gravel-bed streams to make refinements to the Parker 1982 method for more general bedload computation (Bakke et al. 1999). The resulting P-K model may be calibrated for any gravel-bed river and offers a spectrum of analysis options.

6. Suggestions for Evaluation of Transport Thresholds

There is a fundamental need to refine methods of observing transport thresholds so that we may make progress in understanding the dynamic processes involved. Based on modern instrumentation, experiments and sampling systems may be devised that allow accurate measurements of parameters and transport rates. This will provide better knowledge of the dynamic transient conditions involved and may lead to improved prediction equations that include additional parameters.

7. References

Bakke, P.D., P.O. Basdekas, D.R. Dawdy & P.C. Klingeman. Calibrated Parker-Klingeman Model for Gravel Transport. Jour. of Hydraulic Engineering, ASCE. **125**, No.6. 657-660, 1999.

Beschta, R.L. & W.L. Jackson. The intrusion of fine sediment into a stable gravel-bed. Jour. of the Fisheries Resource Board of Canada. **36**. 204-210. 1979.

Diplas, P. Bedload transport in gravel-bed streams. Jour. of Hydraulic Engineering, ASCE. **113**. 277-292, 1987.

Heinecke, T. Bedload Transport in a Gravel-Bottomed Stream. MS Report. Civil Engineering Department, Oregon State University, Corvallis. 45 pp + appendices, 1976.

Helland-Hansen, E..Stability of Gravel at Artificially Created Spawning Beds in Uncontrolled Streams. MS Report. Civil Engineering Department, Oregon State University, Corvallis. 156 pp, 1971.

Helland-Hansen, E., R.T. Milhous & P.C. Klingeman. Sediment Transport at Low Shields-Parameter Values. Jour. of the Hydraulics Division, ASCE. **100**, No.HY1, Jan. 261-265, 1974.

Klingeman, P.C. & R.T. Milhous. Evaluation of Bed Load and Total Sediment Yield Processes on Small Mountain Streams. Civil Engineering Department, Oregon State University, Corvallis. 133 pp., 1970 Combined with other reports and reprinted as Studies on Effects of Watershed Practices on Streams. Water Pollution Control Research Series 13010 EGA 02/71. US EPA.

Klingeman, P.C. & W.W. Emmett. Gravel Bedload Transport Processes. In Gravel-bed Rivers. Hey, Bathurst & Thorne, eds. John Wiley & Sons. 1982. Chap. 7, 141-179, 1982.

Komar, P.D. & Z. Li. Pivoting analysis of the selective entrainment of sediments by shape and size with application to gravel threshold. Sedimentology. **33**, 425-436, 1986.

Matin, H Incipient Motion and Particle Transport in Gravel-Bed Streams. Ph.D. Thesis. Oregon State University, Corvallis,. 268 pp, 1994.

Milhous, R.T Sediment Transport in a Gravel-Bottomed Stream. Ph.D. Thesis. Oregon State University, Corvallis. 232 pp, 1973.

Milhous R.T. & P.C. Klingeman. Bed-Load Transport in Mountain Streams. Abstracts of Papers Presented. Proc. of 1971 ASCE Hydraulics Division 19th Specialty Conference. Ames, Iowa.. pp 85, 1971.

Milhous, R.T. & P.C. Klingeman. Sediment Transport System in a Gravel-Bottomed Stream. Hydraulic Engineering and the Environment. Proc. of 21st Annual Hydraulics Division Specialty Conference, ASCE. Bozeman, Montana. 293-303, 1973.

Mohammadi, A. Flushing of Fine Sediment from a Coarse Streambed. MS Report. Civil Engineering Department, Oregon State University, Corvallis 120 pp + appendices, 1986.

Parker, G., P.C. Klingeman & D.G. McLean. Bedload and Size Distribution in Paved Gravel-Bed Streams. Jour. of the Hydraulics Division, ASCE **108**, No.HY4. 544-571, 1982.

Parker, G. & P.C. Klingeman. On Why Gravel Bed Streams Are Paved. Water Resources Research. **18**, 1409-1423, 1982.

Saluja, H. Oak Creek Bed Load Transport Relations for a Generally Stable Thalweg Armor Layer. MS Report. Civil Engineering Department, Oregon State University, Corvallis 82 pp., 1982

Shih, S.M. and P.D. Komar. Hydraulic controls of grain-size distribution of bedload gravels in Oak Creek, Oregon, USA." Sedimentology. **37**, 367-376, 1990a.

Shih, S.M. and P.D. Komar. Differential bedload transport rates in a gravel-bed stream, a grain-size distribution approach. Earth Surface Proc. Landforms. **15**, 539-552, 1990b.

ADVANCEMENTS IN SEDIMENT TRANSPORT INVESTIGATIONS USING QUANTITATIVE IMAGING TECHNIQUES

MARIAN MUSTE and KWONKYU YU
IIHR – Hydroscience & Engineering, The University of Iowa, Iowa City, IA, 52246-1585, U.S.A., E-mail address: marian-muste@uiowa.edu

1. Introduction

A cursory review of the recent literature and scientific developments regarding sediment transport show a growing awareness that research in this area is stagnating. Moreover, there is critical and enduring need to improve understanding of the fundamentals of sediment transport processes. New approaches, new strategies, and new methods evidently are required attaining that knowledge, especially in coupling between water and sediment motions.

Extensive research efforts in the last few decades have only partially elucidated the fundamentals and complexities of sediment transport. Owing to the lack of adequate theoretical relationships for quantifying the interactions between suspended particles and the carrier fluid, there remains heavy reliance on laboratory flume experiments and field measurements to reveal the mechanisms underlying sediment transport. Laboratory investigations on suspended-sediment transport however, are complex undertakings with respect to attaining a uniform open-channel flow with sediment transport in equilibrium as well as to selecting appropriate measurement instruments in order to avoid error (Muste, 2002).

The advent of the quantitative imaging instruments will significantly accelerate insight into sediment-transport processes. Not only do such instruments possess the requisite temporal and spatial resolution, but, especially important, they are non-intrusive and measure simultaneously two velocity components for both flow phases over a selected flow area. Currently, an increased number of studies of multi-phase flows make increasing use of various imaging instruments. As this paper shows, the instruments and the experimental methods developing with them provide opportunities to investigate the complexities of sediment-laden flows.

Presented herein are measurements and insights obtained with Particle Tracking Velocimetry (PTV) and Particle Imaging Velocimetry (PIV), two very useful image-based instruments for investigating multiphase flows. The full use of these instruments (singly or together) for illuminating sediment transport is still under scrutiny. This paper explores the capabilities of these techniques to provide new insights into the structure of suspended sediment flows, arguably the simplest case of sediment transport. Inferences based on raw images, instantaneous vector fields, and spatial and temporal averages applied to the instantaneous velocity field measured by PTV and PIV also are presented. Special emphasis is given to the potential of these techniques to investigate essential aspects of the particle-fluid interaction in sediment-laden flows as well as in sedimentation and erosion processes.

A. Gyr and W. Kinzelbach (eds.), Sedimentation and Sediment Transport, 237–240.

2. Inferences based on raw flow images

Presented below are images and results obtained in sediment-laden, open-channel flow carrying natural sand (d_{50} = 0.23 mm, and volumetric concentrations from 0.0005 to 0.0015). Flow depth is h = 0.02 m, slope 0.01, and maximum velocity of about 1 m/s. The experiments were conducted at the Research Center for Urban Safety and Security, Kobe University (Japan) using high-speed and high-resolution imaging systems. Details about the setup of the experiments, imaging system, and processing algorithms are provided in Muste et al. (2002).

Sample views of instantaneous and averaged images obtained with the two imaging systems are shown in Figure 1. Quantification of particle images enables distinguishing water tracers from sediment particles. Increased phase discrimination accuracy is obtained by way of a buffer size range (based on calibrations) that is discarded from analysis. Phase separation becomes increasingly difficult in flows with multi-size sediment. A convenient phase discrimination alternative is color-coding of the flow phases.

Raw images allow non-intrusive quantification of particle number and of size distributions over flow depth; whence sediment concentrations can be estimated. Qualitative concentration distributions can be also obtained by superposing several images over appropriate time interval (see Figure 1.b). Raw image analysis also provides free-surface location, its oscillation range, and particle trajectories (see Figure 1.c). These flow and sediment characteristics are difficult to obtain using alternative instruments.

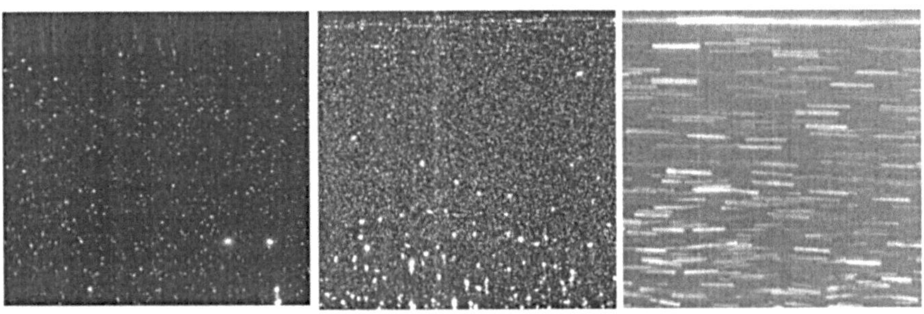

a) HR system (1 frame) b) HR system (75 frames) c) HS system (10 frames)

Fig. 1. Recordings with high-resolution (HR) and high-speed (HS) systems

3. Flow characteristics obtained using PTV-PIV analysis

The PTV processing algorithm locates and sizes individual particles in the image, and it determines vectors associated to each particle type; i.e., water tracers and sediment. Consequently, the raw PTV data consist of instantaneous whole-field velocity vectors. Depending on the type of imaging system used, the velocity vector field can be analyzed using the Eulerian (high-speed or high-resolution systems) or Lagrangian (high-speed system) approach.

Temporal averages applied to instantaneous velocity vector fields using the Eulerian approach lead to important characteristics of the sediment transport governing equations, e.g., mean vector field and turbulence intensities for the flow phases. Sample of such measurements are shown in Figure 2. The mean velocity profiles show that the hydrodynamics of the phases is close, but distinct. There is a lag between the streamwise and vertical velocities of the water and sediment in the mixture. The sediment vertical velocity is not equal to the 0.03 m/s settling velocity of the sediment in still water. The streamwise and vertical turbulence intensities plotted in Figure 2 illustrate turbulence modulation induced by sediment. Turbulence modulation depends on sediment concentration, particle size, and turbulence temporal and spatial scales, hence, it can widely vary over the depth for the same flow.

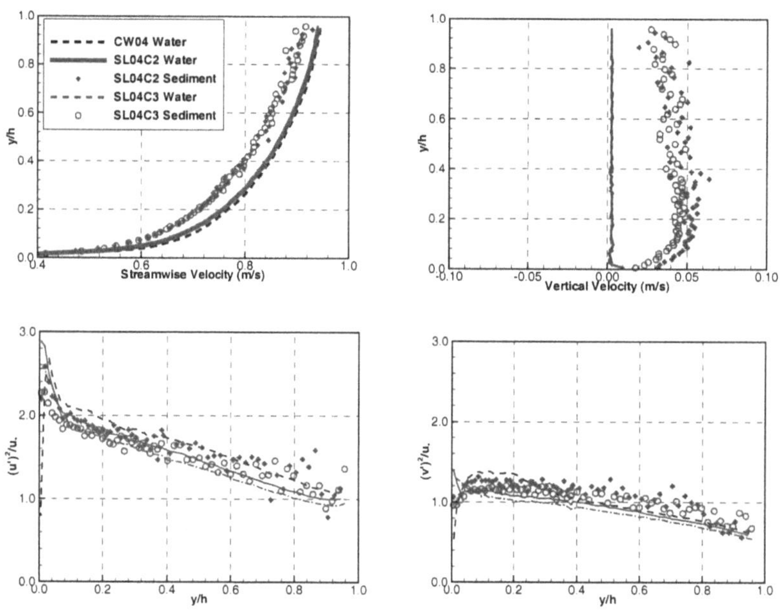

Fig. 2. Water and sediment characteristics in a clear water flow and in two flows of increasing sediment concentration

In addition, the Eulerian investigative approach provides sediment diffusivity, double and triple correlations involved in the energy budget of the mean and turbulent transport equations, as well as stresses and quandrant analysis applied to them. Quadrant analysis is especially relevant for clarification of the energy transport mechanisms through the ejection and sweep mechanisms and for substantiating the connection between the instantaneous and local flow features and the mean flow behavior. Proposed in Figure 3 is a new, spatio- and temporal-averaging method for determining sediment diffusivity using PTV and PIV measurements applied to images collected with a high-resolution imaging system.

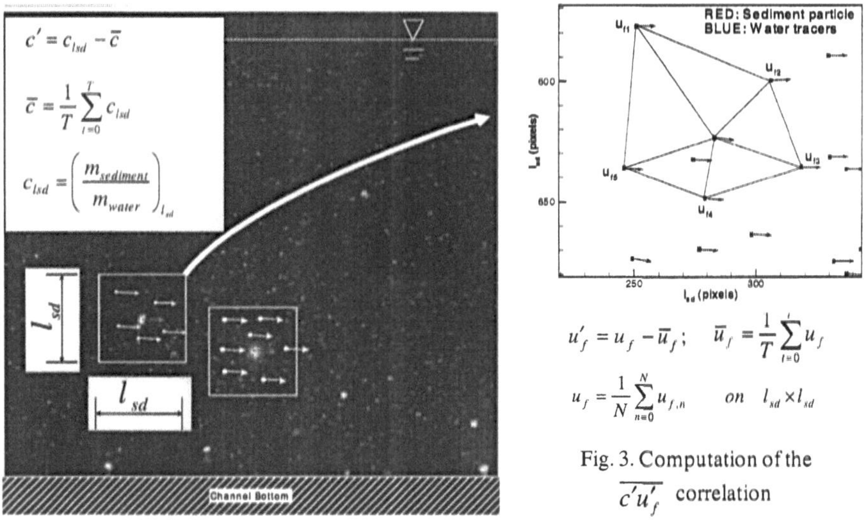

Fig. 3. Computation of the $\overline{c'u'_f}$ correlation

A wealth of information on the kinematics, dynamics, and scales of turbulence in sediment-laden flows that can be obtained using the Langrangian analysis of PTV-PIV measurements collected with high-speed imaging systems. Vectorial (velocity, vorticity) and scalar (sediment concentration) measurements over a range of spatial and temporal scales enable unprecedented insights into the dynamics of the particle-fluid interaction including investigation of the effect of turbulent coherent structures on fundamental processes of sediment transport (i.e., entrainment, deposition, suspension) and of the two-way interaction between flow and sediment.

4. Conclusions

This paper illustrates capabilities of PTV and PIV measurements to illuminate flow and sediment behavior in sediment-laden flows. Of special importance is the ability of these image-based techniques to non-intrusively measure whole-field instantaneous velocities, separately for the two phases of the flows. This ability is unmatched by any alternative measurement methods. Although not very popular in sediment research, the use of the imaged-based instruments with two-phase discrimination capabilities has brought significant advancements in the experimental investigations conducted by other scientific communities. The increased level of complexity associated with the use of the two-phase flow approach might sound overstated for engineering hydraulics, but reliable model development and formulation of predictive relationships for sediment transport require comprehensive and sophisticated experimental approach, well beyond the approaches used in earlier investigations.

5. References

Muste, M. (2002). "Sources of Bias Errors un Flume Experiments on Suspended-Sediment Transport" *J. Hydr. Res.* 40(6), pp. 695-708.

Muste, M. and Fujita, I., and K. Yu (2002). "New Insights into Suspended Sediment Flow Structure Using Particle Tracking Velocimetry," *Proceedings* ASCE-IAHR Joint Conference, Estes Park, CO.

A LID-DRIVEN ELONGATED ANNULAR FLUME (LEAF) FOR THE DETERMINATION OF SEDIMENT TRANSPORT PROPERTIES

Onyx Wing-Hong WAI
Department of Civil & Structural Engineering
The Hong Kong Polytechnic University, Hung Hom, Kowloon, Hong Kong

1. Introduction

Predictions of the fate and transport of sediment are possible only if the properties of near-bed erosion and deposition of the sediment are known. These properties are the erosion rate, deposition rate that is the product of the settling velocity and the sediment concentration, and the critical shear stresses of erosion and deposition. The properties can be determined in laboratories using reconstructed seabed sediments and known flow-generated shear stresses. For cohesive sediment experiments, circular annular flumes with rotating lids are popularly used to avoid disturbing the interaction between particles in the water column. The major drawback of this type of flume is the centrifugal acceleration which causes the cross-sectional velocity distribution to become non-uniform. A Lid-driven Elongated Annular Flume (LEAF), designed to reduce the centrifugal effect, has been built. Figure 1 is a picture of LEAF. The flume has two identical 3-m long straight sections which are meant to create a uniform flow environment. The 2 semi-circle sections have an inner radius of 0.2m and an outside radius of 0.4m. The flow width of the flume is 0.2m. The width of the lid is approximately 0.2m which is almost equal to that of the flume. The water depth and vertical position of the lid can be adjusted freely. The lid is driven by an adjustable speed motor. This paper presents a numerical study of the hydrodynamic characteristics and an experimental study of the sediment transport behavior in LEAF.

Fig.1. LEAF.

2. Numerical study

Since the hydrodynamics in the annular flume is very complicated, the flow patterns in LEAF were studied by a fully 3-D commercial hydrodynamic model—Fluent. The

A. Gyr and W. Kinzelbach (eds.), Sedimentation and Sediment Transport, 241–244.

model solves the continuity and momentum equations for dynamic pressure and 3D velocity field. The turbulent viscosity is determined by the k-ε turbulence model. The lid moves in clockwise direction to drive the water in the surface. The tangential velocity (V_T) of the moving lid in the circular sections is equal to $\Omega \times r$, where $\Omega = 2u_c/(R_1+R_2)$; r is the radius; u_c is the velocity of the moving lid in the straight section; and R_1 and R_2 are the internal and external radius, respectively.

A 3D transformed grid system is used and Figure 2 shows the horizontal and vertical grids in the circular section. The staggering grid and finite volume methods are used in discretization of the transformed governing equations. The SIMPLE numerical scheme is used to solve the discretized equations.

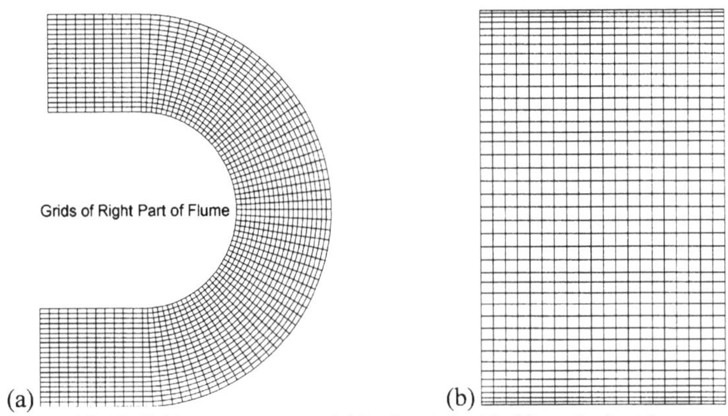

Fig. 2. Grid arrangements: (a) horizontal grid; (b) vertical grid.

In the numerical study, flow depth of 30 cm and lid velocity of 1 m/s were used. The model domain was divided into 421 (in longitudinal direction) × 21 (in lateral direction) × 41 (in vertical direction) grids. The horizontal grid size is $2 \times 1 \text{cm}^2$. In the vertical direction, considering the velocity gradient being greater near to the interfaces, the grid arrangement is denser near the surface and near the bottom. The initial values of velocity component, kinetic energy and kinetic dissipation were set to 0. After 2800 iterations, a converged solution was obtained.

Two ultrasonic velocity profile monitors were installed 1 m upstream from the starting point of the circular reach. The velocity distributions in the cross-section of the flume were measured at different depths. Figure 3 shows the measured and simulated velocity distribution in the cross-section at a water depth 18 cm from the surface. Based on the measurement, the optimal roughness heights of the lid, bottom and sidewall are 0.003m, 0.001m and 0.001m, respectively.

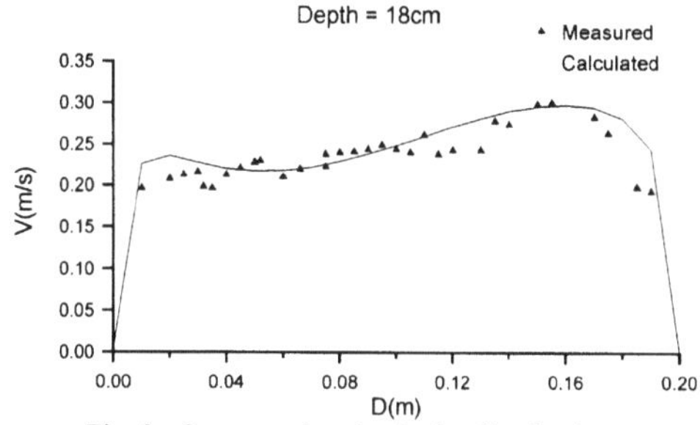

Fig. 3. Cross sectional velocity distribution.

3. Resuspension/Deposition Experiments

Standard test sands with diameter ranging between 90μm~150μm were distributed on the flume bed. The rotating lid penetrated the water surface by about 2 mm to ensure proper contact with the water surface. By rotating the lid, turbulent shear flows can be generated. Also, by changing the rotating speed, it is possible to vary the sediment transport process. Two UVP transducers and two turbidity sensors were used to measure the flow field and suspended sediment concentration, respectively. Turbidity signals with a sampling duration of 125 min at two levels in the flume were measured during the experiment. In the first five minutes, water in the flume was static and the turbidity signals reflected the background turbidity in water before the lid started to rotate. The lid rotating speed was then gradually increased from 0 to 0.8 m/s. Ten minutes later, the speed of the lid increased to 1.2 m/s in order to entrain all the remaining sediment particles from the bottom. This speed was kept for 20 min. After that, the lid speed was slowly lower and the lid was stopped 30 min later. In order to calibrate the turbidity signals, samples close to the turbidity sensors were drawn from the flume at an interval of 10 min during the experiment. The concentration of each sample was determined by filtration, drying and weighing. A laser sensor (LISST1-100) was also used to determine the particle size distribution of the samples. Each time a sample was drawn, the same amount of clear water was added back into the flume. About 15 samples were extracted from each stage of the experiment. The relationship between the turbidity signal and the sediment concentration was estimated by the linear regression analysis. In Figure 4, it is clearly shown that the sediment concentration (turbidity signal) increased and decreased in response to the change of the flow field in LEAF. This indicates that the present flume can be applied to study the sediment erosion and deposition activities near the sediment-water interface.

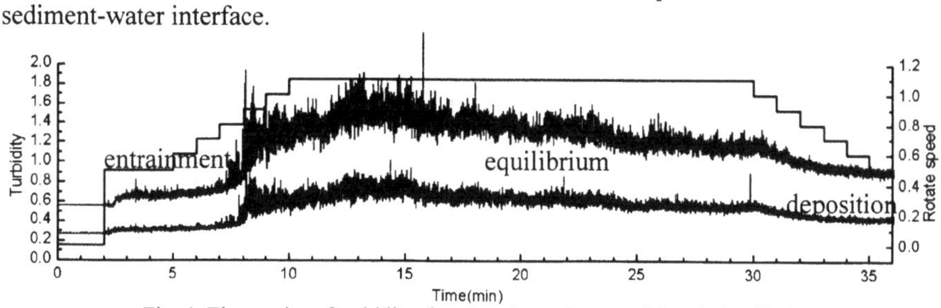

Fig. 4. Time series of turbidity signals and rotating speed in m/s (top line).

4. Concluding Remarks

A Lid-driven Elongated Annular Flume (LEAF) has been built to determine the transport properties of cohesive sediment. A preliminary flow pattern study, based on measurements by ultrasonic velocity sensors and simulations by a computational fluid dynamics software (FLUENT), shows that the flow and bottom shear stress remain fairly uniform in the straight sections. Erosion and deposition experiments of uniform size sediment were also conducted. Two acoustic turbidity meters were installed at two vertical locations to monitor the variation of suspended sediment concentrations. A laser sensor (LISST-100) was used to determine the particle size distribution of samples drawn frequently from a probe located near the bottom. The numerical results and experimental results presented in this paper show that LEAF is reliable to determine the transport properties of sediment such as erosion and deposition rates. This information is essential for accurate prediction of the fate and transport of sediment suspended in the water column which may have large spatial and long term temporal impact on the environment and aquatic life. In addition, LEAF is also suitable for carrying out long-term chemical-sediment sorption experiments because it has a well control environment and is made of non-reactive materials.

Acknowledgements

The funding support for this project was partly provided by the Hong Kong Research Grants Council's CERG grant (grant number: PolyU 5063/99E). Additional funding was sought from a Hong Kong Polytechnic University internal project (grant number: G-T219). The support is appreciated.

References

1. Graham, D. I., James, P. W., Jones, T. E. R., Davies, J. M. and Delo, E.A. (1992). "Measurement and Prediction of surface shear stress in annular flume." *J. Hydraulic Eng.*, 118, pp. 1270-1286.
2. James, P. W., Jones, T. E. R. and Stewart, D. M. (1996). "Numerical and experimental studies of annular flume flow." *Appl. Math. Modeling*, 20, pp. 225-231.
3. Lau, Y. L. and Krisnappan, B. G. (1992). "Size distribution and settling velocity of cohesive sediments during settling." *J. Hydraul. Res.*, IAHR, 30(5), pp. 673-684.
4. Sheng, Y. P. (1989). "Consideration of flow in rotating annuli for sediment erosion and deposition studies." *J. Coastal Research, Spec. Issue*, 5, pp. 207-216
5. Peterson, O. and Krishnappan, B. G. (1994). "Measurement and analysis of flow characteristics in a rotating circular flume." *J. Hydraul. Res.*, 32, 483-494.
6. Yang, Z., Baptista, A. and Darland, J. (2000). "Numerical modeling of flow characteristics in a rotating annular flume." *Dynamic Atmosphere and Oceans*, 31, pp. 271-294.

PETROLEUM PATCH TRANSPORT IN MARINE AND COASTAL ZONES

TATYANA S. KRASNOPOLSKAYA, VYACHESLAV V. MELESHKO
Institute of Hydromechanics National Academy of Sciences of Ukraine, Zhelyabova Str. 8/4, 03680 Kiev, Ukraine; t.krasnopolskaya@tue.nl
Department of Theoretical and Applied Mechanics, Kiev National Taras Shevchenko University, Volodymirska Str. 64, 01033 Kiev, Ukraine; meleshko@univ.kiev.ua

1. Introduction

The transport of pollutants is an important issue in a variety of natural systems. This study deals with distributive mixing of petroleum patches in tidal flows. A good approximation of such flows was suggested by Zimmerman [1]. He deviated from earlier attempts at modeling tidal flows by means of turbulence theory and adopted the idea of chaotic advection, first put forward by Aref [2]. In this pioneering paper it was shown that chaotic mixing of a passive tracer (which in our case will be the petroleum patch) may occur even in deceptively simple flow systems. Zimmerman [1] realized that the macroscale flow in a tidal sea is not much more complicated than chaotic advection model and so should lead to similar phenomena. In particular, he suggested that the many fine striations often observed in scalar dispersal were due to chaotic motion driven by the macroscale, rather than microscale turbulent mixing. By numerical simulations he demonstrated that the patterns formed by dispersed passive particles were very similar to field observations. In the recently published tutorial paper Aref [3] described numerous applications of chaotic advection in fluid mechanics, in general, and in geophysical and geological flows, in particular. It was shown in many other works [4-6] how important it is to identify the parameters and conditions that lead to widespread chaotic advection. In the course of flow evolution an initially designated interface of any petroleum patch may become extremely convoluted, and it appears difficult to follow pattern structures in full detail. In this study we use the special algorithm developed in our previous paper [6], in which extra contour (surface) points are automatically added when needed. Attention will also be given to a quantitative estimation of the transport quality based upon measures first suggested by Welander [7], such as coarse-grained values of density and entropy. For a more complete description of the petroleum patch distribution in a tidal sea we may use the intensity and scale of segregation [6]. The former one presents an average of the size of a clumps of petroleum. Although the transport pattern usually is not clumpy, the coarse-grained representations over an investigation area of the square density, intensity of segregation and entropy demonstrate that this pattern look clumpy, showing a "residence place" for the petroleum material at any instant.

2.Chaotic mixing in the tidal areas

In the present study of the petroleum patch transport in tidal areas we use Zimmerman's kinematical model [1, 4] which is based upon a superposition of a tidal and a residual current flow field. The kinematical model describes a motion of a mathematical point that moves at each instant with the velocity corresponding to its instant position. The

A. Gyr and W. Kinzelbach (eds.), Sedimentation and Sediment Transport, 245–248.

petroleum particle is supposed to be inertialess, not subjected to diffusion or interfacial tension. The two-dimensional advection equations in this model read:

$$\frac{dx}{dt} = \pi\lambda\nu \sin\pi x \cos\pi y + \pi\lambda \sin(2\pi t),$$

$$\frac{dy}{dt} = -\pi\lambda\nu \cos\pi x \sin\pi y, \tag{1}$$

where the parameters λ and ν are not independent. We use the values of these parameters for a tidal area of the Wadden Sea [4], where ν could be less than 0.34 and decreases with increasing of λ ($0.1<\nu<4.0$) to zero. The residual time independent current field is an infinite sequence of clockwise and anticlockwise rotating eddies. Streamlines of the residual velocity field divide the whole area into square cells of equal area, with elliptic points in the centers and hyperbolic points in the corners of cells. The tidal field plays a role of a pertubation of the Hamiltonian system (the stream function is the Hamiltonian for the residual flow). In general this pertubation leads to chaotic dynamics and could be studied in different ways, for example, by means of Poincaré sections. However, in pollutant spreading problems, we are interested in the short term history of pollutant transport. Poincaré sections present the history of motion of points in some area during a long time interval, say, during a thousand periods of a tidal flow. (For the problem of petroleum patch transport this corresponds to the history of one point during almost one and a half year.) On the contrary, ecological considerations demand that disastrous spread of pollution has to be stopped in days or weeks. Therefore, we need to know which part of the Eulerian space will be polluted in a short time and, more importantly, how much petroleum will leak to some specific part of a sea. The orbit expansion method [4], developed for a quantification of the chaotic transport and exploited an assumption that the contributions of tidal and residual currents are of different orders (the tidal is much stronger), does not give answers to those questions. In our case, it is important to know not the mixing region (where presumably mixing is instantaneous) obtained by the long time tool – Poincaré sections, -- or the rate of material exchange (which could be high in a very narrow domain), but how uniformly this mixing region is distributed over the whole area during a specific finite interval of time. We suggest a different approach for an estimation and quantification of pollutant transport.

3. Evaluation of chaotic transport

Here we briefly present a methodology for the quantification of the chaotic transport [6] based on the statistical quantities such as a coarse-grained density. First step is to present a petroleum patch, for example, as a circular blob continuously occupying some part in marine or coastal zone. Then we use a contour tracking algorithm that conserves both area and topological properties (connectedness and non-self-intersection) to find the blob's boundary in Eulerian space at any moment of time. In our algorithm the key idea is the use of a non-uniform distribution of points at the initial contour to present this interface in such a way that (i) the distance between neighbouring points remains between some chosen values (for that points are added when the distance becomes long enough and points are removed when it becomes too short) and (ii) the angle between any neighbouring straight lines is larger than some prescribed value. The principal advantage of our algorithm is that area preservation of the blob enclosed by the contour is guaranteed even after high stretching and complicated folding. Knowing the position of the contour line (the boundary of the petroleum patch) we can construct an Eulerian description of the mixing process, giving an opportunity to quantify mixing at any moment of time.

The next step is dividing the whole area of investigating marine zone S into N square boxes of a side size δ with an area $s = \delta^2$ each, then the marine zone area can be written as $S = Ns$. The conservation of petroleum area enclosed by the contour line permits us to introduce a probability function for the petroleum distribution inside a box with number k as proportional to value of area P_k occupied by petroleum in this box. A ratio of P_k and s denoted as

$$D_k = \frac{P_k}{s} \tag{2}$$

can be called a probability density or a coarse-grained density. If we average this ratio over the whole marine zone by calculating the sum of D_k divided by the number of boxes N, we get the constant value of ratio of the total petroleum area P^* to S. This value does not change in the course of mixing and is the mean density $\langle D \rangle = P^*/S$. Here angle brackets denote an average over the marine zone. If we use the square density, we get the inequality

$$< D^2 >= \frac{1}{N} \sum_{k=1}^{N} D_k^2 = \frac{1}{S} \sum_{k=1}^{N} D_k P_k < \frac{P^*}{S} \ . \tag{3}$$

Welander [7] proved that the uniform ideal mixing (in other words, transport of the initially designated blob material everywhere) is characterized by the minimum of square density $< D^2 >$ and this minimum is $< D >^2$. Thus, the value of the square density may be used as a measure of deviation from the uniform mixing, and the coarse-grained density P_k in each box shows the amount of petroleum transport each moment in time into the concrete box k. If it is zero, there is no pollutant at all in that area, and a box with density P_k equal to one is completely covered by petroleum.

In simulations of system (1) the characteristic parameters λ=2.0 and ν=0.3 for a tidal area like the Wadden Sea were chosen, and the circular petroleum patch with the radius 0.3, initially located at the point (3.0,3.0), was considered. Figure 1 shows the highly ramified petroleum patch with increasingly fine structure and the coarse-grained representation of the mixing pattern after 9 periods of the tidal flow in the zone with 6 cells of the residual current in x and in y direction .

The answer to the question of how much petroleum is transported from one cell to another is obvious from the coarse-grained representation on Figure 1 (b), where the sizes of boxes are equal to sizes of cells, and because initially the petroleum patch occupied 0.07 part of each of the central four boxes shown as the darkest after 9 periods of transport. The dynamics in time of the value of squared coarse-grained density shows also how far from uniform (when petroleum is everywhere) mixing is the present state. At any instant the coarse-grained representation of the petroleum patch distribution clearly demonstrates which boxes are empty, and which have the petroleum material and how much.

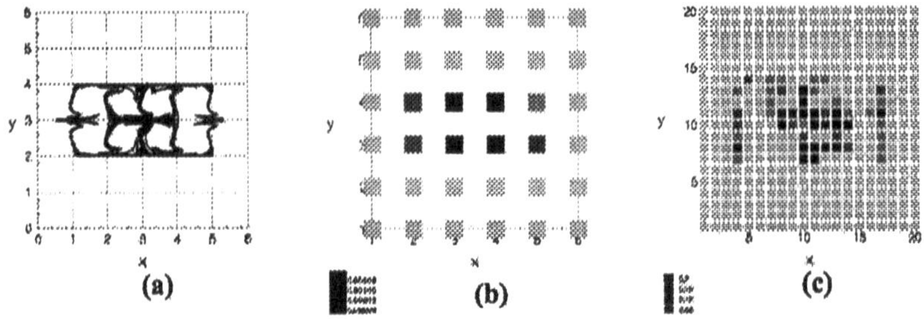

Fig 1. The deformation of petroleum patch (a), and coarse-grained representations in boxes with the sizes (b) δ=1 and (c) δ=0.3. The lightest boxes correspond to areas without petroleum.

It may also be used to show the exchange of material inside boxes, if we study this process and construct a special exchange matrix for the tidal flow in the area under consideration. Then using this matrix we can predict transport of petroleum from any place (any box) in the area to an arbitrary location in the tidal sea and determine the time when it will happen.

4. Conclusion

In this paper, we have decribed how to the study of transport properties of pollutants, such as petroleum patches, in tidal flows. Such study is important for suggesting ecological control strategies leading to effective cleanup. Our investigation was focused on basic characteristics of Lagrangian advection of particles in flows, namely, on mechanisms of the transport, which are provided by hyperbolic points of flows and their unstable manifolds. Attention was also given to a quantitative estimation of transport quality.

References

1. Zimmerman, J.T.F. The tidal whirlpool: a review of horizontal dispersion by tidal and residual currents, Neth. J. Sea Res. **20**, 133-154, 1986.
2. Aref, H. Stirring by chaotic advection, J. Fluid Mech. **143**, 1-24, 1984.
3. Aref, H. The development of chaotic advection, Phys. Fluids. **14**, 1315-1325, 2002.
4. Beerens, S. P., Ridderinkhof, H. and Zimmerman, J. T. F. An analytical study of chaotic stirring in tidal areas. In: Chaos Applied to Fluid Mixing (ed. H. Aref & M. S. El Naschie), Pergamon Press, 267-285, 1995.
5. Meleshko, V.V., Krasnopolskaya, T.S., Peters, G.W.M. and Meijer, H.E.H. Coherent structures and scales of Lagrangian turbulence. In: Advances in Turbulence VI (ed. S.Gavrilakis, L.Machiels and P.A. Monkewitz), Kluwer, 601-604, 1996.
6. Krasnopolskaya, T.S., Meleshko, V.V. , Peters, G.W.M. and Meijer, H.E.H. Mixing in Stokes flow in an annular wedge cavity, Eur.J. Mech.B/Fluids. **18**, 793-822, 1999.
7. Welander, P. Studies of the general development of motion in a two-dimensional ideal fluid, Tellus. **7**, 141-156, 1955.

VIII

Measuring Techniques

VELOCITY DERIVATIVES IN TURBULENT FLOW OBTAINED FROM 3D-PTV MEASUREMENTS

B. LÜTHI AND U. BURR
IHW, ETH Hönggerberg, 8093 Zürich, Switzerland, luethi@ihw.baug.ethz.ch

Ongoing studies show the feasibility of simultaneously tracking two kinds of particles with a three-dimensional particle tracking technique (3D-PTV). Here we report the attempts to use 3D-PTV for studying the field of velocity derivatives and material elements of turbulent flows. Among other things the technique is also expected to provide a better understanding of the basic mechanisms governing particle-laden flows. The procedure for obtaining the tensor of velocity derivatives along particle trajectories from 3D-PTV measurements is outlined. Results related to vorticity, strain and material elements are presented. They are in good agreement - both qualitatively and in many ways quantitatively - with those known from other physical and numerical experiments.

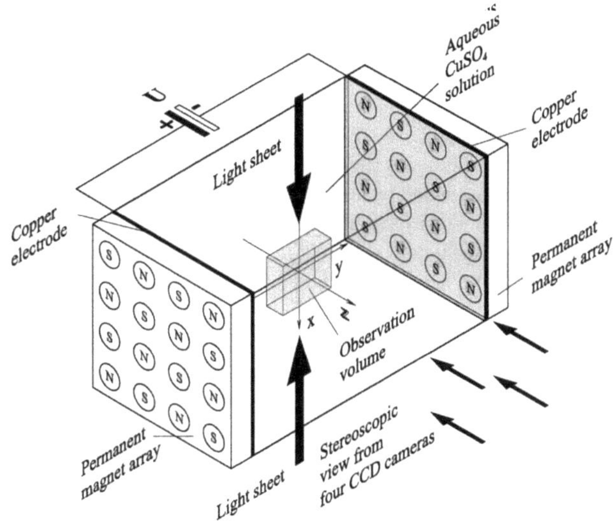

FIGURE 1. Schematic of the experimental facility and coordinate system (x,y,z). Electromagnetic forcing from two opposite walls produces swirls over each magnet. At short distances of the walls the flow becomes three dimensional and towards the observation volume fully turbulent. The flow is recorded by four CCD cameras at a recording rate of 60Hz.

A quasi-isotropic turbulent flow field is produced in a $120 \times 120 \times 140mm^3$ water tank using electromagnetically forcing from two opposite walls. The forcing is accomplished through the Lorentz forces, $\mathbf{f}_L$, $\mathbf{f}_L = \mathbf{i} \times \mathbf{B}$, where $\mathbf{i}$ is the current density and $\mathbf{B}$ the magnetic field. To obtain a current density field, $\mathbf{i}$, high enough a well saturated aqueous copper sulphate ($CuSo_4$) test fluid is used. A DC electric current of $1A$ is a applied. With the $0.12 \times 0.12m^2$ copper plates this is equivalent to a current density of $\sim 70A/m^2$. The flow is seeded with neutrally buoyant particles of $40 - 60\mu m$ in diameter. The seeding density of the detected particles is around $75 particles/cm^3$ which corresponds to a typical particle distance of $\sim 2.3mm$. For illumination of the observation volume a continuous $20Watt$ Argon Ion Laser is used. The laser beam is widened to a laser sheet of $20mm$ thickness with two cylindrical lenses of $f = 80mm$. The flow is recorded with four $JAI\ M10\ 1/2''\ CCD$

A. Gyr and W. Kinzelbach (eds.), Sedimentation and Sediment Transport, 251–254.

progressive scan monochrome cameras with $8bit/pixel$ resolution. Their pixel resolution is 640×480. A recording system operates at $60Hz$ with 72 MB/s over $100s$ which corresponds to $6000 frames$ for each camera. As lenses four *Rodenstock macro lenses* are used. Calibration is done using a calibration block where 36 points are distributed in a volume of $18 \times 15 \times 7mm^3$ leading to a particle position accuracy of $\sim 3 microns$.

The 3D Particle Tracking Velocimetry technique, 3D-PTV, is a flexible non intrusive image analysis based flow measurement technique. It measures particle trajectories from which - if the particles are neutrally buoyant and small enough to perfectly follow the flow - particle velocities and Lagrangian accelerations can be obtained directly. The idea of 3D-PTV was introduced by Chang and Taterson [1], and was further developed by Racca, Dewey and Maas [2,3]. The technique has a history of development for more than a decade at the Institute of Geodesy and Photogrammetry, IGP, at ETH Zurich [4,5]. In cooperation with the Institute for Hydromechanics and Water Resources Management, IHW, at ETH Zurich existing hard and software has been tested and further developed during the course of the work presented [6].

From the derived particle trajectories the flows velocity, $\mathbf{u}$, and Lagrangian acceleration, $\mathbf{a}$, can be derived directly. In the following the details of a procedure to derive velocities, u_i, accelerations, a_i, and the full set of velocity derivatives, $\frac{\partial u_i}{\partial x_j}$, along particle trajectories will be described. For each time step, t, and around each point of a measured trajectory a third order polynom is fitted from $t - 10 \cdot \Delta t$ to $t + 10 \cdot \Delta t$ for each component, x_1, x_2, x_3. This results in a fit to 21 measured trajectory points. This procedure, effectively acts as a low pass filter for the particle position signal with a cut-off frequency of $\sim 10Hz$. From the filtered velocities, $\widehat{u_i}$, spatial and temporal velocity derivatives are interpolated for every particle trajectory point. The interpolation procedures used to obtain $\frac{\partial u_i}{\partial x_j}(\mathbf{x})$ and $\frac{\partial u_i}{\partial t}(\mathbf{x})$ are linear and use weighted contributions from particles according to their distances to $\mathbf{x}$. Again a procedure was applied which acted as a low pass filter to the interpolated velocity derivative signals along their trajectories. This was achieved using Taylor type polynoms of appropriate orders, n, suiting to the specific trajectory lengths, ℓ, and to the quality of the approximations for each component of $\frac{\partial u_i}{\partial x_j}$ and $\frac{\partial u_i}{\partial t}$.

Single and multi point checks based on precise kinematic relations are performed to ensure that the obtained tensor of velocity derivatives, $\frac{\partial u_i}{\partial x_j}$, is of sufficient qualitative accuracy. The performed checks showed that from 3D-PTV measurements in a turbulent flow qualitatively correct velocity derivatives can be obtained and they can be monitored in a Lagrangian way along particle trajectories.

Two selected results on strain and on geometrical statistics with vorticity are shown. They reproduce well known features of turbulence, most of which in a Gaussian flow field would precisely vanish.

In figure 2 the pdf's of $\Lambda_{1,2,3}$ are given. The main feature is the positiveness of Λ_2. It is essentially responsible for 'everything', since without, turbulence would not be able to produce strain and therefore also no enstrophy. Its magnitude however remains relatively small as compared to Λ_1 and Λ_3. To date the shape of Λ_i, that is, Λ_2/Λ_1, remains poorly understood.

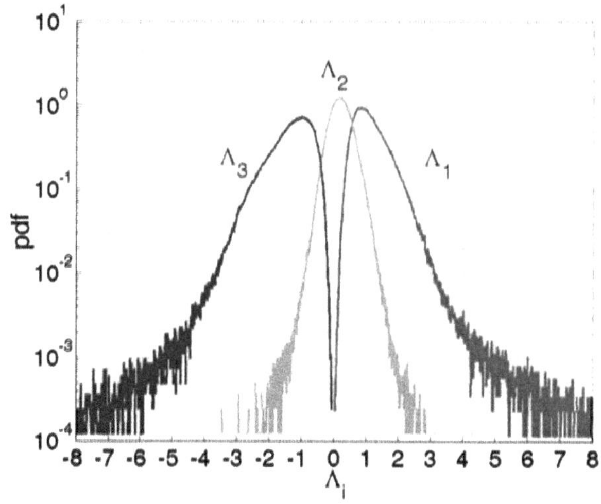

FIGURE 2. Pdf's of the eigenvalues, $\Lambda_{1,2,3}$, of the rate of strain tensor are shwon.

The best known manifestation that turbulence has a dynamically relevant structure even in less intense regions is the orientation of ω relative to the eigenframe of the rate of strain tensor s_{ij}, $\cos(\omega, \boldsymbol{\lambda}_i)$.

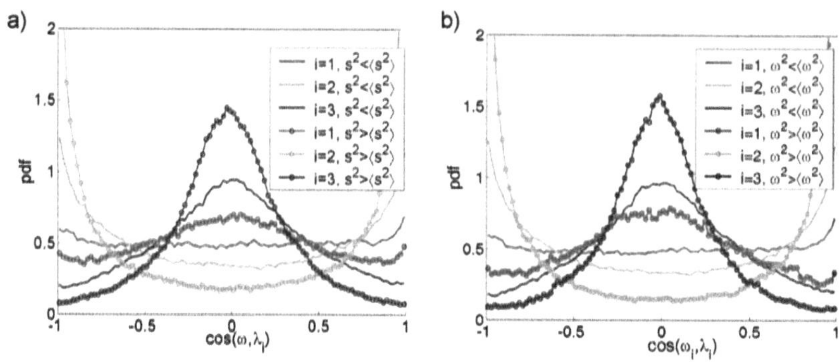

FIGURE 3. Pdf's of $\cos(\omega, \boldsymbol{\lambda}_i)$ conditioned on weak and strong strain (a) and on weak and strong enstrophy (b) are shwon.

Figure 3 shows the pdf's of $\cos(\omega, \boldsymbol{\lambda}_i)$ conditioned on weak and strong strain and enstrophy respectively. The main feature is the predominant alignment of ω with the intermediate eigenvector $\boldsymbol{\lambda}_2$ of the rate of strain tensor. The preferential alignments increase with both, higher strain and higher enstrophy. Even for weak events the preferential alignments persist, suggesting that turbulence has structure also in less intense regimes, see also [7].

The study of enstrophy and strain dynamics is relevant for a deeper understanding of the *dissipative* nature of turbulence. As a consequence of strain and enstrophy and the effect of their combined evolution, the stretching and folding of material elements, is related to the equally important characteristic of turbulent flows - its *diffusivity*. It causes rapid mixing of passive scalars, increased rates of momentum, heat and mass transfer. Following

Girimaji and Pope [8] an infinitesimal material volume deforms under the influence of straining into a volume with length ratios of their principal axes as $(w_1)^{1/2} : (w_2)^{1/2} : (w_3)^{1/2}$, where w_i are the eigenvalues of the Cauchy-Green tensor, $\mathbf{W}$. S, the surface of such a fluid volume, can be defined as, $S = 1/6 \cdot \left(2\sqrt{w_1 w_2} + 2\sqrt{w_2 w_3} + 2\sqrt{w_3 w_1}\right)$.

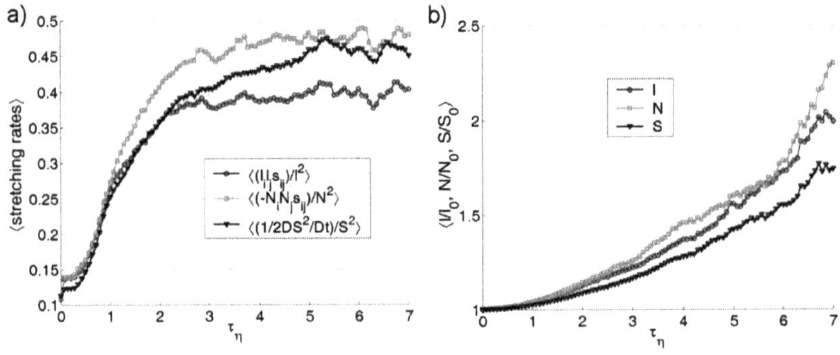

Figure 4. The evolution of stretching rates (a) and absolute stretching (b) are shown for material lines, l, material surfaces, N, and the surfaces of material volumes, S.

Figure 4 shows how the surface S actually evolves and how its growth compares to the growth of material lines, $\mathbf{l}$, and material surface normals, $\mathbf{N}$. The higher growth rates for both types of surfaces, $\mathbf{N}$ and S as compared to $\mathbf{l}$ was already foreseen by Batchelor [9] and is consistent with the findings of [8]. With the same assumptions made he used to estimate the mean rate of material line stretching to be $\sim \langle \Lambda_1 \rangle$ one arrives at a mean rate for surface stretching of $\langle \Lambda_1 + \Lambda_2 \rangle$. Due to non-persistent straining this is significantly over estimating the rates as compared to those obtained from numerical and experimental measurements, but the result that $\frac{1}{2}\frac{DN^2}{Dt} > \frac{1}{2}\frac{Dl^2}{Dt}$ remains correct.

[1] Chang, T. and Taterson, G. (1983) Application of image processing to the analysis of three-dimensional flow fields, Opt. Eng., **23**, 283-287.

[2] Racca, R. and Dewey, J. (1988) A method for automatic particle tracking in a three-dimensional flow field, Exp. Fluids, **6**, 25-32.

[3] Maas,H.-G. (1996) Contributions of digital photogrammetry to 3D PTV, in Three-dimensional Velocity and Vorticity Measuring and Image Analysis Techniques, ed. Dracos, T., Kluwer.

[4] Virant, M. (1996) Anwendung des dreidimensionalen "Particle-Tracking-Velocimetry" auf die Untersuchung von Dispersionsvorgängen in Kanalströmungen, ETH Zürich, Ph.D. Thesis.

[5] Stüer, H. (1999) Investigation of separation on a forward facing step, ETH Zürich, Ph.D. Thesis.

[6] Willneff, J., Gruen, A. (2002) A new spatio-temporal matching algorithm for 3D-Particle Tracking Velocimetry, Proc. 9th Int. Sym. on Transport Phenomena and Dynamics of Rotating Machinery, Honolulu, Hawaii.

[7] Tsinober, A. (2001) An informal Introduction to Turbulence, Kluwer.

[8] Girimaji, S.S. and Pope, S.B. (1990) Material-element deformation in isotropic turbulence, J. Fluid Mech., 220, 427-458.

[9] Batchelor, G.K. (1952) The effect of homogeneous turbulence on material lines and surfaces, Proc. R. Soc. Lond. A, **213**, 349.

LOW COHERENCE TECHNIQUES FOR IMAGING IN MULTIPHASE FLOWS

T. RÖSGEN, R. TOTARO
Institute of Fluid Dynamics, ETH Zurich, Switzerland

1. Introduction

In recent years, quantitative imaging methods have successfully been introduced into many areas of fluids research. Two good examples are particle image velocimetry (PIV) for the measurement of two- and three-dimensional velocity fields and laser-induced fluorescence (LIF) for the determination of concentrations and temperature.
In the study of multiphase flows, some specific problems arise which can complicate the application of such "standard" techniques. On one hand, most flow quantities have to be determined for multiple fluid constituents simultaneously and separately (host fluid, bubbles, particles, etc.) which increases the number of measurement variables. This in turn implies a more complex experimental setup (multiple cameras, multiple illumination sources, higher data rates) which makes measurements more costly and complicated. On the other hand, the generic requirement of unobstructed optical access into the flow under study cannot be met in all cases. This may be due to the geometric complexity of the - often industrial - flow facilities which tend to have a limited number of access ports, but it may also be due to the flow itself. Once too many optically "active" interfaces (fluid-fluid, fluid-solid) are visible, it becomes difficult to image any area inside the bulk flow. The dominant physical effect responsible for the loss of imaging quality is the strong refraction and scattering of light at the phase boundaries which show a rapid change in the optical index of refraction. This makes the flow less transparent so that classical imaging arrangements quickly reach their performance limits. Multiphase flows affected by this problem are for example particle laden and bubbly flows with comparatively large particle concentrations / void fractions or blood flow to be imaged through the surrounding tissue.
(Incidentally, one can have multiphase flows with invisible phase boundaries if the refractive indices of the phases are carefully matched, for example by using glass particles in certain "heavy fluids". At the same time, there are of course also many single phase compressible flows which show optical inhomogeneities that result in light refraction and scattering.)
In very fine suspensions (e.g. in magnetic or electro-rheological fluids) it is not so much the direct scattering that affects the measurements but the absorption of light. These media are essentially opaque and only very little light can penetrate into the fluid so that typical light sheet / camera arrangements become ineffective.
Within this overall context, the present paper will discuss a group of modern imaging techniques, specifically those labelled as "low coherence" methods. While they are being developed mainly for medical applications, their general characteristics could make these techniques also attractive for applications in multiphase flow diagnostics.

2. Low Coherence Light Sources

There are basically two approaches to achieve imaging in highly scattering media, those relying on direct *time-of-flight techniques* and those using *interferometric arrangements*.

A. Gyr and W. Kinzelbach (eds.), Sedimentation and Sediment Transport, 255–267.

In the first case, very short light pulses are sent into the medium. The local effect of scattering at an inhomogeneity can then be described as splitting the incident light into two components. An unaffected part – the so-called "ballistic photons" – travels on in the original direction, whereas the scattered light is deflected into other directions where it experiences further scattering events. Following this model, the un-scattered light components always travel the shortest distance towards an observer. A detector monitoring the light emerging from the scattering medium will consequently record a complex waveform, with the least scattered light arriving first and multiple-scattered components following at a later time. Using suitable gating techniques, one may separate the different contributions and effectively reject the scattered light signal. The method works both in transmission and reflection modes. In the first case, a line-of-sight integration is performed which makes depth resolution difficult. In the reflective arrangement, one observes the single-scattered reflections first and the depth information is preserved in the absolute arrival times.
In order to resolve the very short differences in arrival times, a considerable technical effort has to be made. The light pulses are usually created by mode-locked lasers which are capable of creating relatively strong, ultra-short (sub-picosecond) flashes at high (MHz) repetition rates. On the detector side, high speed imaging devices such as streak cameras and gated microchannel plate intensifiers or auto-correlators have to be employed.
Since the pulsed lasers produce light with a well-defined coherence, it is also possible to use interferometric techniques to simplify the detection process. Here, the incoming scattered light is mixed with an undisturbed replica of the original pulse and, using a square-law detector and a suitable optical delay generator, an optical cross-correlation can be computed. The advantage of interferometric signal detection is that it provides optical gain in the detector and that is permits the use of more sophisticated bandwidth reduction techniques such as heterodyne demodulation and / or lock-in detection.

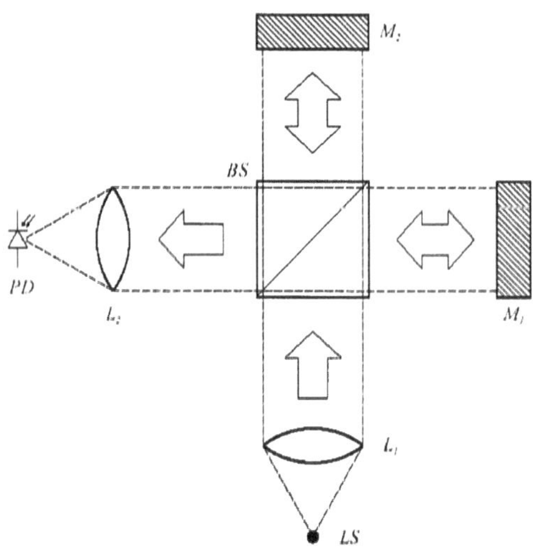

Fig. 1: The Michelson interferometer as optical (auto-) correlator

A typical setup used is the Michelson interferometer (Fig. 1). In this reflective geometry, the light from two propagation paths (light source LS – beam splitter BS –

mirrors $M_{1,2}$ – beam splitter – photo diode PD) is mixed and a modulated intensity signal can be detected. In a simplified mathematical form, one may write for the detector signal

$$I_{PD}(\tau) = I_1 + I_2 + 2\sqrt{I_1 I_2}\,|\gamma_A(\tau)|\cos(2\pi\nu_0\tau + \theta_\gamma(\tau)) \tag{0.1}$$

Here, I_1 and I_2 are the intensities of the light waves arriving from the two arms, ν_0 is the light frequency and τ describes the propagation delay between the two arms. The functions $\gamma_A(\tau)$ and $\theta_\gamma(\tau)$ describe the modulus and phase of the *complex degree of coherence* of the light waves. This degree of coherence is the normalized temporal autocorrelation function of the incident light and captures the coherence properties of the light source. For a perfectly coherent light source, $|\gamma|$ is a constant (=1) and the classical interferometer formula is recovered. The signal I_{PD} varies periodically as the delay τ between the interferometer arms is changed. For a partially coherent light source, $|\gamma|$ has a finite sized, often Gaussian shape which limits the range of delays across which such a signal can interfere. Thus, for delays τ larger than the *coherence time* of the light source, no interference can be observed anymore. This scenario is realized for the ultrashort pulsed lasers mentioned above. Here, the short pulse creates a similarly short autocorrelation function and the coherence is destroyed for delays greater than twice the pulse length. (Obviously, there is no pulse "left" after this time to create an interference signal). Less trivially, the formulation (1.1) points at another alternative to create low coherence signals and their associated interferometric response.

The temporal autocorrelation function $\gamma(\tau)$ can be interpreted as the inverse Fourier transform of a signal's power spectrum, so that any signal with the same power spectrum does have the same autocorrelation function. The spectrum of an ultrashort Gaussian pulse is also Gaussian, with the spectral bandwidth being inversely proportional to the pulse length. A similar spectrum can be tailored by filtering a *continuous*, broadband "random" light source without phase coherence. Thus a "white light" interferometer possesses the same interference behaviour as one based on ultrashort pulsed lasers of equivalent bandwidth, with a significant reduction in the complexity and cost involved.

A third alternative for partially coherent light generation exists in the so-called "superluminescent" laser diodes (SLDs). These are special laser diodes designed to create laser radiation in a broad spectrum (say, about 20 nm wide, see Fig. 2) simultaneously rather than on a single spectral line (Semenov 2000).

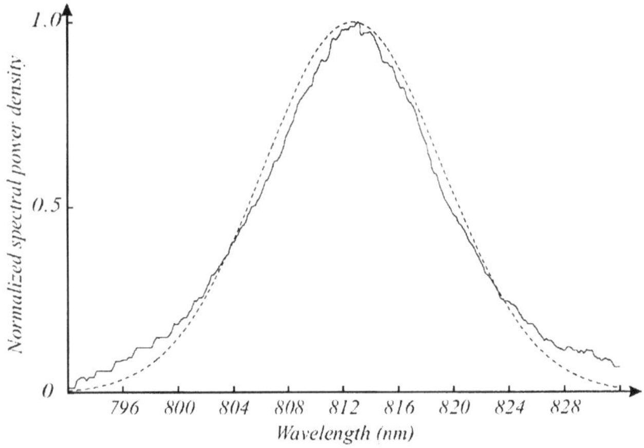

Fig. 2: Power Spectrum of a SLD low coherence light source

Since no mode locking exists, a SLD emits continuous radiation rather than a phase-locked pulse train. Figure 3 shows the interference signal for such a device obtained by moving one of the mirrors in a Michelson interferometer setup.

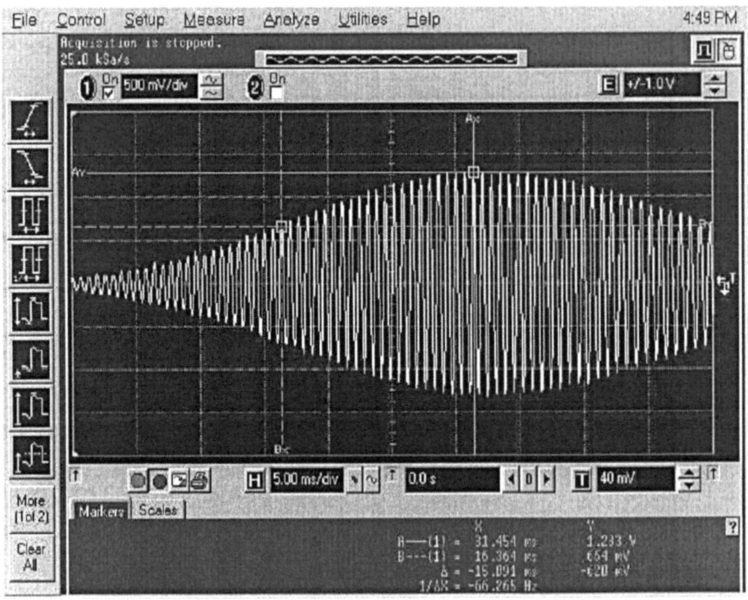

Fig. 3: Interference signal of an SLD measured in a Michelson interferometer. The envelope represents the source's autocorrelation function.

In all cases, the effect is the same: a signal created by a partially coherent light source can create interference within a finite window of temporal or, equivalently, spatial delays.

3. Depth-Resolved Low-Coherence Imaging

The typical low coherence imaging setup is often a variant of the Michelson interferometer. Figure 4 shows a fiber optic version which is used for "optical coherence tomography" (OCT) in biomedical tissue analysis (Huang 1991, Schmitt 1999).

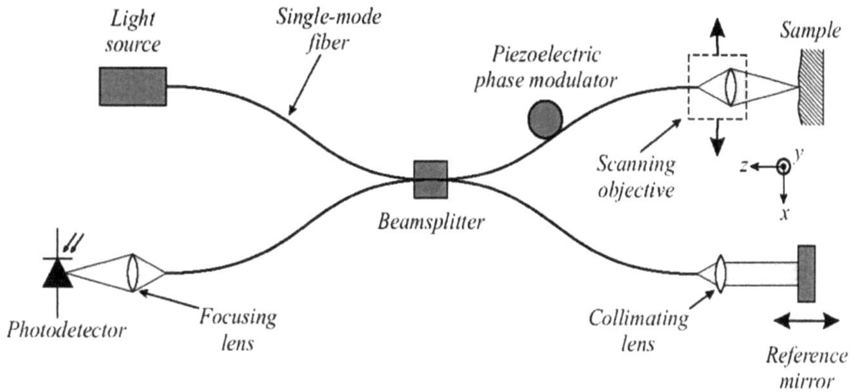

Fig. 4: A typical fiber-optic low coherence interferometer

Two details deserve special mention in this arrangement. The reference mirror is translated to vary the path length of the light travelling in the reference arm. Again, interference is only created when both the reference and sample arms are the same length up to a difference on the order of the light source's coherence length. The reference mirror thus defines the depth plane in which the sample is analyzed. Given typical broadband light sources, the coherence length can be as short as a few micrometers, creating a very high depth resolution. All other light emerging from the sample cannot interfere and creates an incoherent background signal. This feature is very attractive also for the problem of imaging in multiphase flows. No light sheet is required and the flow can be observed and illuminated through a single optical system. The phase modulator is inserted to create an oscillatory signal on the detector even if the sample is stationary. This permits the use of sensitive lock-in detection techniques to extract a (possibly weak) AC signal from the steady incoherent background.
When the frequency shift of the interference signal is analysed explicitly, one may extract information on the relative velocity of the scattering target towards the observer. This variant of OCT, called "optical Doppler tomography" (ODT) is of particular interest for multiphase flow analysis (Chen 1999). Here, the phase modulator allows for a frequency offset to be introduced similar to the one used in laser Doppler velocimetry (LDV) to eliminate the directional ambiguities.

4. Whole-Field Imaging

„Traditional" OCT / ODT configurations use spatial (2D/3D) scanning to build up an image / depth map. They tend to be comparatively slow because the scanning is most often achieved by mechanical means. High resolution flow velocimetry has been successfully demonstrated using this approach, but only for a comparatively slow and laminar microchannel configuration (Chen 1999). For a more complete analysis a direct imaging approach is desirable. For static objects, phase-stepping interferometers can be used to extract a complex interference image from successive camera frames (Beaurepaire 1998).

The situation becomes more complicated for moving (flowing) targets, where for example a whole plane of a flow field has to be processed in parallel so that the spatial velocity correlations can be preserved. This creates the technological challenge of how to extract the interferometric signal from many recording channels simultaneously. Even using a moderate resolution 2D photo detector array, one can easily arrive at tens of thousands of pixels. For each of these pixels, one tries to detect a small oscillatory signal component in a strong steady background. While this can be done with relative ease for a single channel (i.e. a single photodiode, cf. the discussion in sections 2 and 3), the processing for an array requires a different technological approach.

The evolution of CMOS technology has made possible the design of so-called "active-pixel" sensors, where a photodiode array is coupled to analog processing circuitry directly on the silicon chip. Certain processing functions can thus be performed in situ for each pixel, such as gain and offset corrections, rectification, or temporal and spatial filtering. For the detection and pre-processing of interferometric signals, two designs have reached a certain maturity.

In the "photonic mixer" approach, a lock-in or correlation-based detection is performed by rapidly and synchronously switching the photocurrent of each photodiode into a number of local integration capacitors. While this technology has very promising characteristics (including the capability of processing signals with Doppler frequencies of several hundred MHz) it is currently being developed only for single pixel detectors as required for laser radar applications (Schwarte 1997).

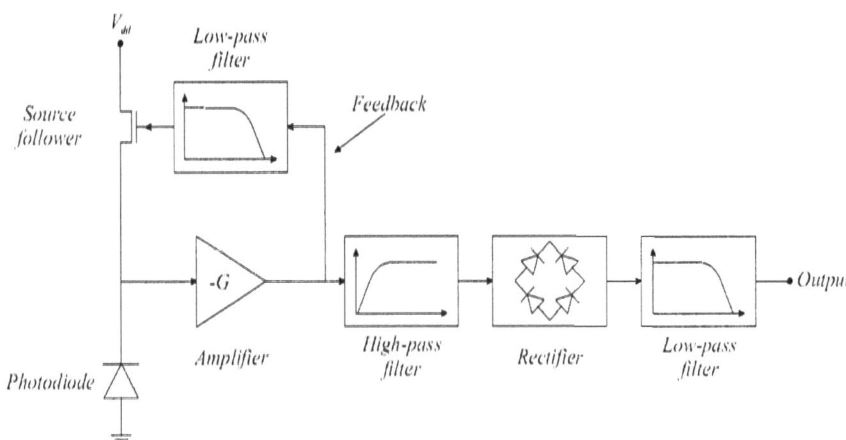

Fig. 5: Processing scheme for each pixel of the FS-49 active pixel sensor

For the present study, an alternative design (FS-49 detector, Bourquin 2000) was chosen based on direct envelope detection. Figure 5 shows the functional processing steps which are implemented *for each pixel* on the chip, which has 58x58 active pixels in total. The resulting signal waveforms at different stages are represented in Fig. 6. The incoming photo detector signal is being demodulated and only the slowly varying envelope of the interferometric signal component is passed to the output multiplexers.

This detector forms the centrepiece of an imaging Michelson interferometer, which is being evaluated at our institute as a planar low coherence flow velocimeter (see Figs. 7,8). One recognizes the low-coherence light source (SLD), the phase-modulator and the FS-49 detector, each designed here to process an expanded beam of approx. 10 mm diameter and thus to look at a similarly dimensioned test volume. The coherence length of the SLD was chosen deliberately to be fairly large (~36 μm) to achieve a certain degree of depth averaging.

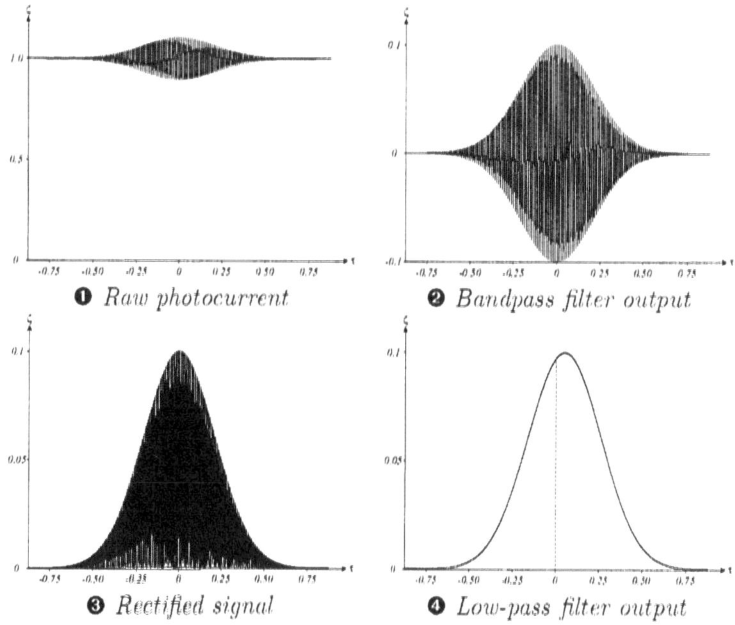

Fig. 6: Analog signal processing steps in each active pixel stage of the FS-49 detector

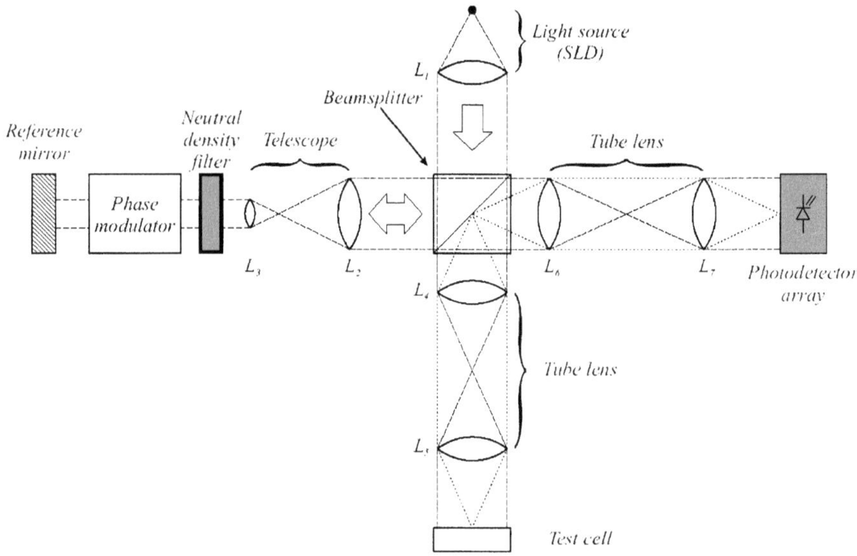

Fig. 7: An imaging low coherence Michelson interferometer for flow analysis

Fig. 8: Experimental setup of the low coherence interferometer in calibration configuration. A,K – SLD, B – beam expander, C – reference mirror, D,J – phase modulator head, E – beam attenuator, F – telescope, G – beam splitter, H,L – tube lenses, I – target (mirror), M – detector array, N,O – auxiliary monitoring equipment

A number of validation tests were performed to evaluate the system's spatial resolution in the depth and lateral directions. For this, hole patterns with controlled diameters and depths were inserted into the imaging volume of the interferometer. The reflections of the holes' bottom surfaces provide a precise gauge for the relative distances imaged in the interferometer. The expected depth resolution of less than 100 μm was verified (see Fig. 9).

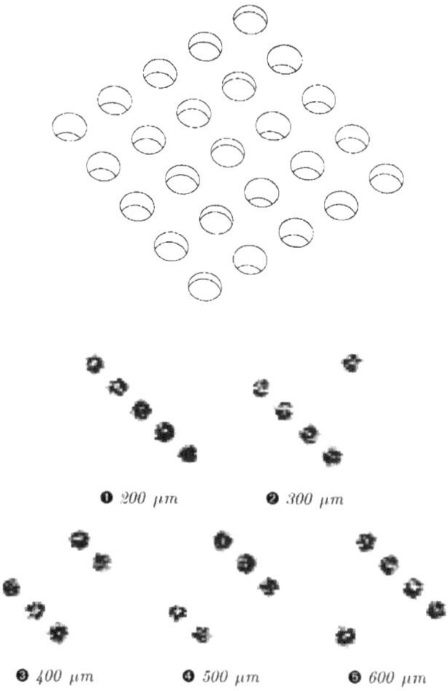

Fig. 9: Test pattern for the verification of the depth resolution (note the alternating depths of the individual holes and their selective visibility)

The lateral resolution was evaluated by imaging patterns with different diameters. Due to the "real estate" required on the chip to implement the active pixel processing functions, the photosensitive area of each pixel was only 10 % (35x35 μm photodiode vs. 110x110 μm pixel pitch). Thus features with a size of less than approx. 200 μm were not expected to be reliably visible. This is confirmed in Fig. 10, were hole patterns with identical depth profiles but different diameters are imaged.

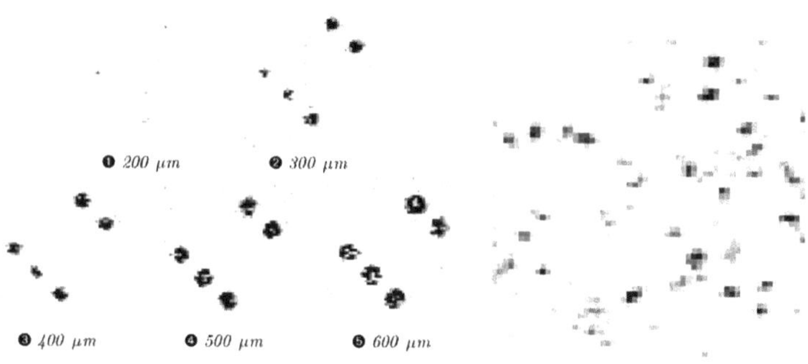

Fig. 10: left – hole patterns with different diameters imaged near the detector resolution limit; right – a random array of glass microspheres imaged on a planar substrate in air

The second image in Fig. 10 depicts the interferometric signature of a random array of glass microspheres that were image on a flat surface in air. The different sphere diameters (150-250 μm) lead to a more varied image. In order to demonstrate the time resolution capabilities of the interferometer, the microsphere sample was translated in the test volume to create a temporal record of the particles' motion. The sampled image pairs were then processed with a standard PIV cross-correlation algorithm to demonstrate the feasibility of in-plane velocity measurements (Fig. 11).

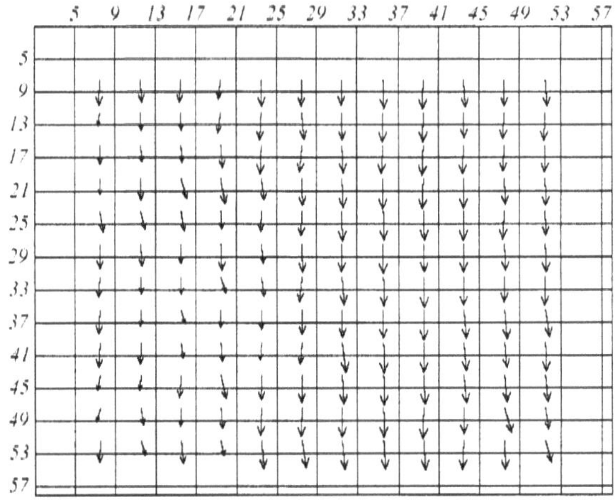

Fig. 11: Displacement field of a moving pattern of random microspheres (PIV processing; spatial coordinates are in pixel units)

These measurement data confirm the design performance of the interferometer. The out-of-plane velocity component of any moving target is reflected in the Doppler frequency of the burst signal. The FS-49 detector does not provide for any direct measurement of the frequency content since the envelope detection eliminates this information. The implementation of suitable extensions of the design (using for example variable frequency shifts in the phase modulator) is being investigated.
A more severe problem arises when the imaging of microspheres immersed in a liquid is attempted (a typical fluid / tracer constellation). Due to the refractive properties of the solid – liquid interfaces, the backscattered light is no longer in the form of an (idealized) spherical wave and the imaging of such a distorted wave front becomes difficult. On the detector surface, a mismatched wave front creates spatial interference fringes which are averaged and reduce the overall signal-to background ratio dramatically. Varying and adapting the optical index of refraction ratio can partially improve the situation but the best match in this respect (identical refractive indices) of course suppresses the backscattering signature altogether.
At present, the functionality of microlens arrays is being evaluated as a solution to the phase mismatch problem. Using an array of microlenses as the light collecting lens system should reduce the interferometer's sensitivity to wave front distortions and at the same time improve the light collection efficiency.

5. Technological Alternatives

While the implementation with the FS-49 envelope detector represents in our opinion at present the most feasible solution, there are several technological alternatives being developed elsewhere which should also be mentioned.
As was briefly described above, low-coherence imaging relies on the detection of an interferometric signal. Due to the design of the interferometer, such a signal is only created in the test volume / plane when the optical path lengths of the reference and

measurement arms are the same. If one replaces the photo detector with a holographic material, the same effect can be utilized to achieve a very efficient background rejection. A hologram requires constructive interference and does not record the incoherent signal contributions. Upon reconstruction, only the scattering targets in the coherence volume are visible (Leith 1992). The technical challenge is to find materials which are suitable for real-time holography, requiring a high sensitivity / low recording threshold and erase capability (Brown 1998, Steele 1998, Tziraki 2000). Motion recording would be limited by the record – readout – erase cycle time. The Doppler frequency information due to motion in the direction of the observer, however, is again lost due to the signal integration during the hologram's exposure.
A different alternative of interest is the use of arrays of so-called "self-mixing" laser diodes. If a laser diode receives part of its coherently emitted radiation back into the cavity a very strong gain modulation can be observed on an (integrated) power-monitoring photodiode. Normally one tries to suppress this phenomenon with suitable optical isolators, but it can be put to use in a low coherence configuration. Again, only the useful signal contributions emerging from the coherence volume can create observable modulations and a very high sensitivity ($\sim 10^6$) is achieved (Rovati 1998, Lacot 1999). Since 2D arrays of laser diodes are currently being developed for telecommunications applications, one may expect to be able to develop imaging systems with this technology soon.
Finally, the emerging field of Terahertz imaging should be mentioned (Mittleman 1999). Here, the parallels with low-coherence imaging are less obvious but the technologies are in many parts quite similar. Ultra-short laser pulses can be used to create optical radiation in the Terahertz regime (between microwaves and very long infrared) which is known to penetrate a number of relevant materials including human tissue. The detection of this radiation is normally achieved by time-of-fight methods using coherent demodulation. So, one may think of such a system as a low-coherence interferometer as well, although operating at very different (and potentially more interesting) wavelengths.

6. Summary

Low coherence techniques may present an elegant solution for imaging applications in multiphase flows. Among the most attractive properties are the capability of imaging in highly scattering media and the possibility of defining precise measurement areas through control of the coherence volume in an interferometric configuration.
In extension of existing scanning systems, a direct imaging low-coherence interferometer was described using a two-dimensional active-pixel sensor as the demodulating detector. The capability of depth resolved in-plane velocimetry was demonstrated using a moving solid target.
The system described can be seen as one alternative among a number of complementary developments aimed at creating a real-time low-coherence imaging system primarily for biomedical applications.

References

Beaurepaire,E., Boccara,A., Lebec,M., Blanchot,L. & Saint-Jalmes,H., *Full-field optical coherence microscopy*, Opt. Lett. 23, 244-246 (1998)

Bourquin,S., *Low-coherence interferometry based on customized detector arrays*, Ph.D. thesis, Swiss Federal Institute of Technology, Lausanne, Switzerland (2000)

Brown,D., Fleurov,V., Carroll,P. & Lawson,C., *Coherence-based imaging through turbid media by use of degenerate four-wave mixing in thin liquid-crystal films and photorefractives*, Appl. Opt. 37, 5306-5312 (1998)

Chen,Z., Zhao,Y., Srinivas,S. Nelson,S., Prakash,N & Frostig,R., *Optical Doppler Tomography*, IEEE J. of Selected Topics in Quantum Electronics 5, 1134-1142 (1999)

Huang,D., Swanson,E., Lin,C., Schuman,J., Stinson,W., Chang,W., Hee,M., Flotte,T., Gregory,K., Puliafito,C. & Fujimoto,J., *Optical Coherence Tomography*, Science 254, 1178-1181 (1991),

Lacot,E., Day,R. & Stoeckel,F., *Laser optical feedback tomography*, Opt. Lett. 24, 744-746 (1999)

Leith,E., Chen,C., Chen,H., Chen,Y., Dilworth,D., Lopez,J., Rudd,.J., Sun,P., Valdmanis,J. & Vossler,G., *Imaging through scattering media with holography*, J. Opt. Soc. Am. 9, 1148-1153 (1992)

Mittleman,D., Gupta,M., Neelamani,R., Baraniuk,R., Rudd,J. & Koch,M., *Recent advances in terahertz imaging*, Appl. Phys. B (1999)

Rovati,L. & Docchio,F., *Low-coherence interferometry using a self-mixing superluminescent diode*, IEEE J. of Photonics Technology Lett. 10, 123-125 (1998)

Schmitt,J., *Optical Coherence Tomography (OCT): A Review*, IEEE J. of Selected Topics in Quantum Electronics 5, 1205-1215 (1999)

Schwarte,R., Xu,Z., Heinol,H., Olk,J., Klein,R., Buxbaum,B., Fischer,H., & Schulte,J., *New electro-optical mixing and correlating sensor: facilities and applications of the photonic mixer device (PMD)*, in Sensors, Sensor Systems, and Sensor Data Processing, Proc. SPIE 3100, 245-253 (1997)

Semenov,A. & Shidlovski,V., *Very High Power, Broad and Flat Spectrum Superluminescent Diodes and Fiber Modules for OCT Applications*, Coherence Domain Optical Methods in Biomedical Science and Clinical Applications IV, Proc. SPIE 3915, 76-82 (2000)

Steele,D., Volodin,B., Savina,O., Kippelen,B. & Peyghambarian,N., *Transillumination imaging through scattering media by use of photorefractive polymers*, Opt. Lett. 3, 153-155 (1998)

Tziraki,M., Jones,R., French,P., Melloch,M. & Nolte,D., *Photorefractive holography for imaging through turbid media using low coherence light*, Appl. Phys. B. 70, 151-154 (2000)

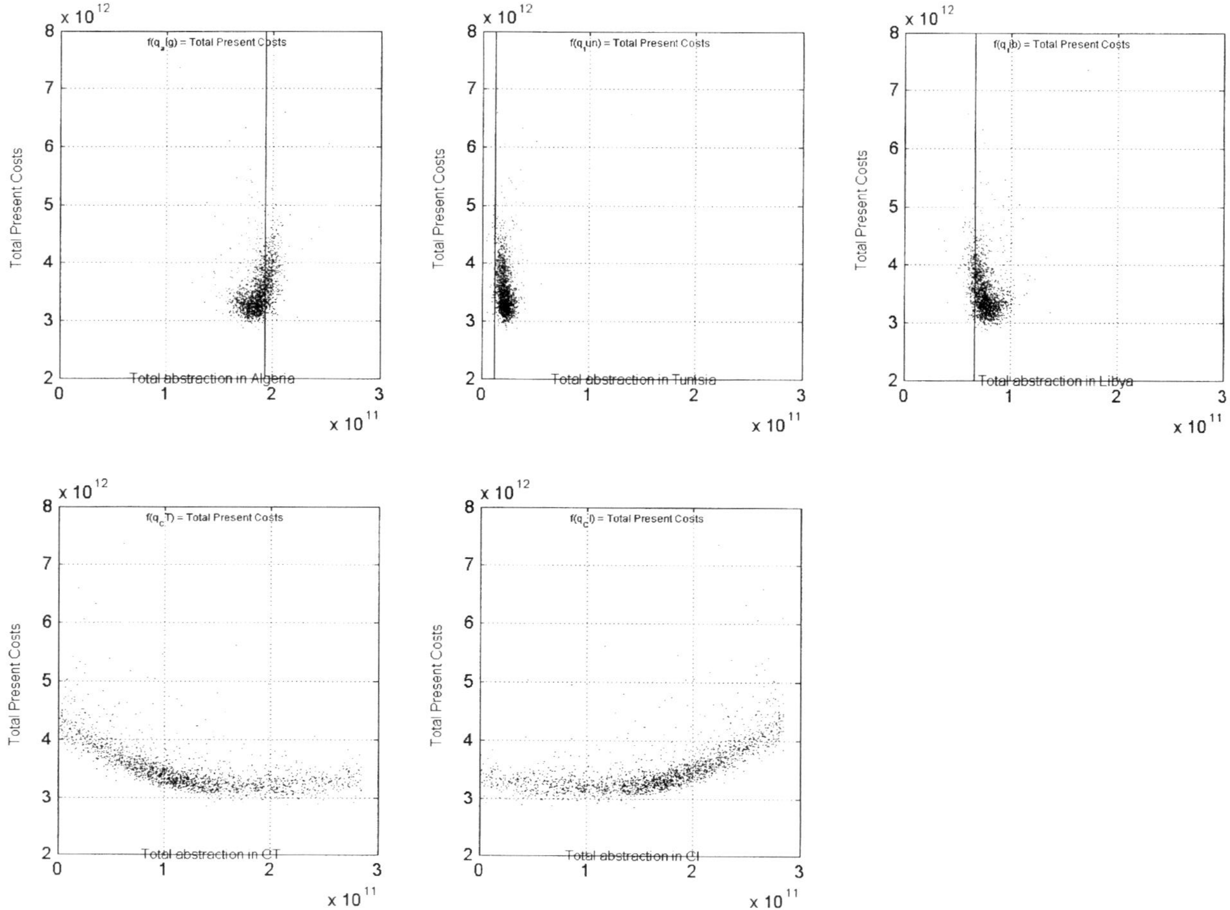

f(q_a lg) = Total Present Costs
Total abstraction in Algeria
f(q_t un) = Total Present Costs
Total abstraction in Tunisia
f(q_l ib) = Total Present Costs
Total abstraction in Libya
f(q_c T) = Total Present Costs
Total abstraction in CT
f(q_c I) = Total Present Costs
Total abstraction in CI
Total Present Costs
x 10^12
x 10^11

KEYWORD INDEX

AUTHOR INDEX

MIX
Papier aus verantwortungsvollen Quellen
Paper from responsible sources
FSC® C105338

If you have any concerns about our products,
you can contact us on
ProductSafety@springernature.com

In case Publisher is established outside the EU,
the EU authorized representative is:
Springer Nature Customer Service Center GmbH
Europaplatz 3, 69115 Heidelberg, Germany

Printed by Libri Plureos GmbH
in Hamburg, Germany